Aviation Noise Impact Management

Laurent Leylekian · Alexandra Covrig ·
Alena Maximova

Editors

Aviation Noise Impact Management

Technologies, Regulations, and Societal
Well-being in Europe

 Springer

Editors
Laurent Leylekian
ONERA
Palaiseau, France

Alena Maximova
Airport Regions Council
Brussels, Belgium

Alexandra Covrig
Airport Regions Council
Brussels, Belgium

European Union's Horizon 2020

ISBN 978-3-030-91196-6   ISBN 978-3-030-91194-2   (eBook)
https://doi.org/10.1007/978-3-030-91194-2

This Springer imprint is published by the registered company Springer Nature Switzerland AG
The registered company address is: Gewerbestrasse 11, 6330 Cham, Switzerland

Foreword

Aviation is a key sector for the European Union's economy, promoting a safe means of transport and connectivity within the Union and globally. Despite the significant progress made in the reduction of CO_2, non-CO_2 and noise emissions due to technological and operational advancements, aviation is still faced with big challenges regarding the reduction of its ecological impact, mainly due to its sustained growth globally. Priorities include, beyond reducing greenhouse gas emissions and improving local air quality, also reducing noise and the number of people severely impacted by it. The European Commission is working on the development of relevant initiatives to address the ecological impact of aviation, including noise, with the European Green Deal, the Fit for 55 package, the Sustainable and Smart Mobility Strategy and the Zero Pollution Action Plan.

The Commission supports thus the Aviation Noise Impact Management through novel Approaches (ANIMA) project, which seeks to directly improve the lives of people affected by aviation noise whilst also supporting policy-makers, researchers and airport managers to make better decisions, which balance economic and regulatory requirements to achieve the best outcomes for everyone. Effective coordination between stakeholders is of the utmost importance to build on existing measures and address the environmental challenges, thus ensuring the long-term success of the aviation sector.

The book on *Aviation Noise Impact Management: Technologies, Regulations, and Societal Well-being in Europe* introduces a panorama of knowledge and expertise on aircraft noise and on the impact of aviation. Building upon past research and covering aspects such as the historical European efforts to reduce aircraft noise, the balanced approach as defined by ICAO and presented in European Regulation, assessment strategies of annoyance and noise metrics and the ways of engaging affected communities, this book gives a contribution to a better understanding of what is at stake.

Rosalinde van der Vlies
Directorate-General for Research and Innovation (RTD)
Director—Clean Planet
Brussels, Belgium

Filip Cornelis
Directorate-General for Mobility and Transport (MOVE)
Director—Aviation
Brussels, Belgium

Preface

Air transport is a key sector for the European Union and for the world at large. If societies and economies are now so deeply connected, and if prosperity is better shared, it may be due to the expansion of information technologies and associated services, but it is also certainly due to the democratisation of rapid, efficient and affordable transports, among which aviation plays a prominent role. In this framework, many European citizens may have some mixed feelings on air transport. On the one hand, we collectively acknowledge how crucial is the associated industry and many of us are therefore unequivocally supporting research and technology that reinforce the competitiveness of the European industry and the subsequent EU wealth and global leadership in this sector. However, on the other hand, we are also very concerned by our quality of life and many of us are therefore keen to lower the negative impact that aviation may have in terms of noise and pollution, especially for communities neighbouring airports.

Competitiveness and impact on people were already the two sources, which inspired Horizon 2020, the finishing Framework Programme for Research and Technology of the European Commission. The Green Deal, for instance, testifies this will put innovation at the service of a better environment. This focus on impact is even reinforced in Horizon Europe, the new Framework Programme for Research and Technology with, for instance, a deep effort to convert our societies to low-carbon economies.

The ANIMA project plays a special role in these overall considerations. It is pioneering an inclusive approach striving to address the impact of aviation noise on people—called the community noise—by associating many different stakeholders from the aviation industry to urban planners, psychologists and sociologists and with the help of SMES and research centres. ANIMA is therefore not focused on reducing noise at source but rather on mitigating and on managing the wide set of noise impacts: annoyance, awakening, social fairness, psychological and physiological health impacts. This project complements more traditional and technological approaches aiming to reduce noise at source (through noise reduction technologies, disruptive aircraft architectures, noise abatement procedures) that we are also supporting. ANIMA is therefore not a stand-alone project but a specific and remarkable one with an ecosystem of other research projects with which it is in constant exchange.

The present book on *Aviation Noise Impact Management: Technologies, Regulations, and Societal Well-being in Europe* thus introduces a panorama of knowledge and expertise on aircraft noise and on the impact of aviation. Beyond ANIMA, it is compiling years of research with a focus on impact and on the latest findings from ANIMA. It illustrates how complex the issue may be in a heavily regulated sector but where regulations come from various levels of authorities; where actors who may know what to do are sometimes not those entitled to do something; where some goodwill initiatives may not meet the communities' expectations; where a huge variety of issues and cases experienced by people may be hidden behind the too vague wording of "annoyance"; where claimed problems may cover implicit or unclaimed ones; where the impact of some initiatives is just not well-known by their promoters or not well-perceived by their supposed beneficiaries or not enough monitored.

Finding a solution is highly challenging in such a context, and indeed, ANIMA along with findings from other projects shows that there is no such thing as a one-fits-all solution. First because what could be a solution around a given airport with some traffic pattern may not be a solution for another place with different traffic. Second because even with similar traffic characteristics, cultural differences may lead to different appreciation here and there. Last, because a solution that may please some stakeholders—for instance, individuals aspiring to respite—will displease some others—for instance, the very same individuals are demanding to have a low-cost flight the day they want or expecting an item they just ordered online to be delivered as soon as possible.

Fortunately, if there is no solution, ANIMA shows that some consensus may however be reached by local stakeholders, providing that they will follow some methodological approach. This methodology is substantiated by several illustrated case studies exemplifying what could be done and what should be avoided. What are the key elements put forward by the project and by this book?

Indeed, they seem to be common sense: first, stakeholders must identify their needs. Why do they want to improve the situation? What is at stake? Which indicators they agree to monitor either to evidence the issue or to follow any evolution? Then they must choose some options to address the identified issue, and they must define how they will practically implement these options. Last, they must agree on the feedback process, which will actually show (or not) any progress and will eventually help to define new objectives. Once again, it may seem common sense but it is probably not blatant as these simple prescriptions are sometimes not obeyed and as so many goodwill initiatives are thereby failing. Indeed, the very first step is about dialogue between all the stakeholders and about the conditions for a fair and fruitful dialogue. Ultimately, it, therefore, appears that the problem of aviation noise is first and furthermost a problem of local democracy.

Certainly, there is room for new studies, new research and new findings. The aviation research community—with the strong support of the European Commission expressed through Horizon Europe—will do its best to initiate new research projects that will put human beings at the core of their objectives. But beyond research, it is certainly time to act yet. Our wishes would be granted if airports and authorities that would like to improve their local situation would consider findings exposed in this book and would adapt methods and tools it exposes. When it turns to aviation, a key sector for the EU global leadership conjugating the high competitiveness of our aviation industry with the most advanced living standards for our citizens is certainly at this price.

Palaiseau, France

Laurent Leylekian
Coordinator of the ANIMA Project

Acknowledgements

The authors of this book and the guest editors would like to express their warmest thanks to Ms. Jordis Wöthge, Mr. Sam Hartley and Mr. Tim Johnson, who performed a careful and fruitful review of the work introduced therein.

This book refers to numerous projects granted by the European Union under its successive Framework Programme for research and innovation. The authors and guest editors of this book would like to acknowledge the contribution of the following projects:

ANIMA has received funding from the European Union's Horizon 2020 research and innovation programme under grant agreement No. 769627.

AERIALIST has received funding from the European Union's Horizon 2020 research and innovation programme under grant agreement No. 723367.

AERONET has received funding from the European Union's Fourth Programme for specific research and technological development under grant agreement No. BRRT975013.

AFLONEXT has received funding from the European Union's Seventh Programme for research, technological development and demonstration under grant agreement No. 604013.

AGAPE has received funding from the European Union's Seventh Framework Programme for research, technological development and demonstration activities under grant agreement No. 625691.

ANNA is a global informal working group interested in non-acoustical factors of aircraft noise. Several meetings of this group were sponsored by X-NOISE and ANIMA projects.

ARTEM has received funding from the European Union's Horizon 2020 research and innovation programme under grant agreement No. 769350.

CALM has received funding from the European Union's Sixth Framework Programme for research under grant agreement No. 516237.

CANNAPE has received funding from the European Union's Seventh Specific Programme "Cooperation": Transport (including Aeronautics) under grant agreement No. 284663.

CIRRUS has received funding from the European Union's Horizon 2020 research and innovation programme under grant agreement No. 886554.

CLEAN SKY and **CLEAN SKY 2** are joint undertakings having received funding from the European Union's Horizon 2020 programme for research and innovation and the aeronautics industry under several grants.

COBRA has received funding from the European Union's Seventh Specific Programme "Cooperation": Transport (including Aeronautics) under grant agreement No. 605379.

COOPERATEUS has received funding from the European Union's Seventh Specific Programme "Cooperation": Transport (including Aeronautics) under grant agreement No. 265211.

COSMA has received funding from the European Union's Seventh Programme for research, technological development and demonstration under grant agreement No. 234118.

DJINN has received funding from the European Union's Horizon 2020 research and innovation programme under grant agreement No. 861438.

DREAM has received funding from the European Union's Seventh Specific Programme "Cooperation": Transport (including Aeronautics) under grant agreement No. 211861.

ECATS has received funding from the European Union's Sixth Research Framework Programme under grant agreement No. 12284.

ENODISE has received funding from the European Union's Horizon 2020 research and innovation programme under grant agreement No. 860103.

ENOVAL has received funding from the European Union's Seventh Programme for research, technological development and demonstration under grant agreement No. 604999.

FLOCON has received funding from the European Union's Seventh Specific Programme "Cooperation": Transport (including Aeronautics) under grant agreement No. 213411.

FLYSAFE has received funding from the European Union's Sixth Research Framework Programme under grant agreement No. 516167.

FORUM-AE has received funding from the European Union's Seventh Specific Programme "Cooperation": Transport (including Aeronautics) under grant agreement No. 605506.

ICARe has received funding from the European Union's Horizon 2020 research and innovation programme under grant agreement No. 769512.

InnoSTAT has received funding from the European Union's Horizon 2020 research and innovation programme under grant agreement No. 865007.

INSPiRe has received funding from the European Union's Horizon 2020 research and innovation programme under grant agreement No. 717228.

INVENTOR has received funding from the European Union's Horizon 2020 research and innovation programme under grant agreement No. 860538.

JERONIMO has received funding from the European Union's Seventh Specific Programme "Cooperation": Transport (including Aeronautics) under grant agreement No. 314692.

NACRE has received funding from the European Union's Sixth Research Framework Programme under grant agreement No. 516068.

NINHA has received funding from the European Union's Seventh Specific Programme "Cooperation": Transport (including Aeronautics) under grant agreement No. 266046.

OPENAIR has received funding from the European Union's Seventh Specific Programme "Cooperation": Transport (including Aeronautics) under grant agreement No. 234313.

OPTI has received funding from the European Union's Seventh Specific Programme "Cooperation": Transport (including Aeronautics) under grant agreement No. 265416.

ORINOCO has received funding from the European Union's Seventh Specific Programme "Cooperation": Transport (including Aeronautics) under grant agreement No. 266103.

PARE has received funding from the European Union's Horizon 2020 research and innovation programme under grant agreement No. 769220.

PARSIFAL has received funding from the European Union's Horizon 2020 research and innovation programme under grant agreement No. 723149.

PropMat has received funding from the European Union's Horizon 2020 research and innovation programme under grant agreement No. 680954.

RECORD has received funding from the European Union's Seventh Specific Programme "Cooperation": Transport (including Aeronautics) under grant agreement No. 312444.

ROSAS has received funding from the European Union's Fifth Framework Programme under grant agreement No. G4RD-CT-2001-00633.

RUMBLE has received funding from the European Union's Horizon 2020 research and innovation programme under grant agreement No. 769896.

SALUTE has received funding from the European Union's Horizon 2020 research and innovation programme under grant agreement No. 821093.

SCONE has received funding from the European Union's Horizon 2020 research and innovation programme under grant agreement No. 755543.

SEFA has received funding from the European Union's Sixth Research Framework Programme under grant agreement No. 502865.

SENECA has received funding from the European Union's Horizon 2020 research and innovation programme under grant agreement No. 101006742.

SESAR is funded by the European Union, Eurocontrol and industry partners.

SILENCE(R) has received funding from the European Union's Fifth Framework Programme under grant agreement No. G4RD-CT-2001-00500.

SOURDINE II has received funding from the European Union's Fifth Framework Programme under grant agreement No. 11011.

SUNJET has received funding from the European Union's Horizon 2020 research and innovation programme under grant agreement No. 640480.

TEAM_PLAY has received funding from the European Union's Seventh Specific Programme "Cooperation": Transport (including Aeronautics) under grant agreement No. 266465.

TEENI has received funding from the European Union's Seventh Specific Programme "Cooperation": Transport (including Aeronautics) under grant agreement No. 212367.

TurboNoiseBB has received funding from the European Union's Horizon 2020 research and innovation programme under grant agreement No. 690714.

VALIANT has received funding from the European Union's Seventh Specific Programme "Cooperation": Transport (including Aeronautics) under grant agreement No. 233680.

VITAL has received funding from the European Union's Sixth Research Framework Programme under grant agreement No. 12271.

X-NOISE has received funding from the European Union's Seventh Specific Programme "Cooperation": Transport (including Aeronautics) under grant agreement No. 265943.

Contents

Understanding the Basics of Aviation Noise

Denis Gély and Ferenc Márki

Abstract This chapter deals with the description of the physical mechanisms of noise, the noise perception and the annoyance induced by air traffic in the aeronautics domain. The authors introduce the basics of aviation noise, describe the main characteristics of the noise emitted by an aircraft in flight, recall the fundamental laws of the audition and noise perception and present the specific context for the annoyance due to aviation noise. This chapter presents and details, as simply as possible, the complex relationships between physical phenomena and noise perception in order to highlight the key notions in aviation noise issues. The readers will find answers to many usual and legitimate questions, for example what is the relationship between the perceived noise and the level of the physical noise related to the European ACARE goals which are expressed sometimes as EPNL reduction in dB or sometimes as perceived noise reduction in percent.

Keywords Physical noise · Source directivity · Annoyance · Noise perception · Audition · Metrics · Loudness · EPNdB · Community noise · Noise contour mapping · ACARE goals · ICAO

Context of Aviation Noise

Over the last decades and since the'60 s, civil aircraft noise has been mitigated significantly: the acoustical energy emitted by average modern civil aircraft—directly related to the noise it is responsible for—is somehow only 10% of an aircraft of the same size in the'60 s. This success was made possible thanks to specific research efforts which focused on reducing the aircraft noise at its source.

Physical Mechanisms, Metrics and Perception

D. Gély (✉)
ONERA, The French Aerospace Lab, Châtillon, France
e-mail: denis.gely@onera.fr

F. Márki
Budapest University of Technology and Economics, Budapest, Hungary
e-mail: marki@vik.bme.hu

© The Author(s) 2022
L. Leylekian et al. (eds.), *Aviation Noise Impact Management*,
https://doi.org/10.1007/978-3-030-91194-2_1

Twenty years ago, at the beginning of the twenty-first century, the International Civil Aviation Organisation (ICAO) promoted the "balanced approach" for more stringent international regulations on air traffic noise. In Europe, the Advisory Council for Aeronautics Research in Europe (ACARE) targeted a further 50% reduction by 2020 and 65% by 2050 of the perceived noise, with respect to the year 2000 as reference.

To understand the basics of aviation noise and its challenges, this chapter aims to describe simply and in short the key knowledge related to it.

Acoustics is a scientific discipline, which combines the physical aspects for describing mechanical phenomena and the physiological aspects for characterising the auditory sensation. Therefore, physical mechanics and neurosciences are inseparable for understanding and interpreting the auditory perception induced by noise and, finally, for evaluating the annoyance. To deal with this last aspect, which goes beyond physics, it is necessary to understand other sciences, human, psychological and sociological.

Many noise sources surround us, like the manufacturing industry, automotive, railway, and, of course, aviation; they all generate annoyance. Air traffic activity is subject to the paradoxical situation: for equivalent noise levels, aviation noise is perceived as the most annoying than railway noise or road traffic noise (Fig. 1).

This comparison illustrates the importance of perception, possibly including the non-acoustics factors (see Chap. 10), and help explain why the ACARE goals for noise were targeted as a mitigation of perceived noise:

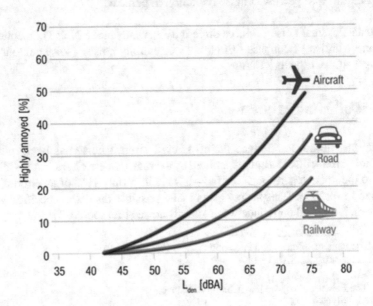

Fig. 1 Comparison of annoyance due to transportation systems. *Credit* European Heart Journal 2014)

> 50% reduction of perceived noise at horizon 2020, equivalent to 10 EPNdB reduction, related to the year 2000
> 65% reduction of perceived noise at horizon 2050, equivalent to 15 EPNdB reduction, related to the year 2000

In the following paragraphs, we will go more into details and explain what the terms *perceived noise* or *EPNdB* mean.

So the key question, related to community noise due to air traffic is:

What is the relationship between the perceived noise and the level of the physical noise?

Basics of Physical Mechanisms, Metrics and Perception

Sound Pressure Level and Loudness

The main physical quantities of acoustics are sound pressure, -power and -intensity. The physical units of these quantities, in the International System, are respectively Pascal (Pa), Watt (W) and Watt per square meter (W / m2). The acoustic power of a small drone is about 10^{-6} W, while the one of a jet-aircraft is about 10^6 W. Such a large range of values to consider brings the acoustician to express the physical quantities using the Logarithm function and a value which serves as a reference because this approach compresses the scales into a meaningful interval.

For the sound pressure specifically, the Sound Pressure Level (SPL) in decibels replaces the pure physical quantity "sound pressure" and is defined as:

$$20 * \log_{10} \frac{p}{p_0} [dB], \tag{1}$$

where p represents the instantaneous pressure, and p_0 the reference level of $2 * 10^{-5}$ Pa.

By convention, the reference pressure was taken equal to $2 * 10^{-5}$ Pascals, because this corresponds to the threshold of average human ear audibility at 1000 Hz. It is interesting to point out that the sound pressure level in decibel is referenced thus to a physiological characteristic. By using this, a sound having a level just "around" the audibility thresholds would result in 0 dB (SPL). Moreover, audition tests have shown that the perception of sound, i.e. the subjective feeling of sound strength (called *loudness* in everyday life) varies with the logarithm of the excitation, so that's one more reason to use it. Finally, decibel was chosenbecause one decibel corresponds to the human discrimination threshold in level, i.e. the just audible amount of change of sound intensity.

If we have to compute the sum of several (independent) noise sources (e.g. L_1 and L_2), we have to compute the resulting sound level (L_{sum}) by adding their (sound) **intensities**, i.e. with the following equation:

$$L_{sum} = 10 * log_{10}(10^{L_1/10} + 10^{L_2/10}). \tag{2}$$

This starts to be a bit too mathematical, but exactly because of this, the following very interesting facts result, worth to note:

$$80\,dB_{SPL} + 80\,dB_{SPL} = 83\,dB_{SPL} \tag{3}$$

Or even

$$0\,dB_{SPL} + 0\,dB_{SPL} = 3\,dB_{SPL} \tag{4}$$

It can be shown that by adding two levels with a significant difference, e.g. of 15 dB, the result is roughly equal to the greater level:

$$80\,dB_{SPL} + 65\,dB_{SPL} = 80\,dB_{SPL} \tag{5}$$

Unfortunately, as you will shortly see, sound reduction sensation does not follow this rule. So when we talk about noise reduction, we always have to specify what we reduce: sound pressure level, sound intensity or the perceived loudness.

The audition characteristics have been studied at the beginning of the twentieth century. In 1933, Fletcher & Munson, engineers at Bell labs, showed that human perception presents a double non-linearity in frequency (i.e. how low or high pitched the sound is) and in level [1]. Figure 2 shows the equal loudness level contours for pure sinus tones, which have been internationally standardised, and were obtained by statistical tests over a large group of young people with normal hearing. Each of the curves represents the necessary SPL of a pure tone at a given frequency to be perceived equally loud as other frequencies on the same curve. The curves are denoted with the metric *phon*, and their value corresponds to the SPL of the 1 kHz tone. (See labelling of the curves at the 1000 Hz vertical line.)

Approaching these curves, from the other side, for differently pitched tones but with the same SPL, generally, the (perceived) loudness falls off at low and at high frequencies (below 500 Hz and above 4 kHz). Please note that the curves are not the same at different levels, although they seem to be quite similar. This means that the "frequency-weighting" nature of our hearing depends on how loud the sound actually is. Let us simply remember that our loudness sensation is very non-linear in relation to both frequency and absolute sound pressure level.

When acousticians have to measure sound events (with the goal to express objectively what people would say about the loudness of the event), they apply—most of the time—the so-called A-weighting. This takes into account the characteristics of the ear, and therefore makes the measuring instruments "listen" more like humans by weighting low and high frequencies to follow approximately (inversely) the equal loudness curve of 40 phons. The A-weighted sound level has been shown to correlate extremely well with a subjective response and is therefore widely used. This is the reason why you can often see dB(A) instead of dB.

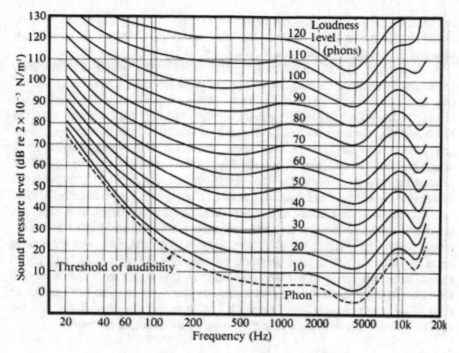

Fig. 2 Standardised equal loudness contours for pure tones

Temporal Behaviour: Peak/instantaneous Levels versus Equivalent Levels

An interesting effect of human hearing is called *masking*. This describes the scenario when a quieter sound cannot be heard in the presence of a louder one. On earth practically, but in cities without any doubt, there is never total silence. Sound sources can be natural, such as dogs barking, birds singing, wind, rain or even human voices, etc., but can also be artificially generated like traffic noise, industrial or construction noise, machine noise from gardening, music, etc. Many of these sounds form together a more or less steady, so-called *background noise*. So when a specific sound event happens, it is either low enough to vanish in the background noise (as this latter masks the quieter sound), or it is high enough to be heard. Our brain is used to living in a noisy world and tries to suppress background noise, i.e. make it unperceived for us (however, this activity certainly causes fatigue...). The more a sound event emerges from the background, the more it attracts our attention. When the sound is welcomed, then we are happy (a nice bird, good music, etc.). If, on the other hand, it is unpleasant, then it causes annoyance. Naturally, high background noise can also cause annoyance, but a disliked sound emerging from the background noise is often more annoying because it cannot be suppressed by the brain. In the long term, if we hear the same sound several times, our brain learns it, and tries to suppress it too, but whether this succeeds depends highly on (1) how strongly it emerges from the background, (2) how fast it becomes audible (slowly or suddenly), (3) how long it lasts and (4) whether it is constant for a while or its loudness fluctuates.

Unfortunately, aircraft noise is hard to suppress, as it often reaches a significantly louder maximum loudness than the background noise (point 1), it lasts for a too-short time to get used to it (point 2), it is by far not constant (point 4), but at least it is not fast increasing/decreasing its level (point 3). The most important negative aspect of this is how high the noise level is. In acoustics, this is called the *maximum sound level* and is denoted by L_{max}. This is completely different from the average sound level, which is called in the literature *equivalent sound level* and is denoted by L_{eq}. (If the measurements were performed with A-weighting then $L_{A,max}$ and $L_{A,eq}$ respectively, or in its simplest text form: LAeq and LAmax.) The use of L_{eq} is reasonable for more or less constant noise sources, like traffic noise, and it is beneficial, as it expresses the overall exposure to noise into just one number. Unfortunately, this became the quasi-standard to measure the *effect of noise* and thus, it is also used for aircraft noise. As you will see in later chapters, this selection is far from optimal, but to defend it, there is no obviously better metric to replace it. Recent research reveals, for example, that the maximum level is one important metric to be used *alongside* the average levels because above a certain level, it cannot distract us from what we are doing during the day, while currently, it wakes us up during the night.

Lden: Day Evening Night Level

The L_{den} (Day Evening Night Sound Level) is the average sound level over a 24h-period. To take into account, the annoyance induced by the noise during the sleeping and the rest periods, a penalty of 5 dB is added for the evening period (usually from 19:00 to 23:00) and 10 dB for the night period (usually from 23:00 to 7:00). The Lden is used for noise contour mapping around an airport allowing the identification of critical areas in terms of community noise (i.e. areas subjected to noise regulations, see Fig. 3).

According to a European Environmental Report, almost 3 million people in Europe are exposed to air traffic noise above 55 dB Lden, which is the EU threshold for excess exposure defined in the Environmental Noise Directive [https://www.eea.europa. eu/airs/2018/environment-and-health/environmental-noise]. For comparison, 125 million people are affected by noise levels from road traffic greater than 55 dB Lden, including more than 37 million exposed to noise levels above 65 dB Lden! [European Environment Agency, www.eea.europa.eu/publications/managing-exposure-to-noise-in-europe].

Fig. 3 Lden contours map around Roissy-CDG Airport (France)

Source Directivity

In everyday life, we are used to teaching our children to turn towards us, when they talk to us. We do this because most sound sources have uneven *directivity*. Directivity expresses how much sound is radiated into one or the other direction. If this wouldn't make our life complicated enough to describe sound sources, we face the fact that directivity is most of the time also frequency-dependent. This means that for example, a source radiates the lower-pitched "parts" of the sound evenly into every direction (behaving as *omnidirectional*), while the higher-pitched content of the sound is radiated strongly into one direction (behaving as *directional*). So we hear low- and high-tuned sound components quieter than mid-tuned ones. This means, in practice, if a directional source is directed towards us, we definitely hear it louder than a source that is turned away from us. And this could happen also with aircraft noise. Noise radiated from aircraft engines is strongly directionally, specifically the direction along the axis of their engines. When the aircraft takes off, the engines exhaust point towards the ground, causing higher sound levels there, generally, but these are also perceived to be as being even louder because of the higher mid-tone content. Additionally, when a taking-off aircraft turns, it rapidly draws a "trace" on the ground with strongly directed sound. For people living there, this means that the sound becomes louder more quickly, and it also reaches a higher maximum sound level than for people who are not affected by the directed part of the engine noise. So let us keep in mind that the directional behaviour of engine noise also causes a measurable sound level difference at some locations, and the perceived noise is even higher. However, related to the overall averaging time (i.e. a year), this more inconvenient fraction of time is too short to appear in numbers. This results in more annoyance without significantly increased long-term noise metrics.

Such a feature must be underlined: annoyance cannot always be described by metrics and especially by time-average metrics.

Effective Perceived Noise Level: EPNL

In aeronautics, the usual metric is the Effective Perceived Noise Level (with units EPNdB), which takes into account the duration and the tonality of a flyover of an aircraft. EPNL is calculated (see Fig. 4) from the sound recording by identifying two instants $t1$ and $t2$ corresponding to the instants for which the sound level is 10 dB lower than the maximum level (PNL max in dB(A)). The equivalent continuous level is calculated for the period between these two instants. If there are tonal components in the sound, then a penalty is added. The equivalent continuous level is then normalised to a period of 10 s to obtain the EPNdB. Note that the duration and tonality of the sound add only a certain amount of decibels to the regular measurements, but overall, EPNL still remains a kind of equivalent level, a "dB" metric. Keeping the characteristics of a flyover but reducing it by a few decibels results in the same reduction in EPNdB.

Generally, we also can consider that 10 dB noise reduction can be achieved by the same amount of EPNdB reduction.

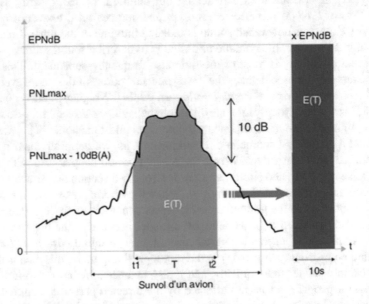

Fig. 4 EPNdB calculation process

Now, to better understand ACARE goals, we have shortly to come back to psychoacoustics and become acquainted with a scale proportional to perceived loudness, the Sone scale. The relationship between loudness in Sones (perception) and loudness level in Phons (physics) is given by:

$$L_{sone} = 2^{\frac{L_{phon}-40}{10}} \tag{6}$$

where L_{sone} and L_{phon} denote the sound level in sone vs phon respectively.

According to this formula, to half the perceived loudness (L_{sone}), it must be reduced by 10 phons. Generally, for changes in sound level, we can neglect the slight deviance of equal loudness curves, and thus say that to **half the perceived loudness, a level reduction of 10 dB is necessary.**

If we apply this knowledge, we can better understand the noise reduction objectives set by ACARE in:

ACARE goal's year	Noise reduction in EPNdB per operations	Perceived noise (L_{sone}) divided by	Perceived noise reduced by %
2020	10	2	50%
2050	15	2,8	65%

Reference

1. Fletcher H, Munson WA (1933) Loudness, its definition, measurement and calculation. J Acoust Soc America 5:82–108

Noise Burden in Europe

Ana Garcia Sainz Pardo and **Fiona Rajé**

Abstract The consequences of noise on the health of the population, as well as the policies and measures that can be adopted to minimise the problem, are a growing concern in Europe. This is highlighted in the recent report prepared in 2020 by the European Environment Agency (EEA), Environmental noise in Europe report (ENER). The main aim of the chapter is to quantify the noise generated by air transport at the EU level and discuss the consequences that this can have on the health of the population exposed to it. The summary of results for air transport contained in the aforementioned report will be presented, as well as those that the EEA presents in more detail for each country in the EEA 2019 Noise country fact sheets (NCFS). All results are derived from the 3rd Environmental Noise Directive (END) round, reported in 2017 and based on 2016 annual traffic data.

Keywords Major airports · Agglomerations · People exposed · High annoyance · High sleep disturbance · Reading comprehension · Ischaemic heart disease · Premature mortality · Cognitive impairment · Years of life lost · Years of life lived adversely

Giving Some Indication on the Actual Impact of Noise on the European Citizen, Mostly by Popularising Data from the Environmental Noise in Europe Report.

A. G. Sainz Pardo (✉)
Airports and Environment, Aeronautical Safety Management, SENASA, Madrid, Spain
e-mail: ana.garcia@senasa.es

F. Rajé
Department of Natural Sciences, Manchester Metropolitan University, Chester Street, Manchester M1 5GD, UK
e-mail: f.raje@mmu.ac.uk

© The Author(s) 2022
L. Leylekian et al. (eds.), *Aviation Noise Impact Management*,
https://doi.org/10.1007/978-3-030-91194-2_2

Introduction

The chapter includes an analysis of the data, summarising the main problems detected throughout the different data delivery phases, as well as exploring future potential challenges. Figure 1 provides an overview of the scope of the noise burden analysis in Europe.

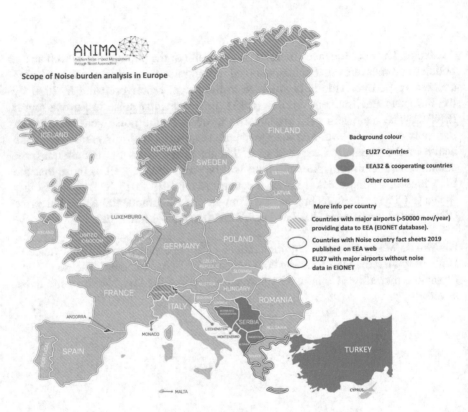

Fig. 1 Scope of noise burden analysis in Europe. The map contains a base colour for the different countries reporting data to the EEA–EU countries, EEA countries and others. Over these colours, there are 3 types of plot: one that indicates the countries that have major airports and the duty to report on them, another for countries that have never reported data to date despite being obliged to do so by the END (EU countries), and another for the countries where the EEA has prepared detailed information on the health condition in the EEA 2019 Noise country fact sheets (NCFS)

The Regulation Behind the Data Collection

The Directive 2002/49/EC, also known as the Environmental Noise Directive (END), aims to "define a **common approach** intended to avoid, prevent or reduce on a prioritised basis, the harmful effects, including annoyance, due to exposure to environmental noise".

To this end, Member States must develop strategic noise maps in order to estimate the level of the population's and/or buildings' exposure to environmental noise using harmonised noise indicators L_{den} and L_{night}. These maps and data are used **to estimate the number of people annoyed and sleep-disturbed** respectively throughout Europe.

The Directive requires the Member States to prepare and publish, every 5 years, the strategic noise maps (SNM) and noise management action plans (NAP) for major airports (i.e. those with more than 50.000 movements a year) and for airports (major and not-major) affecting population agglomerations with more than 100,000 inhabitants.

There have been 3 rounds of SNM thus far: 2007 (showing the noise situation in 2006), 2012 (showing 2011) and 2017 (showing 2016), and 3 NAP rounds in 2008, 2013 and 2018.

The next round (round 4) of SNM has to be delivered by the Member States in 2022, illustrating the situation in 2021. In this round, some changes are expected because of two new directives:

- Directive (EU) 2015/996 establishing common noise assessment methods, to provide complete and homogeneous content to Annex II of the END.
- Directive (EU) 2020/367 establishment of assessment methods for harmful effects of environmental noise, to provide complete and homogeneous content to Annex III of the END.

The Scope of the END Exclusions

Military activities (flights) are excluded from the scope of the Directive 2002/49/CE and from the noise maps, but they are not excluded from the annoyance to residents. There are still quite a few examples of airports with dual-use, civil and military, affecting populations.

The scope of the Directive is major airports and all airports affecting agglomerations. There are many other airports excluded from the scope, but there is still annoyance associated with these. Most countries do not carry out any noise map assessment for airports with less than 50.000 movements/year. Only a few European countries have their own rules about the noise assessment and control for all kinds of airports—large, medium or small airfields, with or without military movements. Moreover, sometimes, training flights with light aircraft—excluded as well from the scope—are the subject of noise complaints because they are flying repeatedly in circles over the same area. Residents may become annoyed and feel helpless because there are no regulations or assessments available to them.

The Method Exclusions

The Annex II of the END, applicable up to 2015, recommended to the Member States, which have no national computation methods, to use ECAC Doc. No 29 "Report on Standard Method of Computing Noise Contours around Civil Airports" (1997 2nd edition) as the method to calculate a noise contour around an airport. Following this, most EU Members have included in their own laws a specific reference to this document and/or to its successive updates. Thus, the first three rounds of SNM were prepared mostly using this "Doc. 29". Where the noise exposure derives mostly from propeller-driven light airplanes or helicopters, **this guidance is not applicable.**[1] **Most countries do not have regulations for light airplane operations or for helicopter operations**. Consequently, most European countries do not include these kinds of operations in their noise contours for major airports. Although not every airport has light airplane and helicopter operations, the associated noise annoyance from these kinds of operations is often overlooked by those that do.

Finally, the aircraft noise calculations in this guidance only take account of noise from aircraft movements. All ground noise sources such as taxiing aircraft, auxiliary power units and aircraft undergoing engine testing are excluded from the guidance. Consequently, most EU countries do not include these kinds of operations in their noise contours for major airports. Once again, not all airports are affected by ground noise, only some of them.

The new Directive (EU) 2015/996 will solve some of these problems. Noise from helicopters, taxiing, engine testing and use of auxiliary power units have to be included with the new methodology, which has to be applied for the next (4th) round of SNM. Accordingly, in some cases, there will be an increase in the number of people exposed as a result of these changes in Annex II. Some other changes are expected because of the method and the way to assign and count people, for example, but in those cases, the noise contours will change homogeneously.

The new Directive (EU) 2020/367 has finally provided a common way to estimate two harmful effects of aviation noise: people who experience high annoyed people (HA) and those with high sleep disturbed (HSD). An additional harmful effect indicator, Ischaemic heart disease (IHD), has been provided solely for road noise. Future revisions of this directive are expected to the extent that research on the matter progresses. Hence, the 4th round will have estimations per country.

---

[1] ECAC.CEAC Doc 29 2nd, 3rd & 4th Editions "Report on Standard Method of Computing. Noise Contours around Civil Airports" Volume 1: Applications Guide.

The Data Collection Exclusions

Annex VI of the END establishes the data to be sent to the Commission by each Member State for each major airport and for each agglomeration affected by airport noise.

Eionet (European Environment Information and Observation Network) is the Central Data Repository for the EEA countries. The END data reporting must adhere to required rules and the EEA has developed Support guidance for reporting under the Environmental Noise Directive 2002/49/EC (END) that the countries must take into account.

The reporting mechanism is clear and well explained. However, checking the reported data per country reveals the following:

- Some countries have been late filing the data. In some cases after 15 years, there is no data.
- The data is not well reported by the countries so that the data are available in some documents but it is not available in the maps presented or in the database. In other words, the number of people exposed was calculated and is available online in Eionet web (or in other official country websites that publish the full documents prepared in accordance with the SNM or NAP), but the digital data to be able to compare and study at European level is not well reported and is not available.

Several significant problems exist with the aviation data and these are highlighted below:

1. Normally the airport/authority in charge of preparing a noise footprint of an airport, following the Annex II methodology, will generate noise contours and the data of people/dwellings/areas affected by these noise contours regardless of whether those people live in agglomerations or outside agglomerations. The SNM contains for both indicators (Lden ¬Lnight) data on the people, area and dwellings affected for each band of decibel levels established. Nevertheless, people exposed to the noise from the same airport have to be reported separately as people inside urban areas and people outside urban areas. Normally this means different authorities in the country are in charge of reporting data internally to the official EU reporter.
2. There is no information about how many and which agglomerations have an airport affecting them. In this case, the agglomeration has to consider all types of airports affecting the agglomeration, not only major airports. The reporting data for each round is entirely inconsistent, and it does not depend on the traffic (unlike the major airports definition). The agglomerations have to present the aviation noise data from major and not-major airports affecting their territories, but in some cases, the data are not reported even for major airports (Italy, Romania, Spain or Sweden). In contrast, the data for major airports in these states and their coverage (tab MAir_list & MAir_coverage from the Noise exposure information under the END Directive -2002/49/EC-

file END_DF4_DF8_Results_2017_190101tabs)were completed and reported. The 4 country examples appeared with 100%

3. The provision of data, concerning the noise sources specified by the END, is reported per round (DF1_DF5) with a description of the location, size and number of inhabitants –agglomerations- or data on the traffic for major airports (> 50,000 movements/year). Nevertheless, not all the countries have adopted the same criteria. As an example, if the list of "major airports" included an airport -and its acoustic data were reported- in the first and second rounds, but in the third round, that airport did not reach 50,000 movements, some countries such as Denmark or Spain report this airport as "−1" -airport without obligation to report-, while others such as Italy report their acoustic data. France presents the same number of movements that they presented in the first round in all rounds. Even if the traffic had dropped below a major airport's definition, they would continue to present data for all of them. Recognition of this disparity of criteria by country is important for global comparisons and by country (Fig. 2).

Notes

- NAR: No major airport to be reported.
- Denmark, Finland and UK expected more airports under the "major airport" definition (column 3 < 100%). They reported only airports complied with the definition (columns 5,7 & 9 = 100%).
- Cyprus, Malta & Slovakia EU27 No major airports to be reported & no NCFS 2019 available.
- Turkey & Liechtenstein EEA32 No data provided.
- Data submitted before 1–1-2019, but for technical reasons, they were not included in ENER.

Health Risk by Aviation Noise in Europe

Aircraft noise exposure is an environmental stressor and has been linked to various adverse health outcomes, such as annoyance, sleep disturbance and cardiovascular diseases.[2]

An updated assessment of the population exposed to high levels of environmental noise and the associated health impacts in Europe for air transport can be extracted from the aforementioned EEA 2020 Environmental noise in Europe report (ENER). Further details by country are also available in the EEA 2019 Noise country fact sheets (NCFS). General –EEA33 aggregated- data exposed come from the publication Health risk caused by environmental noise in Europe (Dec 2020).

---

[2] Further details in Chap. 9 of this book.

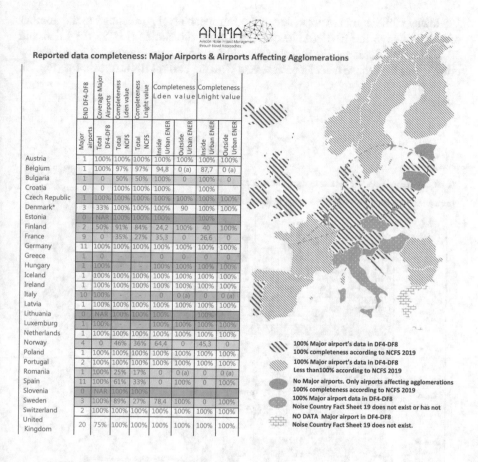

Reported data completeness: Major Airports & Airports Affecting Agglomerations

	Major airports (END DF4-DF8)	Total DF4-DF8 (Coverage Major Airports)	Total NCFS (Completeness Lden value)	Total NCFS (Completeness Lnight value)	Completeness Lden value — Inside Urban ENER	Completeness Lden value — Outside Urban ENER	Completeness Lnight value — Inside Urban ENER	Completeness Lnight value — Outside Urban ENER
Austria	1	100%	100%	100%	100%	100%	100%	100%
Belgium	1	100%	97%	97%	94,8	0 (a)	87,7	0 (a)
Bulgaria	1	0	50%	50%	100%	0	100%	0
Croatia	0	0	100%	100%	100%		100%	
Czech Republic	1	100%	100%	100%	100%	100%	100%	100%
Denmark*	3	33%	100%	100%	100%	90	100%	100%
Estonia	0	NAR	100%	100%	100%		100%	
Finland	2	50%	91%	84%	24,2	100%	40	100%
France	9	0	35%	27%	35,3	0	26,6	0
Germany	11	100%	100%	100%	100%	100%	100%	100%
Greece	1	0			0	0	0	0
Hungary	1	100%	-	-	100%	100%	100%	100%
Iceland	1	100%	100%	100%	100%	100%	100%	100%
Ireland	1	100%	100%	100%	100%	100%	100%	100%
Italy	10	100%	-	-	0	0 (a)	0	0 (a)
Latvia	1	100%	100%	100%	100%	100%	100%	100%
Lithuania	0	NAR	100%	100%	100%		100%	
Luxemburg	1	100%	-	-	100%	100%	100%	100%
Netherlands	1	100%	100%	100%	100%	100%	100%	100%
Norway	4	0	46%	36%	64,4	0	45,3	0
Poland	1	100%	100%	100%	100%	100%	100%	100%
Portugal	2	100%	100%	100%	100%	100%	100%	100%
Romania	1	100%	25%	17%	0	0 (a)	0	0 (a)
Spain	11	100%	61%	33%	0	100%	0	100%
Slovenia	0	NAR	100%	100%	100%			
Sweden	3	100%	89%	27%	78,4	100%	0	100%
Switzerland	2	100%	100%	100%	100%	100%	100%	100%
United Kingdom	20	75%	100%	100%	100%	100%	100%	100%

Legend:
- 100% Major airport's data in DF4-DF8 / 100% completeness according to NCFS 2019
- 100% Major airport's data in DF4-DF8 / Less than 100% according to NCFS 2019
- No Major airports. Only airports affecting agglomerations / 100% completeness according to NCFS 2019
- 100% Major airport data in DF4-DF8 / Noise Country Fact Sheet 19 does not exist or has not
- NO DATA Major airport in DF4-DF8 / Noise Country Fact Sheet 19 does not exist.

NOTES:
- ✓ NAR: No major airport to be reported.
- ✓ Denmark, Finland and UK expected more airports under "major airport" definition (column 3<100%). They reported only airports complied with the definition (columns 5,7 & 9 =100%).
- ✓ Cyprus, Malta & Slovakia EU27 No major airports to be reported & no NCFS 2019 available.
- ✓ Turkey & Liechtenstein EEA32 No data provided.
- ✓ (a) Data submitted before 1-1-2019, but for technical reasons they were not included in ENER.

Fig. 2 Reported data completeness from ANNEX 2 of EEA 2020 Environmental noise in Europe report (ENER), EEA 2019 Noise country fact sheets (NCFS) and END_DF4_DF8_Results_2017_190101

People Exposed to High Levels of Environmental Noise

The Environmental noise guidelines for the European region [9], define long-term noise exposure levels above which a relevant increase in negative health effects occur, expressed in terms of the indicators Lden and Lnight. For aircraft sources, these levels are 45 dB Lden and 40 dB Lnight.

However, the numbers presented at the European level correspond to the number of people above the END (Annex VI) reporting thresholds (i.e. 55 dB Lden and 50 dB Lnight). This means that there could be more people exposed to unhealthy noise levels than those that can be assessed with the current END thresholds (Fig. 3).

Notes

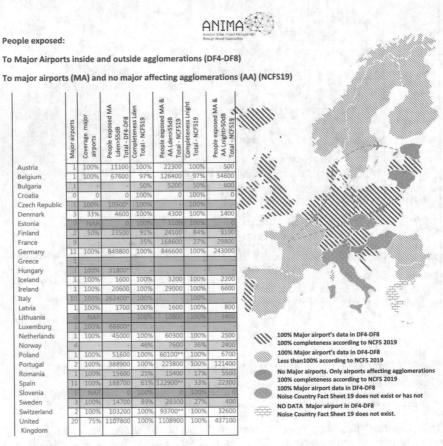

People exposed:

To Major Airports inside and outside agglomerations (DF4-DF8)

To major airports (MA) and no major affecting agglomerations (AA) (NCFS19)

	Major airports	Coverage major airports	People exposed MA Lden>55dB Total - DF4-DF8	Completeness Lden Total - NCFS19	People exposed MA & AA Lden>55dB Total - NCFS19	Completeness Lnight Total - NCFS19	People exposed MA & AA Lnight>50dB Total - NCFS19
Austria	1	100%	11100	100%	22300	100%	500
Belgium	1	100%	67600	97%	126400	97%	34600
Bulgaria	1	-	-	50%	5200	50%	600
Croatia	0	0	0	100%	0	100%	0
Czech Republic	1	100%	10500*	100%	-	100%	
Denmark	3	33%	4600	100%	4300	100%	1400
Estonia	0	NAR	0	100%	3100	100%	0
Finland	2	50%	23500	91%	24100	84%	3100
France	9	-	-	35%	168600	27%	29800
Germany	11	100%	849800	100%	846600	100%	243000
Greece	1						
Hungary	1	100%	31800*	-	-	-	
Iceland	1	100%	1600	100%	3200	100%	2200
Ireland	1	100%	20600	100%	29000	100%	6600
Italy	10	100%	262400*	100%	-	100%	
Latvia	1	100%	1700	100%	1600	100%	800
Lithuania	0	NAR		100%	10800	100%	3400
Luxemburg	1	100%	68800*				
Netherlands	1	100%	45000	100%	60300	100%	2500
Norway	4	-	-	46%	7600	36%	2400
Poland	1	100%	51600	100%	60100**	100%	6700
Portugal	2	100%	388900	100%	223800	100%	121400
Romania	1	100%	15600	25%	15400	17%	5500
Spain	11	100%	188700	61%	122900**	33%	22300
Slovenia	0	NAR	0	100%	0	100%	0
Sweden	3	100%	14700	89%	28300	27%	400
Switzerland	2	100%	103200	100%	93700**	100%	32600
United Kingdom	20	75%	1107800	100%	1108900	100%	437100

100% Major airport's data in DF4-DF8
100% completeness according to NCFS 2019

100% Major airport's data in DF4-DF8
Less than100% according to NCFS 2019

No Major airports. Only airports affecting agglomerations
100% completeness according to NCFS 2019

100% Major airport data in DF4-DF8
Noise Country Fact Sheet 19 does not exist or has not

NO DATA Major airport in DF4-DF8
Noise Country Fact Sheet 19 does not exist.

NOTES:
✓ NAR: No major airport to be reported.
✓ Denmark, Finland and UK expected more airports under "major airport" definition (column 3<100%). They reported only airports complied with the definition (columns 5 & 7 =100%). Less airports than previous rounds. Denmark or Spain did not include these airports in DF1. In all the cases comparations between rounds are not correct because they are considering different number of airports per country.
✓ *No NCFS19 available. Meaning estimated data in global numbers for EU region instead of real numbers.
✓ **The column 6 should be equal or major than column 4 because it includes "no major airports" affecting agglomerations. Poland, Spain have not reported the data of the agglomerations in general, so the air data are not included in the NCFS19, however it was included in DF4-DF8. Meaning estimated data instead of real data in NCFS19 to calculate health impacts.
✓ Switzerland appears in NCFS19 with 93700 people affected by MA in Agglomerations. Nevertheless, in DF4-DF8 appears with 103200.
✓ Bulgaria, Finland, France, Norway, Romania, Spain & Sweden have estimated data in NCFS19.

Fig. 3 People exposed to aviation noise in Europe

- NAR: No major airport to be reported.
- Denmark, Finland and UK expected more airports under "major airport" definition (column 3 < 100%). They reported only airports complied with the definition (columns 5 & 7 = 100%). Fewer airports than previous rounds. Denmark or Spain did not include these airports in DF1. In all the cases comparisons between rounds are not correct because they are considering a different number of airports per country.
- *No NCFS19 available. Meaning estimated data in global numbers for the EU region instead of real numbers.
- **The column 6 should be equal or greater than column 4 because it includes "no major airports" affecting agglomerations. Poland, Spain have not reported the data of the agglomerations in general, so the air data is not included in the NCFS19, however, it was included in DF4-DF8. Meaning estimated data instead of real data in NCFS19 to calculate health impacts.
- Switzerland appears in NCFS19 with 93,700 people affected by MA in Agglomerations. Nevertheless, in DF4-DF8 appears with 103,200.
- Bulgaria, Finland, France, Norway, Romania, Spain & Sweden have estimated data in NCFS19

Health Risk Associated with Noise Exposure

Only those health outcomes that have demonstrated a reasonable causal relationship between noise exposure and adverse human health effects have been estimated. All the health risks associated with noise exposure have been estimated by EEA based on exposure–response functions presented in the Environmental noise guidelines for the European region [9]. Additionally, baseline health statistics, such as incidence of and mortality rates from ischaemic heart disease per country, were used to estimate the number of cases of ischaemic heart disease and the number of premature deaths attributable to noise per year.

- Annoyance (-WHO [9]—exposure–response functions)
- Sleep disturbance (-WHO [9]—exposure–response functions)
- Ischaemic heart disease (-WHO [9] —do not make a recommendation to include these for aviation noise, however, EEA assumed that the cardiovascular effects of road traffic noise can be extrapolated to aircraft noise [10])
- Reading and oral comprehension in children -cognitive impairment- was included following the recommendation of [10]
- Premature mortality due to IHD was included following the recommendation of [10]

The health impacts were estimated using the number of people exposed to levels of noise starting at 55 dB Lden and 50 dB Lnight, as reported under the Environmental Noise Directive (END). The results quantify the concrete health effects of noise in Europe and are easily understood by the public and other stakeholders [10].

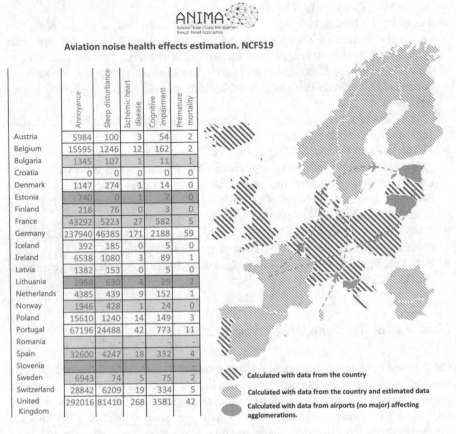

Fig. 4 People health affected by aviation noise in Europe

Nevertheless, uncertainties must also be taken into consideration. The methodology for calculating the burden of disease for noise is not well established and needs more data and further research across the various European countries (Fig. 4 and Tables 1 and 2).

Burden of Disease

Health risks can also be expressed in terms of Disability-Adjusted Life-Years (DALYs). The calculation requires the use of disability weighting as well as data on years of life lost and years lived with disability due to ischaemic heart disease per country. These data were taken from the Institute for Health Metrics and Evaluation.

The DALY estimates how much disease affects the life of the population by combining the burden from mortality, in terms of Years of Life Lost (YLL) because

Table 1 Relationships between noise and the health effects used by EEA in all the sources consulted

Health effect	Relationship
High annoyance (all people exposed)	Guski et al. [7] $(-50.9693 + 1.0168 \times Lden + 0.0072 \times L^2den)/100$
High sleep disturbance (all people exposed)	Basner and McGuire [1] $(16.7885 - 0.9293 \times Lnight + 0.0198 \times L^2night)/100$
Reading comprehension (children exposed)	Clark et al. [2] and van Kempen [11] $1/(1 + exp(-(ln(0.1/0.9) + (ln(1.38)/10 \times (Lden-50))))$ if $Lden \geq 50$ dB and 0.1 if $Lden < 50$ dB
Ischaemic heart disease incidence (all people exposed)	van Kempen et al. [12] relative risk (RR) derived from road noise (applied to aviation with not significant evidence) $RR = exp(ln(1.08)/10 \times (Lden - 53))$ if $Lden \geq 53$ dB, and $RR = 1$ if $Lden < 53$ dB
Premature mortality due to ischaemic heart disease (all people exposed)	van Kempen et al. [12] RR derived from road noise (applied to aviation with not significant evidence) $RR = exp(ln(1.05)/10 \times (Lden-53))$ if $Lden \geq 53$ dB, and $RR = 1$ if $Lden < 53$ dB

Table 2 Estimated number of people health affected by aviation noise, EEA-33 (Turkey not included)[3]

	High annoyance	High sleep disturbance	Ischaemic heart disease	Premature mortality [a]	Cognitive impairment in children
Inside urban areas	848 300	168 500	700	200	9 500
Outside urban areas	285 400	82 900	200	50	2900

[a]Refers to mortality due to ischaemic heart disease.

of premature death due to disease, and morbidity, in terms of Years of Life Lived adversely affected by Disease (YLD). One DALY corresponds to one lost year of a healthy life, attributable to morbidity, mortality or both. The disability weighting used are described in the WHO [9] (Fig. 5).

We argue in ANIMA for better communication and engagement, especially with the public and affected communities. The use of a term such as DALY can make such communication challenging: it is not readily comprehensible to the general public and this is undesirable. The fact that it needs further explanation could lead to

---

[3] EEA 2020 Environmental noise in Europe report (ENER), Table 3.5.

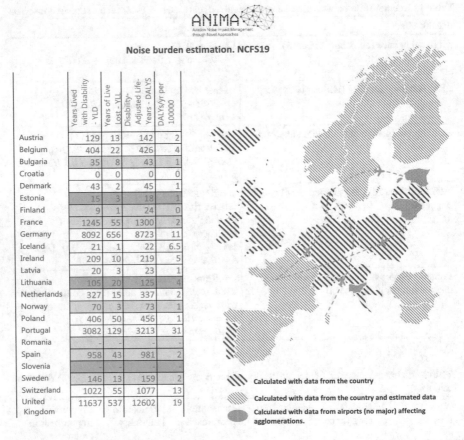

Fig. 5 Aviation noise burden estimation in Europe

an unintended element of obfuscation which would not be acceptable for clear and succinct communication. Also, DALYs have considerable data requirements which may not be economically justifiable in increasingly financially straightened times.

Uncertainties in the Estimation Data

Uncertainties in the exposure level. Data collected in the 3rd round under the END do not cover all aviation sources as discussed earlier, nor do they cover populations exposed to levels of noise below 50 dB Lnight and 55 dB Lden -although Lden/Lnight 45/40 were suggested by WHO [9]. Noise data from different countries, regions or cities may be modelled using different methodologies, and are often reported late or in different ways (due to a lack of internal coordination in some countries). All of these issues introduce some inconsistencies into the EU wide combined dataset. The completeness of the aviation data reported for the 2017 round of noise mapping is about 66% -75%. Therefore, gap-filling data were used to estimate the total area covered by the END following a clear methodology to estimate missing data. ETC/ATNI Report 1/2019: Noise indicators under the Environmental Noise Directive. Methodology for estimating missing data.—Eionet Portal (europa.eu).

Uncertainties in the population exposed.[4] Different countries use varied methodologies to assign the population to dwellings, which creates inconsistencies. In addition, the exact distribution of the population per decibel level is not known, as reporting is aggregated into 5 dB bands.

Uncertainties arising from the baseline and health data.[5] Baseline morbidity and mortality data are based on national statistics. Therefore, using national health data for other sub-national units (e.g. urban areas) may bring about uncertainties, as health baseline data may not apply to the territory uniformly.

Uncertainties arising from the exposure–response functions used.[6] Using the generalised exposure–response functions from the World Health Organisation may introduce uncertainties, for example, if the magnitude of the exposure–response estimate in different countries varies with age and gender distributions [10]. The health impact depends on the 'baseline' prevalence (frequency) or incidence (new cases per year) of health effects. These differ between countries and were taken into account in the calculations. The calculations in this assessment include a non-uniform distribution across noise bands, which was estimated using a 1-dB resolution for calculating the average exposure in each band. The methods employed for this Health Impact Assessment (HIA) are described in more detail in ETC/ACM [6].

The results of acoustic insulation interventions or a quiet façade implemented in the surroundings of one airport **are not reflected** within the conclusions of noise exposure maps provided in the 3rd round. Noise insulation measures or façade interventions might have significantly reduced the indoor exposure and thereby annoyance and sleep disturbances.

Challenges for the Near Future

It is evident that there need to be movements towards more reliable data collection and improved development of knowledge about the noise burden and how it can be addressed effectively and efficiently.

The chapter describes the data collected and the directives that govern its acquisition. There are evident inconsistencies across the piece and it is important that solutions be considered to address the exceptions from the scope, the method and the data collection. In addition, there need to be efforts made to reduce the number

---

[4] EEA 2020 Environmental noise in Europe report (ENER).

[5] EEA 2020 Environmental noise in Europe report (ENER).

[6] EEA 2020 Environmental noise in Europe report (ENER).

of data missing from various countries. This needs to be complemented by improved national and EU-country coordination.

Also, there would seem to be a need to reduce the time spent on obtaining better and updated results with higher quality. For example, the estimation of health impacts from the 2016 data were not made available until the end of 2020, yet data from some countries was still missing.

As the ANIMA work has revealed throughout, effective communication of information to the public is vital for progressive policy-making, community learning and intervention development to address noise and health concerns. To this end, the public should be presented with clear horizons and noise objectives for both individual airports and countries.

It is equally important to tackle differences in the estimations of health effects by country and within a country. Thus, for example, it is also essential to investigate the application of interventions like noise insulation within the estimation of harmful effects, especially in light of the newly published Annex III.

ANIMA research on noise and health [8] calls for efforts to build consensus on what constitutes a socially acceptable response to the challenge of health risk reduction in the EU. It underlines that addressing sleep disturbance and annoyance is most important as research moves forward, indicating that these two elements are associated with annoyance in local communities and that persistent annoyance has been linked to adverse health effects through the stress mechanism. By reducing both sleep disturbance and annoyance, it follows that there should be a decrease in adverse health effects of aircraft noise. (Please see Chap. 9 for further details.)

Comprehensive approaches to lessening the burden of aircraft noise should be based on sound frameworks that are built on consistent and reliable data that clearly communicates the noise picture across the EU. With consistent, improved and timely data availability that is open to comparison across and within territories, the emerging extent of the noise burden can be more easily understood. This would allow, for example, for the assessment of noise interventions (e.g. sound insulation) across different geographical areas and increased knowledge of how they may be applied and adapted to better suit local conditions, with associated lowering of adverse health impacts. While, at the same time, this would enable shared learning for airport operators which may preclude unnecessary and expensive trials if good practice is revealed and built upon instead. In this way, financial and other resources may be released from investment in insulation and may be better used to address health impacts of noise through effective engagement with affected communities to co-design alternative and/or improved interventions and offers that optimally serve their experience and needs.

References

1. Basner M, McGuire S (2018) WHO environmental noise guidelines for the European region: a systematic review on environmental noise and effects on sleep. Int J Environ Res Public Health 15(3). https://doi.org/10.3390/ijerph15030519
2. Clark C et al (2006) Exposure-effect relations between aircraft and road traffic noise exposure at school and reading comprehension The RANCH Project. Am J Epidemiol 163(1):27–37. https://doi.org/10.1093/aje/kwj001
3. EEA (2020) Environmental noise in Europe report—2020, EEA Report No 22/2019, Accessed 10 April 2021
4. EEA (2020) Noise country fact sheets 2019—Published in 2020, accessed 10 April 2021
5. EEA (2020) Health risks caused by environmental noise in Europe—Dec 2020, EEA Briefing no. 21/2020, Accessed 10 April 2021
6. ETC/ACM, 2018/10, Implications of environmental noise on health and wellbeing in Europe, Eionet Report –Feb 2018, accessed 10 April 2021
7. Guski R et al (2017) WHO environmental noise guidelines for the European region: a systematic review on environmental noise and annoyance. Int J Environ Res Public Health 14(12):1539. https://doi.org/10.3390/ijerph14121539
8. Kranjec N, Gjestland, T, Vrdelja M, Jeram S (2018) Slovenian standardised noise reaction questions for community noise surveys. https://zenodo.org/record/2582860#.YZykKtBBxnJ
9. WHO Europe (2018) Environmental noise guidelines for the European region. World Health Organisation Regional Office for Europe, Copenhagen, Accessed 7 December 2018
10. van Kamp I et al. (2018) Study on methodology to perform environmental noise and health assessment, RIVM Report No 2018–0121. National Institute for Public Health and the Environment (Netherlands)
11. van Kempen EEMM (2008) Transportation noise exposure and children's health and cognition. Dissertation, Utrecht University. http://dspace.library.uu.nl/handle/1874/25891
12. van Kempen E et al (2018) WHO environmental noise guidelines for the European region: a systematic review on environmental noise and cardiovascular and metabolic effects: a summary. Int J Environ Res Public Health 15(2):379. https://doi.org/10.3390/ijerph15020379

Regulating and Reducing Noise Today

Balanced Approach to Aircraft Noise Management

Oleksandr Zaporozhets

Abstract ICAO Balanced Approach (BA) to aircraft noise management in airports is reviewed in accordance with historical and technological challenges. All four basic elements of the BA are subject to noise exposure control with dominant emphasis on reduction of noise at source and compatible land usage inside the noise zoning around the airports. Noise abatement procedures and flight restrictions are used at any airport due to its specific issues and should be implemented on a basis of cost–benefit analysis. Noise exposure reduction is an intermediate goal, a final goal—to reduce noise impact, which is mostly represented by population annoyance as a reaction to noise exposure, is discussed also.

Keywords Aircraft noise certification · Reduction at source · Land-use planning · Aircraft operational measures · Aircraft operating restrictions

From Noise Exposure to Noise Annoyance—Introduction Issues

Noise has always been a dominant environmental factor in the area of air transportation, first of all affecting residential communities close to airports. Environmental noise, in general, is an obvious example of unwanted technological and social outcomes in continuous human development, with obvious negative health and behavioural aspects for exposed the population. Particularly in Europe, one-third of the reported problems with noise exposure (ranging between 14 and 51% in the particular EU States) are observed in urban conglomerates mostly connected with population annoyance as a reaction to noise [1]. Among other sources, aircraft noise can also be a substantial source of annoyance. The International Civil Aviation Organisation (ICAO), in line with its own key role, is aware of and continues to

---

Small introduction explaining the legal framework of ICAO chapters and the difference between standard for certification and research objectives

---

O. Zaporozhets (✉)
National Aviation University, Kyiv, Ukraine
e-mail: zap@nau.edu.ua

© The Author(s) 2022
L. Leylekian et al. (eds.), *Aviation Noise Impact Management*,
https://doi.org/10.1007/978-3-030-91194-2_3

address the adverse environmental impacts (aircraft noise is one of them) and strives to limit or reduce the number of people affected by significant aircraft noise [2]. The EU regulations require from airports to inform the authorities and the population on extent of aircraft noise exposure and the number of people affected by this exposure.

Aircraft noise is still the most significant adverse stressor for community-related to the number of environmental factors in the operation and expansion of airports. It is expected to remain the case worldwide for the foreseeable and even remote future. Any new technological changes in the aviation sector may contribute to further increased noise annoyance, especially the introduction of faster possible new generation of supersonic aircraft for business and scheduled air transportation. Even the introduction of electric aircraft, which are expected to be considerably less noisy in comparison with today's aircraft with similar flight performances, will continue to be scrutinised due to their very close operation to residential areas. Thus, the subsequent reduction of annoyance for neighbouring communities will not necessarily be proportional to their actual noise reduction at source. Thus, limiting or reducing the number of people significantly exposed and impacted to aircraft noise is one of the main priorities of ICAO and one of its key environmental objectives. Division on exposed and impacted population is quite important because not every exposed is obligatorily assessed as impacted by noise. There is only around 20% of the population, which is exposed to significant noise levels over 65 *LDN*, usually reacts adversely to it.

Aircraft noise exposure can lead to more than one effect, and the community impacts (usually health effects, which can be chronic) depend on multiple effects [3]: the primary recognised health consequences of community noise exposure are sleep disturbance during night-time and annoyance during composite daytime. The cardiovascular disease and cognitive impairment of children are also major consequences [4]. WHO Reports [1, 5] studied the link between environmental noise and diseases (such as cardiovascular disease, sleep disturbance and tinnitus, etc.), and they probably provide the most pertinent evidence of noise impact on human health. Efforts to reduce noise exposure should concentrate on diminishing the annoyance and sleep disturbance, improving the learning conditions for children and lowering the prevalence of cardiovascular diseases and other risk factors [3]. However, the efforts recommended by these guides usually have different coping abilities for all these types of health consequences. All these notions will be extensively addressed by other sections of this book.

In general, the severity of any hazard impacts depends on the level of exposure inside the affected area. If there is no exposure, the impact is absent too. But evidence exists to show that risk has increased worldwide not only due to increases in hazard exposure of population and/or its assets but the vulnerability of the population to hazard is also fundamental to our understanding of general risk for a population to be damaged by the hazard [6] and of the risk to be annoyed by noise in particular.

Human response to noise is varying differently to different environmental noise sources with the same acoustic levels. Therefore, human sensitivity to noise levels depends on the source of noise. This feature is accounted for by the original Schultz interpolation curve, used in Standard ISO 1996-1 (with small modifications from the initial one) [7] to define the proportion of people highly annoyed by noise independence with day-night noise index L_{DN} (*DNL*). The standard recommends using such dose–response curves for assessing the annoyance. Among the three basic modes of transportation inside urban environments, aircraft noise is the most annoying for the same given L_{DN} value. For example, at 65 dBA *DNL*, the proportion of the population who are highly annoyed by aircraft noise is 10–20% higher than for road traffic and railway noise.

For aircraft noise, any flight event (or any aircraft engine run-up at an airport) is leading to scenarios of noise exposure and impact. The same is valid for aircraft engine emission and air pollution, but the probability of hazard exposure and especially following impact due to any scenario is dependent on the specific location of the point of control relative to the flight path. People are impelled to complain when some burden factor in the environment gives rise to any effect and when this stressor reaches a lower limit value (for example, for noise index $L_{DN}/L_{DEN} = 45$ dBA or for $L_{night} = 40$ dBA [1, 4, 5]). Exposure, in general, is defined as "the people, property, systems, or other elements present in hazard zones that are thereby subject to potential losses" [8].

Today ICAO [2] and ACARE [9] targets and goals are not confined to reducing noise levels: intended noise level reduction (it is level of stressor only) receiver point s not the final result, but it is just an instrument to achieve the real final goal, which is the noise effects mitigation (just an evaluative tool to achieve the real end goal of reducing noise effects). For ICAO, this effect is currently defined as 'a reduction of the number of people affected by aircraft noise', measured through the number of exposed people by noise over given values defined by guidance or through the number of highly annoyed people. This rationale led currently to new approaches and concepts, even realised measures, to reduce the human annoyance (sleep disturbance and other effects of noise impact), for given levels of aircraft noise [10].

Compared with the traditional noise management approach defined by physical effects of sound generation and propagation, annoyance relies therefore on psychological elements. Up to recently, attempts to explain annoyance relied only on non-acoustical factors like sound intensity, peak levels, duration of time in-between sound events, number of events, etc. [11]. The non-acoustical factors ("moderators" and/or "modifiers" of the effect) have still received empirical attention, without a deep theoretical approach, despite the fact that various comparative studies reveal that they play a major role in defining the impact on people [12].

Therefore, protecting residents from aircraft noise appears now to be a dynamic process: The evaluation limits must be repeatedly tested in view of new scientific findings and adapted, if necessary. Besides traditional approaches of the ICAO balanced approach (BA) [13], which includes reducing aircraft noise at source, implementing operational procedures and restrictions (mostly in airports), new or changed existing flight routes and other forms of mitigation, it is needed to embark on a dialogue with neighbouring communities [14, 15]. The main objective is to address the issue in an environmentally and economically comprehensive way, to preserve potential benefits of air transport for all categories of stakeholders.

The Global Footprint of Aviation Noise

An individual's perception of noise makes it clear that sound levels become an unwanted experience when they occur at the wrong place or at the wrong time. The aircraft noise exposure in and around an airport depends upon a number of factors including the types of aircraft operating the airport, the overall number of daily ground and flight operations, the time of day that they occur, the runways and flight routes that are used for departures and arrivals, airport-specific flight procedures, weather conditions, topography, and other operating conditions. The main effect of noise on population, usually caused by aircraft operations in airport, is dependent of noise exposure, but somewhat subjective at the same time. It may depend on a number of non-technical factors related to the cultural, socio-economic, psychological, and physical issues of the exposed by noise individuals, and may vary from no effect to severe annoyance. Thus, the noise impact is more complex and harder to define the exposure. It is therefore a significant challenge to fully understand, predict and characterise the noise exposure and impact of aviation.

The number of people exposed (which covers a number of impacted and other people not reacting on noise) to aircraft noise is the metric normally used to estimate aircraft noise influence few decades ago. For that a MAGENTA model was developed by ICAO's Committee on Aviation Environmental Protection (CAEP), it was used for assessing global (and regional for comparison) exposure to the noise of transport aircraft. Estimates from the ICAO MAGENTA program have shown an improvement in the global noise situation with a reduction in the size of the population within the 65 dB *DNL* contours of about 30% in 2006, relative to the 2000 level. More recently, ICAO developed a range of scenarios for the assessment of future aircraft noise trends [16]. The noise indicators used in these scenarios are the total contour area and population inside (simply talking—exposed population to specific for contour noise level) the yearly average day-night level *DNL* at 55, 60 and 65 dBA contours. Such indicators were assessed for over three hundred airports worldwide, covering over 80% of the global air traffic. Aircraft noise exposure was modelled for four scenarios (baseline and low, moderate, and advanced technology [16]):

- Scenario 1 (baseline) assumes the growth of air traffic without any further aircraft technology or operational improvements after 2015.
- Scenarios 2, 3, and 4 assume the same air traffic rise that all new aircraft delivered by manufacturers into the market after 2015 will reduce their noise levels on 0.1, 0.2, and 0.3 EPNdB per annum (so called low, moderate and advanced technology improvements), respectively.

Figure 1 shows the total 55 dBA *DNL* noise contour area from 2010 to 2050. In 2020, this area was 18,400 km^2, and covering the population inside that contour approximately 31 mln people [16]. By 2050, the global 55 *DNL* contour area is expected to grow up to 2 times, compared with 2020, depending on the technology scenario. In 2050, the total population exposed increases to 34.2 million people due to scenario 4 (quite small—only 10% of growth—comparing with the baseline value

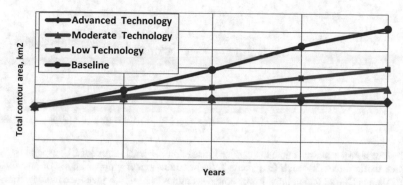

Fig. 1 Total aircraft noise exposure assessment for over three hundreds major airports worldwide, 2010 to 2050: **a** contour area above 55 dBA DNL (km²); **b** global population exposed to noise above 55 dBA DNL [16]

in 2020), and to 66.9 million people with scenario 2 (near to 100% growth from 2020).

EASA provided some similar analysis at the European level (EU28 + European Free Trade Association—EFTA) [17]. Current growth in air transportation inside European region is certainly supported by strong demand and represents an average annual rate of 1.5% over this period. Comparing with ICAO CAEP global trend, the number of yearly flights from and to this EU28 + EFTA area attaining 13.6 million in 2040 with ~2% growth of air transportation annually, compared to value in 2017 (Fig. 2a) under the most-likely future scenario. It has brought traffic back to the most likely scenario— "Regulation and Growth"—from the 2013 forecast [17] (Fig. 2b), which is close to moderate scenario in ICAO forecasting (Fig. 1).

The total exposed population to noise levels $L_{den} = 55$ dBA and over in vicinity of the 47 major European airports was assessed 2.58 mln people in 2017 (Fig. 3). Due to European regulations the noise exposure at night is also important—for the same scenario the 50 dBA L_{night} contours may disturb 0.98 mln of residents at night. The

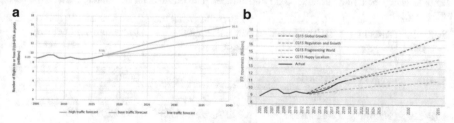

Fig. 2 Number of flights increases by 42% between 2017 and 2040 under the base traffic forecast: **a** forecasting from [European Aviation Environmental Report 2019. European Aviation Safety Agency (EASA), European Environment Agency (EEA), EUROCONTROL, 2019. ISBN: 978-92-9210-214-2 https://doi.org/10.2822/309946. Catalogue No.: TO-01-18-673-EN-N [www.easa.europa.eu/eaer]]; **b** forecasting from [European Aviation in 2040. CHALLENGES OF GROWTH. EUROCONTROL - 2018 (Edition 2)]

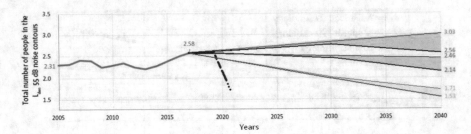

Fig. 3 Latest fleet renewal is able to stabilise noise exposure levels at major EU airports by 2030: for three traffic forecasts—high (red), base (blue) and low (green); upper bound of any forecast reflects the unchanged technology, lower bound—employment of advanced technology for noise control (forecasting from [17])

increase of exposure is equal 12% and 13% for L_{den} and L_{night} respectively comparing to 2005. nd at the same time the average noise single flight event exposure went down by 14% over this period [17].

If the latest manufactured aircraft types now entering the European fleet deliver their expected noise reductions (look the Table 1 in text below), the total exposed by noise population (to $L_{den} = 55$ dBA and $L_{night} = 50$ dBA) around these 47 major airports should be stabilised for the scenario of Global Growth (mostly optimistic) and possible to decrease by 2030 (in comparison with continuous increase globally). These forecasts assume that airport expansion will be absent, and the population rise will be absent around these airports. However, around 110 airports in Europe could handle more than 50,000 annual aircraft movements [19] by 2040, although there was 82 airports in 2017, thereby new affected by noise population should be expected. Taking in mind the current trend—close to the "Regulation and Growth" scenario—the total number of people exposed to aircraft noise should stabilise in Europe much earlier, likely at the beginning of the 2020s. Of course, such forecasts did not take into account the dramatic effect of the COVID-19 pandemic.

Table 1 New generation aircraft recently introduced or about to enter the market (before the end of 2020) [24]

Seat category	Aircraft category	2010 reference	New generation (examples)	Entry into service	Fuel saving reference (%)
51–100	Regional jet	ATR/CRJ	MRJ	2020	20
		E-Jet	E-Jet E2	2020	24
101–210	Narrow body	A320/B737	A220/A320neo/B737 MAX	2016/2017	20
211–300	Wide body	B767	A350/B787	2015/2011	20–25
301–400		A330/B777	A330neo/B777X	2018/2019	14–20
401–500		A380/B747-8	A330neo/B777X	2018/2019	14–20

The ICAO Noise Certification Procedure for Aircraft

Standards and recommended practices (SARPs) cover ICAO noise requirements to allow civil aircraft to operate safely for population and environment as a whole. They are contained in Annex 16 [20] to the Convention on International Civil Aviation (in Volume I *Environmental Protection—Aircraft Noise*). It should be noted that the first generation of aeroplanes with jet engines was not covered by SARPs of Annex 16 and were hencereferred to as non-noise certificated airplanes (among them the aeroplanes Boeing 707 and Douglas DC-8). The noise certification scheme of Annex 16 [20] considers the overall noise produced by the operation of an aircraft, the engine plus the airframe. Through subsequent chapters to Annex 16, these SARPs have been subsequently updated, becoming stricter, since then to reflect and to motivate improvements in aircraft and engine technology (see Subchapter on *Reduction of Aircraft Noise At Source* below or Chap. 5 and 6 of the book).

Chapter 2 of Annex 16 [20] was the first Standard for aircraft noise and set the limits as a function of maximum take-off mass recognizing the fact that heavier aircraft would be essentially noisier than lighter ones. Those limits in Effective Perceived Noise Level (*EPN*) dB were set for three measurement points (Fig. 4): at the side of the runway on take-off, under the flight path on climb after take-off, and under the flight path on the approach to landing (this set is still used for limiting noise levels in subsequent chapters of the Annex 16).

Nowadays, newly manufactured aeroplanes must comply with Chap. 14 requirements in ICAO Annex 16 Volume I [20]. In difference with Chaps. 2 and 3, currently the ICAO aircraft noise limits are defined as cumulative values equal to the arithmetic sum of the certification levels at each of three points, which are still the same. Annex 16, Volume I [20] also contains provisions (noise limits and procedures to assess them) for the certification of propeller driven aeroplanes and helicopters.

Each successive chapter of the Annex 16 Volume I [20] has set and enacted higher stringencyin certification conditions, aircraft have become quieter and areas affected by its noise significantly reduced. However, the air traffic is constantly increasing on growing demand for passenger and cargo transportation. ICAO and

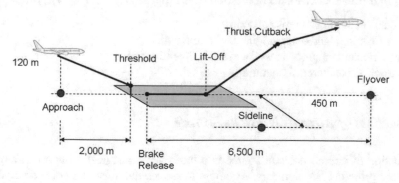

Fig. 4 Aircraft noise certification reference points [16]

IATA are both assessing and further forecasting during closest decades around 5% of air traffic increase worldwide due to this constant increase in the community demand for flights (Fig. 2 confirms this trend for European aviation sector, which is going along the most wishful scenario Regulation and Growth). The overall aircraft noise exposure is increasing due to this existing air transportation trend, but overall population exposure by noise may increases also due to constant population desire to provide the great number of activities, including residential, closer to airports. So, additional ways to control the noise exposure and accordingly the impact are required in the vicinity of the airports to provide the sensitive to noise human activities as far as possible. This is organised through noise zoning and appropriate land usage in vicinity of the airports. Further, to keep the size of the zones constant, or even to reduce them, is an important task of noise management in any airport—it is of the same strategic importance as an aircraft noise control in source. Reduction of noise zones benefits those living and working close to the airport.

In 2001 the 33rd Session of the ICAO Assembly adopted a new policy for aircraft noise control globally, referred to as the "balanced approach" to noise management. This new SARPs provided the Member-States with an agreed approach for addressing aircraft noise management components in an effective and economically-responsive way. Because of existent specifics at any region and even at any airport it is eventually the responsibility of any State to implement the various noise protection measures by developing appropriate combinations between them to provide balanced solutions in their programs for managing the noise exposure at airports. The highest effectiveness and economical efficiency are reached with due regard to ICAO provisions and policies while recognizing that States have responsible lawful obligations, including various agreements, laws, and policies on noise management around the airports. The Balanced Approach guidance was developed by ICAO and contained the explanation of all elements in general details [13].

The goal also is to identify the efficient noise abatement measures that achieve the maximum environmental benefit, using objective and measurable criteria, at any specific airport most cost-effectively. At the first step, the noise problem at an airport should be identified and then a comprehensive noise protection program must be developed analyzing the mostly available measures to reduce the noise exposure using four principal elements of the Balanced Approach (BA), namely [13]:

- reduction of aircraft noise at source;
- noise zoning, land-use planning and management;
- noise abatement procedures for aircraft operation; and
- restrictions for aircraft operation.

Reduction of Aircraft Noise at Source

Reduction of aircraft noise at source is a basic and strategically important among the four principal BA elements. Attempts to reduce the noise the noise emitted by the aeroplane are understandable and fundamental in setting new, more stringent standards for noise radiation, so ICAO pays special attention to this element.

Among the principal factors defining the sound level from aircraft flyby over the point on a ground surface are the noise radiated by acoustic sources of the aircraft and the sound propagations factors like the local topography and atmosphere state and both are dependent on weather conditions. Aircraft is a complex acoustic source consisting of engine and airframe components (Fig. 5), and the principal aircraft elements for dominance of any specific acoustic source at a specific flight stage are the following: type of aircraft and engine type, the engine installation (over or under the wing, at the tail, etc.), the aerodynamic configuration of the aircraft, flap and airspeed management procedures being used during the flyby, distances from the aircraft to point of noise control. The airframe and engine contribution to overall aircraft noise are shown in graphs of the Fig. 6, during the principle flight stages of aircraft take-off and landing. The contribution of any acoustic source is dependent of flight mode and it is an important factor for technology improvements in aircraft noise reduction.

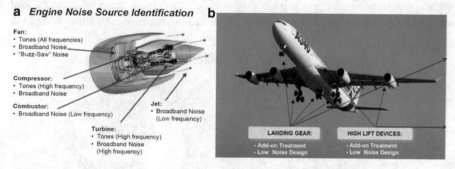

Fig. 5 Engine and airframe noise source identification: **a** by engine component and sub-component; **b** airframe component contributions to total aircraft noise for a modern turbofan powered aircraft. [xx]

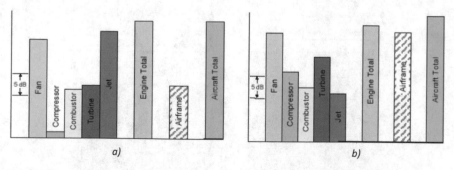

Fig. 6 Sources of acoustic emission during: **a** takeoff; **b** landing [Leylekian, L., Lebrun, M., Lempereur, P. 2014. An Overview of Aircraft Noise Reduction Technologies. AerospaceLab Journal, Issue 7, pages AL07-01 (https://doi.org/10.12762/2014.AL07-01)]

Noise from a single aircraft is primarily produced by the engine, but the total engine noise is the sum of broadband and discrete noise radiated by fan, compressor, combustor, turbine and jet as shown in Fig. 5a). Noise is also created by the airframe as it moves through the air, which is complex also like the engine. It includes the contributions of aerodynamic noise from the high lift devices and the landing gear in air flows (Fig. 5b)).

More detailed information on aircraft noise redaction at source is described in this book in Chap. 5 for existing aircraft and Chap. 6 for expected their future generations. The noise from the first jet aircraft was completely defined by noise radiated by the exhaust jets—it was a loud roar or rumble, very difficult to be reduced by technical means. For the first turbofans (with small bypass ratio $m = 1$) this dominance of exhausted jet noise became less because the high speed of primary jet was sufficiently reduced and a new acoustic source—a fan—is appeared (Fig. 7a). The achievements of the first technological solutions to reduce engine noise were the basis of the first international standards—Chap. 2 of Annex 16 "Aviation Noise" to the ICAO Convention.

The next generation high-bypass-ratio turbofans (with bypass ratio $m = 5$–6), which characterised by much higher fuel efficiency, again reduced primary and secondary jet velocities and generated noise in consequence. For these turbofans

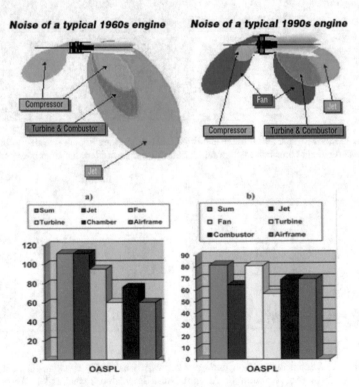

Fig. 7 Engine noise source contributions high bypass benefits and changes in source ranking

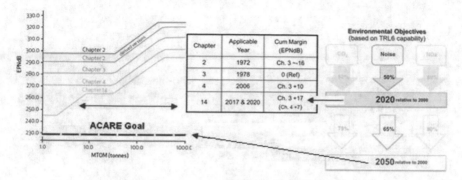

Fig. 8 Comparison between the ICAO requirements and ACARE 2050 goal to aircraft noise

the exhaust jet is still a significant noise source, but not absolutely dominant to total aircraft noise as at take-off so as at landing stages of flight, fan noise became the same significance contributor to overall aircraft noise, especially for flight in approach and landing conditions (Fig. 7b). High bypass technology reduced the engine noise on a value which was fixed in the second ICAO aircraft noise standards—Chap. 3 of Annex 16, Volume 1 "Aviation Noise".

Decreasing the bypass ratio of the turbofans to high ($m = 6$–8) and very high ($m = 10$ and beyond) provides new noise reduction benefits (a specific fuel consumption continues to be reduced also) for the aircraft. They were complementarily accomplished with other measures of noise reduction at the engine: reducing the fan noise, optimising the engine cycle, using the contra-rotating rotors, using the sound-absorbing structures and reducing the level of turbulence in the inlet section. These benefits stimulated new changes to aircraft noise standards—Chap. 4 became applicable in November 2008. Chapter 4 includes also provisions related to noise measurement conditions during noise certification testing and began to combine the levels measured at separate three points into a single summary value, used also in the following Chap. 14 requirements to aircraft noise, Fig. 8.

Due to the introduction of quieter aircraft (in accordance with putting more stringent requirements of the ICAO Annex 16 standards at the relevant Chaps. 2, 3, 4 and 14, Figs. 7 and 8) and the global phasing out of Chap. 2 aircraft between 1998 and 2002 as agreed upon in the 28th Assembly of ICAO noise levels from the separate flyovers have been decreased at most airports worldwide. As a result, a modelled overall 65 dBA L_{DN} noise contour at airport has shrunk in size consistently since 1970-ies (Fig. 9) regardless a rise in total air traffic because of the replacement of noisy ICAO Chaps. 1 and 2 aircraft with much quieter Chaps. 3 and 4 aircraft, further implementation of Chaps. 14 aircraft in operation will decrease them much more.

Aircraft produced today are 75% quieter than the first civilian jets appeared in operation 50 years ago (Figs. 8, 9 and Table 1). The newly manufactured aircraft typically produce around a half the noise of the aircraft they are replacing, so with this advance the air traffic movements can double without increasing the total noise exposure output. In more detail, The British "Sustainable Aviation Noise Road-Map"

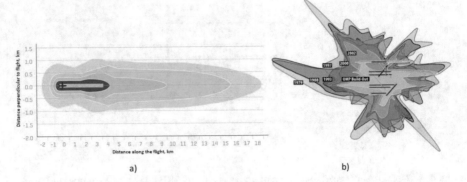

Fig. 9 Aircraft noise exposure reduction by ICAO influence through Annex 16 noise certification norms: **a** single aircraft departure footprint: Chap. 2 aircraft (represented by B737-200)—yellow contour; Chap. 3 aircraft (MD80 or B737-200 Hush Kit)—blue contour; Chap. 4 aircraft (A320 or B737-800)—green contour; current Chap. 14 aircraft (A320neo)—red contour; **b** overall 65 dBA L_{DN} noise contour changes at airport O'Hare (Chicago, USA) due to implementation of quieter (with more stringent noise standard requirements to them) aircraft regardless flight traffic growth: till 1979—the Chap. 2 fleet; till 2000—the Chap. 3 fleet evolution; at 2002—a fleet with Chap. 2 phase-out; after 2002—a Chap. 4 fleet evolution

[23] predicts that as current aircraft are replaced by *'Imminent'* and *'Future'* aircraft, the noise exposure from UK aviation reduces by around 20%, which is close to EU forecasting (Fig. 3).

The noise reduction of engines is provided by further increasing the by-pass ratio of the engine, by the use of a low-speed advanced fan (reduction of the dipole acoustic noise source). Wide-chord fan blades have a twist in height and are made of composite materials. Reduction of the acoustic interaction between the fan impeller and the outlet guide vanes is provided by inclined and specially profiled blades, the number of which is selected to provide the "cut-off" effect. The development of optimised Outlet Guide Vane (OGV) made it possible to significantly reduce discrete and broadband noise.

To achieve ACARE noise reduction goals, engine manufacturers are developing engines under the Advance (bypass ratio $m = 11$) and UltraFan ($m = 15$) programs, which are planned to be completed in 2020 and 2025, respectively. These engines feature for instance: a three-shaft modular design of a high-bypass engine; improved aerodynamics of the impeller machines and a high-pressure ratio up to 60 for Advance and 70 for UltraFan; multiparametric optimisation of aerodynamic and acoustic characteristics, use of the "Intelligent Engine" concept; application of 3D printing technology; an inlet section of a special design contributes to the reduction of the turbulence level at the engine inlet, which in turn reduces the level of vortex noise at the fan inlet; sound-absorbing materials to reduce the tonal noise of a low-pressure turbine.

Effective noise reduction technology—acoustic liners in the nacelle and inside engine ducts—are important in lowering the noise from engine internal sources as it propagates along and out of the intake, bypass duct or core duct (Fig. 10).

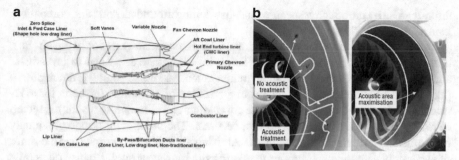

Fig. 10 Engine noise reduction technologies [27]: **a** acoustic liners inside the high bypass turbofan engine; **b** maximising inlet acoustic liner area

To reduce the noise level, it uses a noise active control system, which includes acoustic resonators installed on blades of the fan outlet straightener and devices based on MEMS technology and installed along the inner passage of the engine nacelle. Besides, special porous materials are under development, the use of which can reduce the acoustic emission from the source. The application of such a system makes it possible to reduce the noise level in a source by ~50–60%, or overall aircraft noise reduction achieved by fan forward and fan rearward noise attenuation through acoustic liners was in the range of 10–12 *EPN*dB (cumulative) on recent aircraft types [21].

The proposed measures allow to reduce engine weight and increase fuel efficiency by 20% for Advance and 25% for UltraFan, as well as to reduce noise level and harmful emissions level [24]. The price to pay for very-high-bypass-ratio engines is increased size, weight, and drag, which result in more mission fuel burn. Higher bypass ratio engines require much higher pressure-ratio and temperature gas generator cores, which may have a negative effect on emissions, especially nitrogen oxides (NOx).

During the 50 years of aircraft noise standardisation from ICAO (1st Edition of Annex 16—Aircraft Noise was published in 1969) and continuous strengthening of the requirements from Chap. 2 till current 14 the cumulative reduction was gained up to ~35 dB (Fig. 5), close to this value is necessary to be reached till the ACARE noise goal at 2050. In 2014, ICAO adopted a new (latest) standard that resulted in a 7 EPNdB decrease compared to the Chap. 4 Standard, so in-production aircraft are prohibited to be manufactured currently with noise higher than Chap. 14 requirements.

Land-Use Planning and Noise Impact Management Issues

Land-use planning and management is a necessary means to ensure that the human activities nearby airports are consistent with aviation activities [30]. Its main goal is to minimise the population, usually the residents in vicinity of the airport, affected by aircraft noise by introducing specific land-use zoning around airports and compatible land usage inside them.

Noise reduction at source alone is not sufficient concerning not only the aircraft, but other transportation sources also. To all of them there is a need to investigate additional measures for noise problem management and noise zoning is among them [1x]. Residential developments near transportation infrastructure (airports for aviation sector) are the places generating the complaints and community reaction directed to reduce noise due to transportation operations. The conflicting trends previously mentioned in designing quieter aircraft and rising air traffic only emphasize the sense of accurately assessing and further managing the impact of aircraft noise on people at their residence site. On the other side of the problem, his may put airport capacity constraints and significantly increase the costs of providing transportation services.

If the main goal in aircraft noise control to reduce noise level at source of its generation, the main goal for noise zoning and land use management to prevent the people from the levels, which are inconsistent with their health and welfare. These levels are defined as sanitary norms (limits) for the population doing any kind of activity—residential, educational, rehabilitation, etc. For all the existing human activities the noise limits are defined because for all of them the noise is a disturbing factor.

The first task of noise zoning is to define the area around the noise source—an airport in our case—where any of the human activities (land usage) are not possible be permitted. The second task—to define the area without any prohibition for human activities. The territory between them is an area of noise management with any possible measures for noise protection.

Any airport development should be accompanied by a program of aircraft noise management in its vicinity. Otherwise, any noise-sensitive human activities (land usage), introduced without attention to noise from the airport, will eliminate the aircraft operation by restrictions and reduce the airport capacity. In the same way, that aircraft noise must be controlled to prevent existing developments from being further exposed.

Airports are usually located within or close to the limits of large urban areas, in better case a distance to existing noise-sensitive land usage (residential or recreational) may provide human protection from noise exposure and minimise the adverse impacts of their operations. Inside the zone of noise management, it is necessary to organise a set of plans (a program for noise protection) that govern urban planning and management with respect to the airport activities. In reality each airport is different in its operational, social, economic and political situation, as well as in the type of land use in its vicinity. That is why, the airport noise protection program should include a land-use control system to assure that all the prescribed measures comply not only with the airport development plan but also with the plan of urban development and the goals of the communities involved.

ICAO is continuously developing a policy on land use planning and management. Current version is contained in Assembly Resolution A40-17, Appendix F [25], it urges States, complementarily with the development and introduction of quieter aircraft, to minimise aircraft noise impact problems through preventive measures of noise zoning and land use management:

(a) locate new airports (or new infrastructure in existing airports during reconstruction) away from noise-sensitive residential areas;

(b) take into account the appropriate protection measures for existing and planned land use at the earliest stage of any new airport development or of an existing airport;

(c) define protection zones around airports associated with different noise limits taking into account population size and growth as well as forecasts of air traffic growth and establish criteria for the appropriate land use (as recommended by ICAO guidances [13, 30]);

(d) enact legislation with supporting guidance and other means to show and recommend how to achieve compliance with those criteria for land use; and

(e) ensure that information on aircraft operations and their environmental effects are available and understandable to communities near airports.

Airport Planning Manual for Appropriate Land-Use Inside Noise Zone

ICAO's *Airport Planning Manual* (APM) in Part 2 [30] covers three key issues for compatible coexistence of the airport and urban environment: land-use, land-use planning, and land-use management. For that, the APM describes a variety of possible land-uses with an appropriate estimation of their relative sensitivity to aircraft noise exposure. Also, the guidelines indicate the compatibility or incompatibility of these land uses to noise exposure and also to airport operations themselves.

Land-Use

Human activities inside the area of land management (other words—land uses) such as natural, agricultural and recreational are usually considered as the most compatible with noise since they are realised outdoors and normally don't involve constant human use.

Commercial and industrial land uses are also considered compatible with aircraft noise because they are normally carried out during the daytime and they do not touch the problem of noise at nighttime—sleep disturbance does not occur, as usually happens in residential areas.

On the other hand, the development of residential and institutional land uses, which include single and multi-family dwellings and community support facilities such as schools, hospitals and churches should not be encouraged in airport surrounding areas, since they are extremely incompatible with noise.

The development of residential and institutional land uses, including all necessary urban components such as schools, hospitals, and churches are extremely incompatible with noise, they should not be encouraged in the airport vicinity, where the noise levels exceed the limits.

Fig. 11 Example of noise contours Kyiv/Zhulyany international airport—Kyiv, Ukraine

Land-Use Planning

The problem of airport noise inside the surrounding area can be solved successfully by implementing all possible measures and means, and the proper land-use planning can contribute significantly to the final solution Land-use planning to existing airports is especially important because their possibilities for prompt land-use transformations are limited usually. A trend for continuous reduction in aircraft noise in the source can cause the noise contours to approach closer to the airport boundary. In such a case, the withdrawn area from noise exposure should be prevented from immediate additional encroachment of incompatible land uses. Especially substantial benefits can be acquired from the appropriate land-use planning for new airports under the development without or with quite small constraints for human activities in the surrounding area.

Figure 11 depicts typical noise contours for an airport. In red a noise zone with complete prohibition for any type of human activities is shown, in yellow—a noise zone with specific noise control measures to be implemented. The overlap of urban areas within the NPZ exists and it indicates that a population inside the zones is affected by noise and needs for protection. Noise control measures can be included in program for noise management in airport, adopted by airport and urban authority to ensure that future developments in airport and inside residential areas will be compatible with aircraft noise.

Land-Use Management

Among the alternatives to regulate land developments inside the surrounding area affected by the airport a number of the modification or restriction of land-uses exists to achieve greater consistency between aviation and human activities or other words—compatibility between the airport and its environs. These control measures may be divided into three categories, as follows: Planning Instruments, Mitigating Instruments, and Financial Instruments. There are only some examples of these instruments listed below.

Planning Instruments

Planning instruments	Land-use planning and control authority is covered by local governmental bodies, as a rule, which must take into account aviation noise exposure and measures for its reduction. They must account for existing development, first of all residential and rehabilitation types, and ensures that future planned development is also consistent between aviation sector and society, and compatible with various community goals including the changes in aviation noise exposure
Noise zoning	Noise zoning regulations should specify land development depending on the level of noise exposure and use restrictions, based on certain noise levels, incompatible with human activities. The regulations should protect both—the airport and the residents in their mutual developments
Easement acquisition	Easements should restrict the use of land to that which is compatible with aircraft noise levels. They should not violate the right of flights over the property and to create noise

Mitigating Instruments

Building codes	Building codes are the legal instruments of requiring the inclusion of proper sound insulation in new and existing construction inside areas exposed by noise
Noise insulation programmes	Sound insulation is used to reduce interior noise levels especially for buildings that are not possible to remove out the areas exposed by noise. Effective noise insulation may be reached with closed-window conditions, which may impose additional costs to homeowners or an airport noise protection fund
Land acquisition and relocation	The acquisition of land through purchase by the airport operator (or another responsible subject in case of new developments under their authority) and the relocation from the acquired area of all incompatible with noise levels residences, administrative and business structures

Financial Instruments

Capital improvements planning	Capital improvements can be planned in order to locate or support existing infrastructure networks in areas where industrial and commercial growth would be compatible with airport noise
Tax incentives	Tax incentives can be provided to occupants of existing incompatible use facilities in order to encourage structural improvements to reduce interior noise levels or to encourage their relocation to quieter areas or otherwise the expansion of industry as a means to diversify the local economy
Noise-related airport charges	Noise reduction and prevention around the airports require financial investments which may be recovered through airport noise charges

Aircraft Operational Measures for Noise Reduction

The noise of aircraft is subject to ICAO certification SARPs that are intended to apply to the manufactures of aircraft worldwide. This does not prevent local authorities from applying stricter noise limits at specific airports because the land usage at any location may require for that. Due to their unique nature, the certification procedures will not capture the detail and variations that do occur during real operations and, as a result, actual noise levels will not necessarily be the same as those measured during certification.

Local airport rules can include noise limits, curfews and penalties on excessive noise levels. These measures are considered mostly as constraints, they may limit the operational capacity of airports (for example, by restrictions for flights during night) and they may affect the economics of air transportation by limiting the take-off weight, payload and consequently reducing the economic benefit of specific flight. When analysing operational measures to arrive at an optimum result, it is important to involve all the stakeholders to ensure that interdependencies between the various aspects are fully identified and that any unintended consequences are avoided or minimised to the extent possible. This subchapter discusses the use of aircraft operational measures as a noise reduction method, one of the elements of ICAO's Balanced Approach. It presents a discussion of aircraft procedures for both departures and arrivals/approaches, and their potential effect on noise levels.

Departure Procedures

Noise abatement operational procedures are being applied in airport operation management to provide locally effective noise reduction to communities active in airport surroundings, from both arriving and departing aircraft. ICAO PANS-OPS, Volume 1, contains a general guidance for the development of a maximum of two noise abatement departure procedures (NADP's) designed to eliminate noise over the normative limits either close in (NADP 1) to the airport, or further out (NADP 2) along the departure path (Fig. 12a). By ICAO guidance both NADPs terminate at 1000 m altitude, however there may be noise reduction benefits above this, but also the deterioration is possible in some areas, and these should be taken into account as well. Review [31] contains a list of current NADP's with operational and noise mitigation analysis in use by air carriers for a wide range of aircraft types.

At the beginning, the ICAO PANS OPS guidance [32] contained only two recommended noise abatement departures procedures (NADP) for aircraft—ICAO A and ICAO B. This guidance prescribed the choice of either of these two NADP without regard to aircraft acoustic (or noise attenuation technology realised in aircraft design) and flight performance.

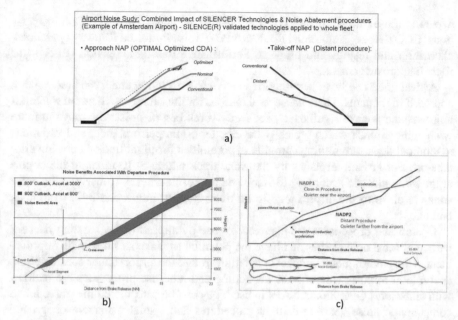

Fig. 12 Departure and approach noise abatement procedures: **a** SILENCE® validated technologies for NAP at Schiphol airport; **b** noise reduction efficiency beneath the flight path; **c** noise contour reduction efficiency

After first revision, the ICAO PANS OPS guidance changed the prescriptive ICAO A and ICAO B flight procedures including the criteria for development of NADP. These criteria are the following:

- Engine thrust reductions till minimum safe aircraft climbing value is prohibited below 800 ft above the runway (Fig. 12b);
- The thrust reduction in aircraft operation cannot be below the thrust level required by the certificated aircraft flight manual if not approved in addition by manufacturers' operations manual;
- The current guidance on NADP and the similar on noise abatement arrival procedures (Fig. 12a) is contained in PANS OPS, Doc 8168 [33], Part I, Sect. 7.
- Designing noise procedures must account for elementary aviation criteria such as regulatory, safety and operability objectives; and it is these criteria that will limit what is ultimately achievable.

NADP 1 and 2 provide air carriers standardized departure profiles for worldwide flight operations, enhancing safety, as well as providing airports and air navigators with a predictable departure metering of aircraft into the en-route airspace structure. These procedures should be the same for all aerodromes. The procedural differences between the NADP 1 and 2 are the thrust or power reduction altitudes and the high-lift device retraction—acceleration segments. Today, the operational opportunities for departure noise mitigation also (besides NADP) include: Continuous Climb Operations (CCO), Noise Preferential Routes (NPR), Noise Preferred Runway operations, alternation and respite, and usage of Performance Based Navigation (PBN)—for flight track management.

Arrival noise can be of key importance to residents living under the predominant approach flight paths. There are several reasons for this change in focus: new aircraft noise standards have significantly reduced overall engine noise; the move to more two-engine aircraft which typically have better climb performance; and variations in terminal departure routings provide opportunities for flight track dispersion after take-off. For arrival, airplanes fly the same track within 5–10 miles of the airport at the same altitude according to stabilised approach criteria. Consequently, noise contours close to the airport tend to be more heavily influenced by arrivals than before.

Operational procedures are intended for use by aircraft of the existing fleet for particular local noise issues and have the potential to make an immediate improvement in the environmental impact of aviation around the airport under consideration, as a rule at airports where the noise zoning and land use procedures are realized with omissions. Operational NAPs in use today can be categorised into three broad components: noise abatement flight procedures and spatial management; ground movement management. Figure 13 shows schematically the operational opportunities of the NAPs [23]. It also gives an indication of the areas (distances to runway at departure and arrival) benefiting from some of any procedures outlined.

Given the above, the following guiding principles should be adopted when considering operational opportunities to reduce noise:

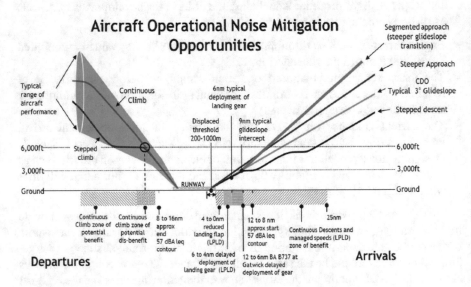

Fig. 13 Aircraft operational noise mitigation opportunities—illustrative, not to scale. [23. Sustainable Aviation Noise Road-Map, 2018 [www.sustainableaviation.co.uk]]

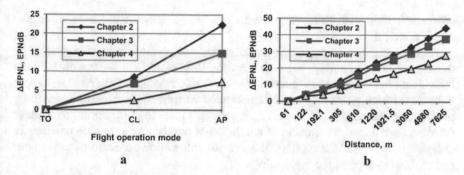

Fig. 14 Aircraft noise standard stringency influence on NAP ability to reduce noise level: **a** at point of noise control; **b** along with flight path distance [10]

(a) Safety must not be negatively affected;

(b) Operational procedures should be developed in accordance with relevant ICAO provisions or regulatory guidance, while allowing for implementation of new procedures as that guidance evolves;

(c) Changes to operational procedures must consider aircraft and operator capabilities and limitations with appropriate approval by the regulator;

(d) Appropriate assessment tools and metrics to support decision making and post-implementation review of conformance should be maintained;

(e) Interdependencies should be considered between other environmental and non-environmental impacts and disproportionate trade-offs should be avoided.

Of course, any progress in designing low noise aircraft would therefore lead to relax the stringency of the NAP to be used (Fig. 14).

Arrival/Approach Procedures, the Continuous Descent Arrival/Approach (CDA) can lead to significant reductions in noise, emissions and fuel burn [34]. The CDA concept was elaborated during long time to exclude (or to minimise them) from the approach flight the constant altitude segments and to fly along the descent segments only. The descent flight of an aircraft is made usually with thrust at the flight idle (or very close to it) setting, noise level from descend segments is much less than produced during the level flight segment. The development of the standardised CDAs takes into account environmental aspects (noise and emissions), as well as fuel efficiency issues while ensuring safe and efficient operations of the ATM operations.

Engine noise tends to be lower as the thrust is reduced to slow down or descend, and as technology has developed, engine noise has reduced significantly. However, airframe structures and aircraft configuration contribute to the overall aircraft noise levels, especially during arrivals. They contribute more than during departures for two reasons. Engine thrust is higher during departure and the undercarriage (landing gear) is retracted soon after take-off, after which the high-lift devices (such as flaps and slats), are retracted on schedule as the aircraft climbs, whereas during approach, the engines tend to be at a low power setting. As engines have become quieter, the balance of noise sources has changed and, although engine noise still is the major component during departure, there is more of a balance between airframe and engine noise on arrival.

Aspects of flight safety, which are especially important during the descent of the aircraft before landing, come to the fore. First of all, the separation between individual aircraft should be control by Air Traffic Management (ATM)—simply a distance which previously in flight operation was much more easily reached by providing speed changes to aircraft flying along a level flight segment. To manage the separation between a number of continuously descending aircraft is much more complex task for ATM and pilots, it needs for automation on board of the aircraft and in ATM to support the flight safely.

A landing runway may be nominated for noise abatement to provide preferred final approach routes and landings, similar to preferential take-off runways. The first consideration must be safety, but within operational constraints, consideration can be given to community noise concerns.

PBN procedures are being increasingly deployed at airports for departure, arrival, and final approach procedures with measureable benefits on airport capacity, efficiency and emissions reduction as well as reduced noise exposure. There is a general understanding that as a consequence of PBN, procedural predictability and accuracy will be increased with a corresponding concentration of flight over the defined path. This concentration of flight tracks will, in turn, increase the frequency of noise events and hence noise exposure in close proximity to the PBN flight path. As a result, it is important that PBN procedure design takes population impact into account.

Aircraft Operating Restrictions to Reduce Noise Exposure

An operating restriction is defined in ICAO's Balanced Approach guidance [13], as "any noise-related action that limits or reduces an aircraft's access to an airport". The guidance recommends to avoid applying any operating constraints as a first measure to eliminate noise exposure, but after considering the exposure reduction to be obtained from the other three BA elements. If the total efficiency of the first three is not enough to reduce noise at any location in the vicinity of an airport, operating restrictions may be implemented, even to exclude it at all.

Usually implementation of any operational restriction is a subject of impossibility to fulfil the requirements to environmental noise at location of interest. In fact, such requirements are defined by the national (State Sanitary Rules) or international (EU Directive) rules and they are used as fundamental basics for establishment the rules for noise zoning around dominant noise sources including the ones in the vicinity of airports. So, noisy type of the aircraft, which is appropriate for doing efficiently transportation work at any route may not be able to fly over the location with installed specific noise limit without violation, even with realised noise abatement procedure. The problem may be due to inaccuracies in the establishment of zoning rules around the airport, or land use (for example, by permitting an activity that is more sensitive to noise than required by the rules of the current noise zone) within the zone.

Banning of certain noisy operations at noise-sensitive airports was a first type of implemented noise restrictions in aviation sector 30–40 years ago. Evidently, they limited the operational capacity (by installing so-called environmental capacity) of the airports decreasing the economic efficiency. Their number was growing up so quickly that an Extraordinary 28th Session of the ICAO Assembly must consider a consensus on a global framework for the eventual phase-out of aircraft compliant with Volume 1, Chap. 2 of Annex 16, but unable to comply with Chap. 3 Standards. During a period 7 years beginning 1 April 1995 all the States obliged to phase out operations of Chap. 2 aircraft, which had completed 25 years of service on this date (if they were not immediately affected by this requirement). By 2007 97% of Chap. 2 aircraft were withdrawn from operation worldwide. This is a type of consensus which was grounded by environmental and economic benefits of the approach.

At the European level, policy-makers reinforced the legislative framework endorsing the "ICAO BA to noise management and sustainable development of air transport". From 2002 to 2014, this regulation was enacted through directive EC 2002/30 providing the Member States with a range of possibilities about its practical implementation. Since 2014, the new EU Regulation 598/2014 [35] has superseded the directive and gives clear and mandatory guidance on how to implement it.

Today hundreds of airports worldwide are implementing aircraft operating restrictions for noise management purposes (Table 2) on a case-by-case basis, whilst limiting capacity, but improving the noise climate around airports (Fig. 15, it is built on data of Boeing data base Airport Noise and Emissions Regulations [36]). Any of them may fall into one or more of the four of the below-described categories, depending on how they are applied:

- *Global*—restrictions adopted worldwide or inside large regions to be applied at any airport. ICAO and EU decisions on Chap. 2 aircraft phase-out from operation are the examples.
- *Local*—restrictions adopted by airport authority or by the State to eliminate the operation of noisy aircraft types, for example Chap. 3 aircraft, other way the environmental constraints in a specific airport may reduce its efficient work
- *Aircraft-specific*—restrictions applied to a specific type based on individual aircraft noise performance, usually at specific route of departure or arrival at airport.
- *Partial*—restrictions applied for specific flight directions (because aircraft noise becomes inappropriate with new conditions for land use in this direction) or/and for certain runways at the airport, during noise-sensitive time periods (evening and night) of the day, on specific days of the week (weekend).
- *Progressive*—restrictions which provide for a gradual decrease in the maximum level of traffic or noise energy used to define a limit over a period of time, for example an installation of quota for night-time traffic in airport.

The decisions of the 40th Session of the ICAO Assembly regarding operating restrictions are contained in Assembly Resolution A40-17 [25], Appendix E "Local noise-related operating restrictions at airports". It is a policy document and the ICAO still discourages the application of operating restrictions as the first option to mitigate

Table 2 Aircraft operating restrictions for noise management purposes in airports

Operating restrictions	Explanations
Restriction rules	The rules should define the number of operations not to be exceeded at an airport usually during the whole day or noise-sensitive time of the day. They can be global and/or partial measures, i.e. applicable to all or specific (due to certified level, certified margin, cumulative margin) operations during an identified period of the day on specific or all runways of an airport
Noise quota or budget	It is generally used to limit the overall noise exposure from aircraft operations within a given area to established by airport total value over a given period of time taking in mind that a human reaction to this noise is defined by the exposure level. The operators begin to use quieter aircraft, to increase the traffic not violating the noise limit
Non-addition rules	The aircraft-specific measures are aimed at prohibiting the operation of specific aircraft (usually new aircraft) based on their acoustic performance (certification noise levels)
Nature of flights	The non-scheduled flights (also non-maintenance based flights, check flights, and training flights are consistent with this group) may be forbidden during a specified noise-sensitive time period
Night-time restrictions	Sleep disturbance is a dominant noise effect for people in the EU, so night-time restrictions are of special concern in aircraft noise protection programs
Curfews	Overal or partial operating restrictions, somewhere aircraft-specific, in airports that prohibit take-off and/or landing during an identified time period

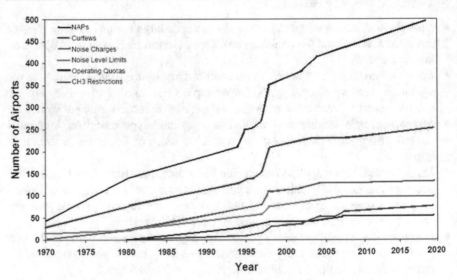

Fig. 15 Growth in aircraft noise restrictions at airports worldwide

noise exposure around a specific airport. As stated above it is limiting the operational and economic efficiency of the airport work. If the benefits from the first three BA elements are not enough to fulfil the environmental requirements to noise, operating restrictions should be considered in the following way:

(a)   be based on the acoustic performance of the aircraft, which should be determined as the noise certification results, consistent with procedures of Annex 16, Volume I;
(b)   be fitted to solving the noise problem of the airport concerned in accordance with the balanced approach principles;
(c)   be mostly of a partial nature, not the complete withdrawal of operations at an airport;
(d)   take into account the consequences for air transport services, especially if the suitable alternatives are absent (for example, long-haul flights);
(e)   conditions of competitiveness should not be violated; for example, exemptions may be granted exemptions for carriers of developing countries;
(f)   be introduced stepwise, taking into account the possible economic burden for operators and, if reasonable, to give operators a time period of advance notice for preparations.

Operational constraints can provide a significant reduction in the impact of aircraft noise around airports immediately, but they can increase the financial burden on both airport operators and airlines.

Conclusion: From Noise Exposure to Noise Impact Management

Until now, all the existing BA elements have been assessed by changes in the noise exposure, mostly via noise contour modelling, and in some cases via monitoring. This allows for the evaluation of noise control measures to determine the most cost-effective and benefitial for environmental protection [37]. In the best cases, the process is performed with public notification and consultation procedures, supplemented with mechanisms for dealing with disputes and complaints. This important approach is recognised in the European Environmental Noise Directive [19]. It requires developing noise action plans with obligatory participation of the public, especially if their residential/rehabilitation area or substantive environmental aspects are impacted by aircraft noise.

Besides the technical elements, which are based on noise intensity metrics completely, the noise annoyance (and other types of outcomes of aircraft noise exposure to neighbouring residents) must now be addressed. This evolution may lead to a new vision of the balanced approach to aircraft noise control in very near future.

It is principal to differentiate between noise exposure and the resulting noise nuisance (annoyance first of all) in different communities and to manage each appropriately. The protection of the residents from aircraft noise exposure is understood as a dynamic process, meaning that the evaluation criteria (both for exposure and nuisance) must be repeatedly tested and—if necessary—adapted to new scientific findings [38]. In comparison to the traditional ICAO BA elements, which are defined by physical phenomena of sound generation and propagation, involves non-acoustical factors must now be included to reduce the annoyance. Up to now, annoyance was mainly explained through acoustical factors like sound intensity, peak levels, duration of time in-between sound events, number of events [11]. The non-acoustical factors ("moderators" and/or "modifiers" of the effect) have still received empirical attention but without a deep theoretical approach, despite the fact that various comparative studies reveal that they play a major role in defining the impact on people [12].

Addressing such human-centric concerns, encompassing fear, negative health effects, and other environmental issues may lead to adding a fifth element to the ICAO BA to aircraft noise management around the airports.

It should be a primary objective of future research into environmental noise impact to investigate the interaction of sound level management and perceived noise management. New and additional measures to mitigate noise impact may result from the redirection of attention from sound to noise and to noise annoyance.

Strategies that reduce noise annoyance, as opposed to noise, may be more effective in terms of protecting public health from the adverse impacts of noise and its interdependence with other environmental, operational, economic and organisational issues of airport and airlines operation and maintenance.

References

1. Environmental noise guidelines for European region. World Health Organisation Regional Office for Europe, Copenhagen, Denmark, p 160. ISBN 978 92 890 5356. Available online https://www.euro.who.int/__data/assets/pdf_file/0008/383921/noise-guidelines-eng.pdf. Accessed date: 30.03.2021
2. ICAO Resolution A39-1 (2016) Consolidated statement of continuing ICAO policies and practices related to environmental protection—general provisions, noise and local air quality ICAO resolutions adopted by the assembly, Provisional edition, Oct 2016, p 138
3. Basner M (2014) Auditory and non-auditory effects of noise on health. Lancet 383(9925):1325–1332
4. Berglund B et al (1999) Guidelines for community noise. World Health Organisation, revised version, Geneva, p 141. Available online https://apps.who.int/iris/handle/10665/66217. Accessed date: 30.03.2021.
5. World Health Organisation (2011) Burden of disease from environmental noise. Quantification of healthy life years lost in Europe. World Health Organisation Regional Office for Europe, Copenhagen, Denmark, p 128. ISBN 978 92 890 0229 5. Available online https://www.who.int/quantifying_ehimpacts/publications/e94888.pdf
6. UNISDR (2016) Exposure and vulnerability short concept note: work stream 2, working group, 27–29 January 2016, Geneva International Conference Centre, p 12
7. ISO 1996-1 (2016) Acoustics—description, measurement and assessment of environmental sound—part 1: basic quantities and assessment procedures, International Standard ISO 1996-1. International Organisation for Standardisation, Geneva

8. UNISDR (2009) Terminology on disaster risk reduction. UN Office for Disaster Risk Reduction, p 30. Available online https://www.undrr.org/publication/2009-unisdr-terminology-disaster-risk-reduction. Accessed date: 30.03.2021

9. Strategic Research and Innovation Agenda 2017. Update, vol 1. https://www.dlr.de/dlr/Portaldata/1/Resources/documents/2017/acare-strategic-research-innovation-volume-1-v2.7-interactive.pdf

10. Zaporozhets O, Blyukher B (2019) Risk methodology to assess and control aircraft noise impact in vicinity of the airports. In: Karakoc TH, Colpan CO, Altuntas O, Sohret Y (eds) Sustainable sviation. Springer International Publishing, Springer Nature Switzerland AG, pp 37–79. Print ISBN 978-3-030-14194-3. https://doi.org/10.1007/978-3-030-14195-0_3

11. Janssen SA, Vos H, van Kempen EE, Breugelmans OR, Miedema HM (2011) Trends in aircraft noise annoyance: the role of study and sample characteristics. J Acoust Soc Am 129(4)

12. Job RFS (1988) Community response to noise: a review of factors influencing the relationship between noise exposure and reaction. J Acoust Soc Am 83:991–1001

13. ICAO (2004) Guidance on the balanced approach to aircraft noise management. ICAO Doc 9829, AN/451, Montreal

14. Woodward JM, Lassman Briscoe L, Dunholter P (2009) Aircraft noise: a toolkit for managing community expectations. ACRP report 15, Washington, DC

15. ICAO Circular 351 (2016) Community engagement for aviation environmental management. ICAO Cir. 351-AT/194, 2017

16. ICAO Environmental Report 2019. Aviation and environment. Destination green. 999 Robert-Bourassa Boulevard, Montreal, QC, Canada, H3C 5H7. www.icao.int

17. European Aviation Environmental Report 2019. European Aviation Safety Agency (EASA), European Environment Agency (EEA), EUROCONTROL, 2019. ISBN 978-92-9210-214-2, https://doi.org/10.2822/309946. Catalogue No.: TO-01-18-673-EN-N. www.easa.europa.eu/eaer

18. European Aviation in 2040. challenges of growth. Eurocontrol, 2018 (Edition 2)

19. Directive 2002/49/EC of the European Parliament and of the Council of 25 June 2002 relating to the assessment and management of environmental noise, OJ L 189, 18.7.2002

20. ICAO Annex 16 to the Convention on International Civil Aviation—Environmental protection: Volume I—Aircraft noise; Volume II—Aircraft engine emissions; Volume III—Aeroplane CO_2 emissions; Volume IV—Carbon offsetting and reduction scheme for international aviation (CORSIA), 2019

21. ICAO, Doc 10127, 2019. Final Report of the Independent Expert Integrated Technology Goals Assessment and Review for Engines and Aircraft

22. Leylekian L, Lebrun M, Lempereur P (2014) An overview of aircraft noise reduction technologies. Aerospace Lab J (7):AL07–01. https://doi.org/10.12762/2014.AL07-01

23. Sustainable Aviation Noise Road-Map, 2018. www.sustainableaviation.co.uk

24. IATA Fact Sheet, 2019. Technology Roadmap for Environmental Improvement Fact Sheet, June 2019

25. ICAO Resolution A40-17, 2019. Consolidated statement of continuing ICAO policies and practices related to environmental protection—General provisions, noise and local air quality

26. X-NOISE, 2015. Evaluation of Progress Towards ACARE Noise Targets. Aviation Noise Research Network and Coordination, X-NOISE EV, Project Number 265943, Deliverable D06.31, Date of preparation: June 2015

27. Astley RJ (2014) Can technology deliver acceptable levels of aircraft noise? Inter-noise 2014, Melbourne, Australia, paper 369, p 12

28. Dobrzynski W, Chow LC, Guion P, Shiells D (2002) Research into landing gear airframe noise reduction. AIAA 2002–2409

29. Hall CA, Schwartz E, Hileman JI (2009) Assessment of technologies for the silent aircraft initiative. AIAA J Propulsion Power 25(6):1153–1162

30. ICAO Document 9184, Airport Planning Manual, Part 2—Land Use and Environmental Control

31. ICAO Document 9888, 2007. Review of Noise Abatement Procedure Research and Development and Implementation Results

32. Operation of Aircraft. Annex 6 to the Convention on International Civil Aviation Part I International Commercial Air Transport—Aeroplanes. Adopted on 25 February 201, Published by ICAO, p 240
33. Doc 8168 Aircraft Operations (PANS OPS) Volume I Flight Procedures, 6th ed., 2020. Published by ICAO, p 214
34. Continuous Descent Operations (CDO) Manual. Doc 9931 AN/476, 2020, 1st ed., 2010. Published by ICAO, p. 76
35. EU Regulation 598/2014 OF THE EUROPEAN PARLIAMENT AND OF THE COUNCIL of 16 April 2014 on the establishment of rules and procedures with regard to the introduction of noise-related operating restrictions at Union airports within a Balanced Approach and repealing Directive 2002/30/EC. Available online https://eur-lex.europa.eu/legal-content/sv/ALL/?uri=CELEX:32014R0598. Access date: 02/04/2021
36. Airports with noise and emissions restrictions. Database. Available online https://www.boeing.com/commercial/noise/list.page. Access date: 02/08/2021
37. Zaporozhets O, Tokarev V, Attenborough K (2011) Aircraft noise: assessment, prediction and control. Glyph International, Taylor & Francis
38. Chyla A, Bukała M, Zaporozhets O et al (2020) Portable and continuous aircraft noise measurements in vicinity of airports. In: Systemy s SrodkiTransportu. Bezpieczenstwo I MaterialyEksploatacyjne. Red. Naukowa Leida K., Wos P. Monografia, WydawnictwoPolitechniliRzeszowskiej, Rzeszow, pp 69–80

Perspective on 25 Years of European Aircraft Noise Reduction Technology Efforts and Shift Towards Global Research Aimed at Quieter Air Transport

Eugene Kors and Dominique Collin

Abstract This article provides a perspective on 25 years of European aircraft noise reduction technology efforts as well as the gradual shift towards a more global research effort aimed at quieter air transport activity. It covers the following aspects:

1. Introduction - Background and general context of noise from air transport operations
2. European context – ACARE Strategic Research Agendas and establishment of the 2020 and 2050 aviation environmental goals
3. Phased strategy towards 2020 targets and beyond
4. A coordinated European aviation noise research effort
5. Assessment of progress relative to ACARE noise reduction targets
6. Addressing the longer-term objectives – Noise and the ACARE SRIA
7. Community building
8. Lessons learned

Keywords ACARE · XNOISE · ICAO · Roadmap

Introduction—Background and General Context of Noise from Air Transport Operations

The White Paper, "European Transport Policy for 2010: time to decide", has highlighted the noise issues associated with aviation. "Air transport is having growing problems gaining acceptance, particularly from local residents who suffer from the

---

Browsing the panorama of noise sources and technologies employed to reduce these noises as well as the trend for future progress. Introducing in this regard the ACARE Agenda and the X-Noise network.

E. Kors (✉)
Safran Aircraft Engines, Paris, France
e-mail: eugene.kors@safrangroup.com

D. Collin
Retired, Paris, France

© The Author(s) 2022
L. Leylekian et al. (eds.), *Aviation Noise Impact Management*,
https://doi.org/10.1007/978-3-030-91194-2_4

noise generated by airports. Introduction of measures to reduce noise caused by air traffic is a sine qua non if the industry is to continue to grow…."

Despite very significant technology improvements over the past twenty years, and, despite the attention being paid to other environmental impacts, aviation noise remains a major problem in Europe, which has to be solved by the air transport industry as a whole, to deal with the expected growth. Stakeholders and policymakers are faced with the particular challenge that, while noise reduction at source has generally been progressing by leaps, in particular by the evolution of engine concepts, there is a need for an economically viable, but continuously quieter airline fleet to accommodate the expected traffic growth without adverse environmental impact.

In practice, this calls for new, more encompassing systemic approaches, implying that, associated with the successful development of novel technology by manufacturers, additional elements have to be taken into consideration for noise source reduction to meet its goals and play its full role in the face of further air transport development.

International standards governing the noise of newly manufactured aircraft are developed by the International Civil Aviation Organisation (ICAO). In view of the long cycles (research, design, development, production, operation, evolution of infrastructures) involved in the air transport business, its purpose is to provide the needed stability, supporting a global long term view for the manufacturers to anticipate future needs through development of affordable technologies. Within ICAO, standards are being developed by the Committee for Aviation Environmental Protection (CAEP). In 2001 CAEP approved the definition of more stringent noise limits (Chap. 4) effective as of 2006. In the process, to contain any increase in noise exposure beyond the benefits provided by Chap.2 aircraft phase out (2002), recommendations were made in favour of a Balanced Approach, challenging the ICAO member states to "study and prioritize research and development of economically justifiable technology", besides complementary actions on airport land-use planning and noise abatement operational procedures. A comprehensive Balanced Approach guidance document was completed by ICAO, describing the various steps and instruments, including in last resort such as operational restrictions that could be used to cope with specific airport situations.

The task of monitoring noise technology research programmes has been a particular focus of the International Civil Aviation Organisation (ICAO) since the 6th meeting of its Committee on Aviation Environmental Protection (CAEP/6) in 2004. Over the last ten years, it has been working to develop a broader view of worldwide research activity and place the ambitious goals established for the wider initiatives in perspective.

The first dedicated CAEP Noise Technology Workshop was held in Sao Paulo in December 2001 and information on worldwide noise research efforts has been regularly updated since then. This has included contributions to the Noise Technology Independent Expert Reviews held in 2008 and 2011, as well as the status reports provided to CAEP meetings every 3 years.

International Noise Technology Research Programmes(2018)

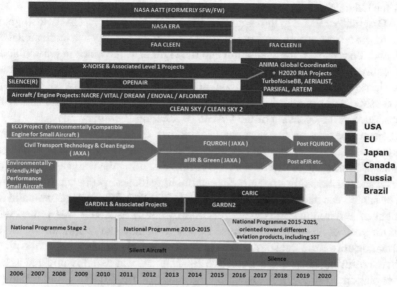

Fig. 1 International Noise Technology Research Programmes as of end of 2018

The latest report presents an overview of noise research efforts up until the end of 2018, covering known national and regional research initiatives and providing an up-to date view of on-going and planned efforts with respect to their technical scope as well as their set objectives (see Fig. 1).

It should be noted that the major initiatives reviewed in 2001 (in the US, EU, and Japan) at the time of the first workshop have been either maintained or expanded, while new significant efforts have been initiated over the years in Canada, the Russian Federation and Brazil, giving us a picture of a truly worldwide effort.

This overview of the research situation demonstrates a significant commitment of all research stakeholders (manufacturers, research establishments, and funding agencies) to investigate and develop novel technology solutions aimed at reducing noise at source. However, it should be noted that beyond the stated research goals, anticipated progress trends will remain conditioned by several factors such as the capability to ensure viable industrial application for promising technology break-throughs as well as the commitment to maintaining steady funding over a significant period of time.

As part of its global forecasting effort, developing a forward view on technological developments was deemed essential for CAEP and in 2007 a process was initiated to develop mid and long term ICAO Noise Technology Goals by means of an Independent Experts panel reviewing potential achievements from research programmes worldwide. Dedicated Noise Technology Reviews were organised in 2008 and 2011 for the Experts Panel to establish a full set of CAEP Noise Technology goals (reported at the CAEP/9 plenary meeting of February 2013).

Originally related to the modeling and forecasting effort, CAEP also got focused on the issue of Impacts, inviting in 2007 worldwide experts to a dedicated workshop. Within this context, the evaluation and measure of Annoyance in particular attracted attention, the registered situation at main airports leading to question the precedence of aircraft noise levels over the frequency of operations as the critical factor. A second workshop was organised in 2015, paving the way for a white paper in noise effects produced at the occasion of the CAEP/10 meeting (February 2016).

Focusing now on the European regulatory context, the European Commission issued in 2002 a directive (DG TREN 2002/30/EC «Noise Related Operating Restrictions at Community Airports» aimed at implementing the Balanced Approach at EU level and harmonising strategies for establishing local airport operational restrictions. The same year, another directive (DG ENV 2002/49/EC « Assessment and Management of Environmental Noise», referred to below as END) was issued within the framework of an overall EU Noise Policy to achieve an evaluation of the baseline situation across the transport modes and subsequently define local action plans based on the resulting noise maps. Directive 2002/30/EC was recently updated as European Regulation 2014/598/EU, extending the scope of aircraft that could be submitted to local operational restrictions.

European Context—ACARE Strategic Research Agendas and Establishment of the 2020 and 2050 Aviation Environmental Goals

Since 1995, through the early work of the Aeronautics Task Force on "The Environmentally Friendly Aircraft" and subsequent work by the Advisory Council for Aeronautics Research in Europe (ACARE), there has been a definite will to develop a consistent research strategy aimed at addressing Aviation Environmental issues on a problem-solving basis.

This priority was reflected in the 2001 report of the Group of Personalities "European Aeronautics—a Vision for 2020+ " on meeting society's needs and winning global leadership. In particular it addressed the goals of reducing perceived noise to one half of current average levels eliminating noise nuisance outside the airport boundary by day and night through quieter aircraft, better land-use planning around airports and systematic use of noise reduction procedures.

Taking up on these goals, the first edition (2002) of the Strategic Research Agenda (SRA) issued by the Advisory Council for Aeronautics Research in Europe (ACARE) promoted the development of an appropriate strategy encompassing:

- The elaboration of technology development strategies aimed at a new generation of noise reduction means for both fixed wing aircraft and rotorcraft, including the associated adaptation of research infrastructures, in particular tests and computing facilities, and covering potential synergies with national efforts.
- The elaboration of an action plan aimed at taking advantage of technology advances in aircraft and air traffic systems to favour implementation of environmentally friendly operational practices such as noise abatement procedures (NAPs).
- The elaboration of a development plan for impact assessment tools and instruments aimed at improved airport noise planning and environmental management practices.

The proposed approach clearly mirrored the Balanced Approach concept and aimed at setting the conditions for a successful implementation of ICAO's recommendation from the early steps of research. This was further substantiated by way of a quantified target addressing the first noise objective of Vision 2020, translated in quantitative terms as an average reduction of 10 decibels per aircraft operation (departure or landing), taking into account technology benefits (Source noise Reduction) as well as operational improvements (Noise Abatement Procedures).

Moving further along these lines, in 2011, the report "Flightpath 2050—Europe's Vision for Aviation" issued by the High Level Group on Aviation Research did set a new target for 2050, stating that by then "the perceived noise emission of flying aircraft is reduced by 65% relative to the capabilities of typical new aircraft in 2000".

To address the targets set by Flightpath 2050, the 2012 ACARE Strategic Research and Innovation Agenda (SRIA) aimed at the 2035/2050 timeframes confirmed the importance of addressing the impacts aspects as part of a coordinated research strategy, stating that the targeted 65% noise reduction relative to the 2000 situation "should be achieved through a significant and balanced research programme aimed at developing novel technologies and enhanced low noise operational procedures, complemented by a coordinated effort providing industry, airports and authorities with better knowledge and impact assessment tools to ensure that the benefits are effectively perceived by the communities exposed to noise from air transport activities".

The general approach to ACARE 2020 and 2050 environmental objectives is summarised in Fig. 2.

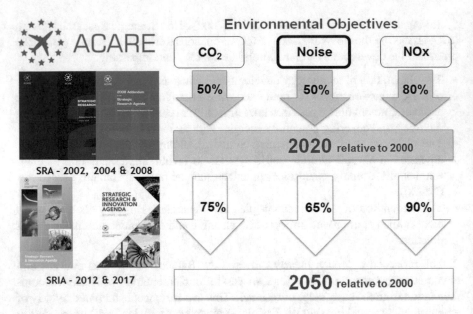

Fig. 2 ACARE environmental objectives

Further on, the 2017 update version of the SRIA maintained these noise reduction goals while stating the ever stronger focus on emissions reduction. This emphasized the need to look after and understand the issue of environmental interdependencies whether on technology, operational measures and overall impacts.

Finally, in its 2021 document "Time for change—the need to rethink Europe's FlightPath 2050", ACARE is re-affirming its environmental commitment stating that "societal expectations on CO_2 mitigation have strongly increased, especially in the field of aviation. Aviation has been pinpointed as a potential major contributor to CO_2 emissions and global warming, although it is currently estimated that the aviation industry represents only approximately 2% of global human-induced CO_2 emissions. The ratified Green Deal objectives demand that the European aviation sector achieves drastically reduced emissions by 2030 and climate neutral aviation by 2050. These targets include emissions, air quality and noise around airports, and ECO-design and end-of-life recycling. This societal change demands disruptive technological solutions; conventional technologies are not enough. New energy sources need researching, integrating, and deploying as new generation aircraft types enter airline fleets".

Phased Strategy Towards 2020 Targets and Beyond

The first noise 2020 objective defined by the ACARE SRA-1 aimed at reducing noise emission of flying vehicles by half, which can be translated in quantitative terms as an average reduction of 10 decibels per aircraft operation (departure or landing), taking into account technology benefits (Source noise Reduction) as well as operational improvements (Noise Abatement Procedures).

The two contributors identified to achieve the reductions associated with the −10 dB were further described in terms of associated technical and operational solutions as shown below:

- Source Noise Reduction associated solutions: Noise Reduction Technologies (NRT generation 1 and 2), Novel aircraft and engine/powerplant architectures
- Noise Abatement Procedures associated solutions: Improved Operating Practices with Current Concepts/Optimised Operations with New Technology/ATM-ATC Integration

The second noise objective defined by the SRA-1 aimed at ensuring that the 10 dB benefit in noise emission anticipated for fixed-wing aircraft effectively led to there being no impacted people outside airport boundaries, provided the appropriate management practices are in place.

As shown in Fig. 3, to address the −10 dB noise target, a phased approach was developed aimed at meeting an interim 2010 target of 5 dB with the help of

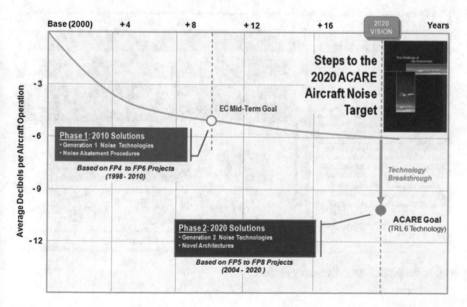

Fig. 3 Steps to ACARE 2020 noise target

more readily (higher TRL) available solutions, paving the way for the technology breakthroughs needed to achieve the full target in 2020.

To lay out the foundations of the following phase, namely addressing the 65% noise reduction target set by Flightpath 2050 and the 2012 ACARE Strategic Research and Innovation Agenda (SRIA), a complete set of recommendations were expressed aiming at the 2035/2050 timeframes, identifying solutions capable by 2050 of reducing noise at departure and arrival by 15 dB per operation relative to Year 2000.

On top of expected 2020 achievements, anticipated solutions would involve the development of a 3rd Generation of Noise Reduction Technologies (NRT), relying in particular on active and/or adaptive techniques to reduce the noise of engines, landing gears and high-lift devices. The emergence of novel aircraft configurations was also considered an essential factor in ultimately delivering the needed source noise reduction. In the shorter term, advanced tube and wing concepts associated with ultra-high by-pass ratio propulsion concepts should provide through masking effects an anticipated 2 dB contribution to the ACARE target. In the longer term, wider options associated blended-wing body concepts such as embedded nacelles or distributed propulsion systems should also significantly contribute to further noise reduction. A broad view of the anticipated solutions is presented in Fig. 4.

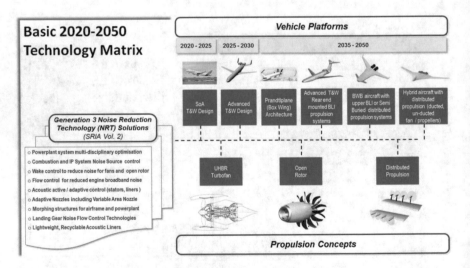

Fig. 4 Basic 2020–2050 Technology Matrix

Moreover, in order to exploit new technology and low noise operations developments and to enable integrated impact mitigation solutions, it was considered of utmost importance to:

- improve and continuously update the understanding of how noise from air transport operations implemented through new Air Traffic Management solutions affects people
- provide the technical support to successful implementation of planning policies compatible with traffic growth for the long term benefit of the communities.

In line with this comprehensive strategy, a number of "Enabling Factors" foreseen as key contributors to the 2050 noise goal achievement were pointed out, namely:

- improved numerical simulation capabilities, together with test facilities incorporating advanced measurement techniques, in order to support further noise reduction at source level as well as the implementation of multi-disciplinary optimisation techniques and aircraft/engine integrated design practices contributing to lower noise through efficient integration of noise reduction solutions, reduced weight, decreased drag, improved powerplant efficiency and flight path design,
- stimulated advances in related technology areas, such as materials and electronics, to allow the introduction of novel low noise technologies, including active/adaptive techniques,
- updated, internationally recognised, Annoyance and Sleep Disturbance models, taking into account the evolution of aircraft noise signatures and traffic conditions (multiple events), also considering airport specificities,
- tools supporting transparent communication policies covering relevant indices, flight path / operations on-line forecast and tracking as well as comprehensive assessment of environmental interdependencies and monetisation of impacts.

Detailed recommendations were provided along these lines by the noise research community to ACARE and included in the Volume 2 of the SRIA which included the foreseen solutions for all areas. These are summarised in Fig. 5.

In fact, the ACARE SRIA confirmed the importance of addressing the impacts aspects as part of a coordinated research strategy, stating that the targeted 65% noise reduction relative to the 2000 situation "should be achieved through a significant and balanced research programme aimed at developing novel technologies and enhanced low noise operational procedures, complemented by a coordinated effort providing industry, airports and authorities with better knowledge and impact assessment tools to ensure that the benefits are effectively perceived by the communities exposed to noise from air transport activities".

This successive set of strategic recommendations shaped the overall noise research effort implemented to this day. In face of the diversity of anticipated solutions, it also called for efficient coordination from the early stage such an effort to ensure achievement of the noise targets.

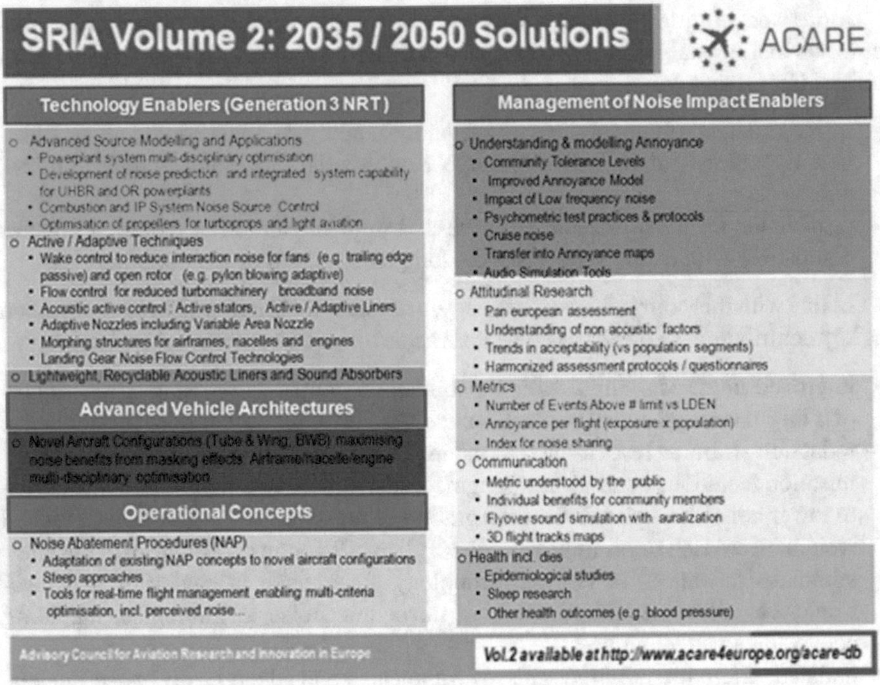

Fig. 5 ACARE SRIA Vol. 2035–2050 solutions from air transport noise mitigation

A Coordinated European Aviation Noise Research Effort

Overall Approach to Coordination

To implement the phased strategy described in the previous section, a coordinated approach was established. The basic concept of the European Aviation Noise Research Network (X-NOISE) emerged in a similar timeframe as the ACARE SRA.

From 1998 to 2015, based on the 3-Pllar approach described in Fig. 6, the network concept demonstrated its ability to accommodate the evolution of the broader context and provide key support in the definition and implementation of a research strategy aimed at reducing the impact of noise from air transport. It established well recognized dissemination and communication features and developed an active research community covering a vast majority of EU member states. Working around a common set of priorities and objectives, it also favored a better exploitation of innovative upstream research developed at national level into larger European projects aimed at downstream research.

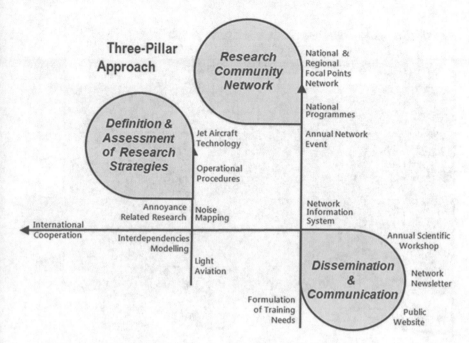

Fig. 6 Three-Pillar approach to networking and coordination activities

Within the specificity of the European general framework supporting collaborative research, the effective implementation of the phased strategy is best represented by the roadmap of European research projects directly contributing to the 2020 noise target achievement and supporting the early steps towards the 2050 target (Fig. 7).

As can be seen, over 25 years, more than 35 noise dedicated projects aimed at implementing the ACARE agendas were launched and complementary efforts on operational procedures and noise impacts initiated. Noise reduction was also supported through significant participation in large architecture-oriented multidisciplinary projects such as VITAL, DREAM, ENOVAL and CLEAN SKY.

This resulted from a significant mobilisation of research actors, achieving a well balanced participation between Industry and Research Organisations from a large majority of EU and Associated States. Definite steps was also taken towards wider international cooperation, in particular with countries actively involved in discussing international aviation noise standards within ICAO/CAEP such United States, Canada, Russia, Japan and Brazil.

Following the last X-NOISE project completed in 2015, a new phase in research coordination was then implemented in 2017 building on the X-NOISE legacy, supported by the ANIMA project with the following objectives:

- Establish and update through a scenario-based approach a common strategic research roadmap for aviation noise reduction, addressing the development of new technologies and methodologies.

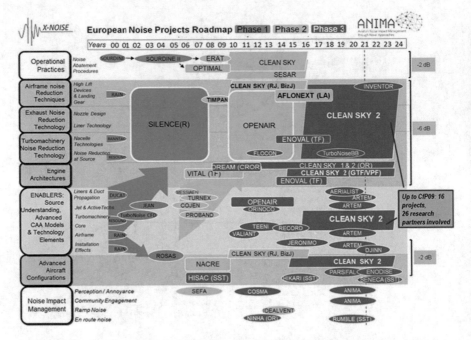

Fig. 7 Roadmap of European Union funded projects relating to aircraft noise

- Coordinate a European wide network of experts and research actors in support of the roadmap process, stimulating the emergence of novel approaches to overcome gaps, ensuring minimal duplication and fragmentation in the course of future research efforts.
- Explore possibilities, facilitate and establish practical conditions for targeted international collaboration in line with the strategy for EU international cooperation in research and innovation as well as the needs put forward by the strategic roadmap for aviation noise, including aspects related to international regulatory discussions.

Aspects pertaining to the common strategic research roadmap are further developed in Section "Establishing the Common Strategic Research Roadmap for Aviation Noise Reduction"

The networking aspects, lessons learned and possible ways forward will be further developed under Section "Community Building" and "Lessons Learned"

Noise Reduction at Source—Technological Achievements and Perspectives

A key contributor in impulsing the whole research effort was the SILENCE(R) project, funded under the 5th Framework 'Growth' programme. SILENCE(R) remains to this day the largest project devoted to aircraft noise ever supported by the European Commission for a total budget of 111 Million Euros. Completed in 2007, the 6-year project involved a consortium of 51 partners and focused on the development of aircraft noise reduction technologies (NRT) Generation 1. Research activities were being carried out in various fields, such as Engine Source Noise, Nacelle Technologies, and Airframe Source Noise and Active Control Applications. More than 35 prototypes were to be tested during the SILENCE(R) programme.

In a number of ways, Silence(R) established a blueprint for larger EU funded research projects to come through its effective risk management approach and the first implementation of a dedicated Technology Evaluation process, which proved to be both a valuable tool in decision making and a key asset to assess the progress made to this day relative to ACARE 2020 noise target.

As can be seen on Fig. 8, at project end, thanks to two dedicated flight tests on the A320 and the A340 as well as a number of engine full scale tests, ten new

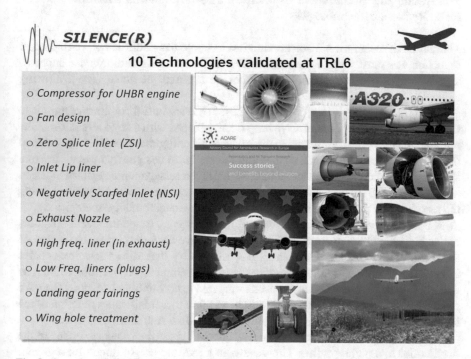

Fig. 8 Overview of SILENCE(R) key technologies

technologies were validated from the noise reduction standpoint. These technologies were considered mature enough for further work aimed at successfully addressing through industrial development work the design tradeoffs issues pointed out by the technology evaluation process.

A significant effort was subsequently dedicated to technology enablers throughout the 6th Framework programme, focusing on advanced methods for prediction of fan noise, jet noise and nacelle liners efficiency together with low TRL airframe noise reduction concepts. Further maturation of Generation 2 NRT solutions aimed at all significant noise sources was then achieved through the OPENAIR and AFLONEXT projects.

In terms of achievements, the noise reduction at source effort aimed at both 2020 targets and 2035 horizon can be divided into three headlines:

- Maturation and validation of Generation 2 powerplant and airframe noise reduction technologies
- Enabling development of Generation 3 solutions (prediction tools and advanced technology concepts)
- Contribution from novel propulsion concepts and novel aircraft configurations.

Maturation and Validation of Generation 2 Powerplant and Airframe Noise Reduction Technologies

Maturation of Generation 2 noise reduction technologies dedicated to engine noise reduction was mostly performed in the OPENAIR project including noise suppression techniques to cover the key sources of fan and jet noise. The following technologies were successfully matured to TRL4.

MDO Outlet Guide Vanes (OGVs) and Lined OGVs: Multi-disciplinary Optimised (MDO) OGVs and the Lined OGVs have been designed with a large reduction of the number of vanes (~10) compared to the traditional configuration (~40). This enabled thin blade designs with special shapes and thicker blade designs that permit internal space for acoustic liners. Design objectives balanced aerodynamic performance and acoustic design. Both broadband and tonal noise sources were reduced, whilst leaving the aerodynamic performance unchanged as demonstrated on large scale fan rig.

Intake Technologies: Various new inlet liner concepts, based on recent advancements in Computational Aero Acoustics (CAA) have been developed and tested on a fan rig. The "Folded Cavity Liner" has a geometry that allows low frequencies to be damped through a large space that is folded behind the conventional liner, so that lower nacelle thickness can be used compared to conventional designs. Forward fan noise reductions have been achieved, in addition to a significant reduction of the buzzsaw noise, which is an annoying noise source also audible in the cabin.

Highly Curved Bypass duct: The highly curved bypass duct seeks to open out the duct earlier to a higher radius. This results in a reduced height duct with a greater liner area per unit length. These attributes allow a shorter nacelle to be used giving significant reductions in weight and drag.

Acoustically lined splitters and fins: Configurations of supplementary liner area in the bypass duct were designed and tested on a large scale fan rig. Both "splitters", which fully cover the height between the inner and the outer wall, as well as so-called "fins", which protrude from the outer wall up to about halfway of the duct height were investigated. These may provide a solution to the acoustic area loss when shorter nacelles are desired in the future.

Negatively Scarfed Nozzles: The scarfed shape changes the directivity of the rear-ward radiated fan noise to higher angles and provides a general reduction in the total engine noise. Scarfed nozzles have been designed for the secondary nozzle of both short and long cowl nacelles and tested in an anechoic wind tunnel.

Active Stator: Research efforts on this active noise control technology have been pursued with improved actuators, sensors and algorithms. Objectives were extended to include also rearward fan noise control, in addition to the already demonstrated forward fan noise control. The new system was validated on a large scale fan rig. Integration aspects were matured in parallel through a full scale composite OGV demonstrator.

Active Nozzle: Various active flow/noise control concepts have been explored before selecting the "Microjet" technology for large scale testing wind tunnel testing. Extensive integration studies have been performed in parallel on the air supply through the nacelle to the primary and secondary nozzles.

Through OPENAIR, CLEAN SKY and AFLONEXT, technologies were developed to reduce the noise of landing gears and high lift devices. While CLEAN SKY focused on solutions aimed at regional aircraft, OPENAIR and AFLONEXT investigated and validated techniques for larger commercial models.

Low Noise Landing Gear: Wind tunnel tests at full and reduced scale have been carried out on both wing-mounted and fuselage-mounted landing gears for large airliners and regional aircraft. Amongst all tested configurations, showed the highest noise reductions:

- For the large aircraft fuselage-mounted gear: Torque link mesh fairing, deceleration plate
- For the large aircraft wing-mounted gear: Low noise dressing routing; Forward location for cardan pin; Torque link mesh fairing; Door aligned with flow; H-shape side-stay with spring and large side-stay mesh fairings.
- For the regional aircraft main and nose landing gears: Low Noise Concepts including landing gear bay acoustic treatments, strut fairings and wheelpack fairings (covers, hubcaps, wind shields).

Adaptive Slats: In the adaptive slat concept, the trailing edge of the slat is a flexible morphing structure that can fully close the gap between the wing and the slat. In normal operation with low angle of attack, the gap is closed and quiet.

Droop Nose: This solution consists in a smart hingeless, gapless, leading edge highlift device, based on a rotational drive concept assisted by SMA actuators. The device extends up 67% wing span due to structural/system installation constraints. The "skewed" configuration is aimed at avoiding streamwise longitudinal gaps in the retracted position.

Porous Flap Side Edge (PFSE): Various acoustically porous materials were evaluated for their environmental (certification) requirements before a final selection was manufactured for large scale acoustic testing at the DNW The PFSE proved an effective means of reducing noise without any significant aerodynamic penalty.

Lined Flap: Innovative, acoustically treated, low noise trailing edge flap conceived as a sound-absorption integrated multilayer liner structure with micro-perforation on the external facing sheet.

Morphing Flap: An advanced structure has been conceived to match an aerodynamically optimised "target shape". Morphing performance of two architectures was demonstrated on 2D mechanical prototypes: Smart Actuated Compliant Mechanism (SACM) and Deeply Embedded Smart Actuator (DESA).

Despite being only a third, budget wise, of the Silence(R) project, more than 50 engine and airframe oriented technology developments took place in OPENAIR, all executed around a number of key technologies identified at project start. From this set, the technologies described above achieved TRL4/5 through large scale testing in wind tunnels and/or dedicated engine fan or exhaust rigs.

Part of the 50 technology developments were as risk-reduction, backup or long term solution related to the key technologies. A total of 144 low TRL ideas were collected before project start in an "innovative concepts" database from which the project selected 19 for inclusion in OPENAIR. This approach was perceived as very successful for OPENAIR and other related projects and proposed for best practice in the future.

On the aircraft side, the project AFLONEXT successfully advanced further maturation of airframe noise reduction technologies, flight testing in 2018 a series of flap and main landing gear solutions.

The scope of technology development achieved in OPENAIR, CLEAN SKY and AFLONEXT is represented in Figs. 9, 10 and 11.

Enabling Development of Generation 3 Solutions (Prediction Tools and Advanced Technology Concepts)

These activities are aimed at supporting advanced noise reduction technologies (Generation 3) for industrial application in the 2030 + timeframe. Recent developments are reported for each of the main areas of research and projects featured in the Fig. 7 roadmap.

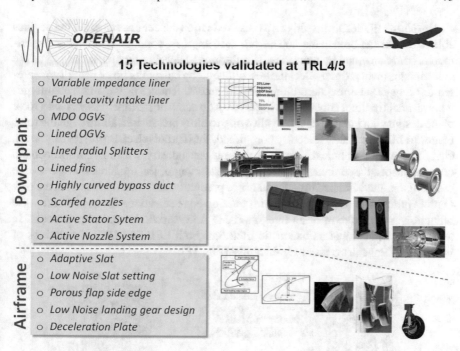

OPENAIR

15 Technologies validated at TRL4/5

Powerplant
- o Variable impedance liner
- o Folded cavity intake liner
- o MDO OGVs
- o Lined OGVs
- o Lined radial Splitters
- o Lined fins
- o Highly curved bypass duct
- o Scarfed nozzles
- o Active Stator Sytem
- o Active Nozzle System

Airframe
- o Adaptive Slat
- o Low Noise Slat setting
- o Porous flap side edge
- o Low Noise landing gear design
- o Deceleration Plate

Fig. 9 Overview of OPENAIR Key technologies

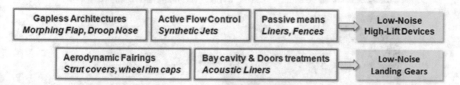

Gapless Architectures Morphing Flap, Droop Nose	Active Flow Control Synthetic Jets	Passive means Liners, Fences	⟹	Low-Noise High-Lift Devices
Aerodynamic Fairings Strut covers, wheel rim caps	Bay cavity & Doors treatments Acoustic Liners		⟹	Low-Noise Landing Gears

Fig. 10 Scope of investigated airframe noise solutions in CLEAN SKY

DLR's A320 Advanced Technology Research Aircraft (ATRA) fitted with noise-abatement technologies pre-flight

Brake covers and partial fairing on main landing gear (Safran LS)

Porous Flap Side Edge (Airbus)

Fig. 11 AFLONEXT key technologies

Fan Noise: FLOCON investigated noise reduction concepts and associated devices able to reduce fan broadband noise from aero engines, conducting lab-scale experiments, complemented by numerical simulations to develop an understanding of the mechanisms involved and select the best concepts by balancing noise benefit and integration impact on a real aero-engine environment. Among promising technologies were an adaptive liner concept serving an overtip acoustic treatment as well as a rotor blade design with internal channels allowing to blow pressurised air through the rotor blades to fill their wake. Recently completed, the TurboNoiseBB project took over, enabling a major technical leap in providing the industry with low fan broadband noise concepts, based on an improved understanding of the broadband noise source mechanisms and validated broadband noise prediction methods. This included a new Outlet Guide Vane (OGV) design with leading edge serrations for broadband noise reduction and validation of an integrated 3D Aero/Mech/Noise design process. In terms of computational aeroacoustic methods, both LEE and DES simulations of rotor–stator stage progressed significantly (Fig. 12).

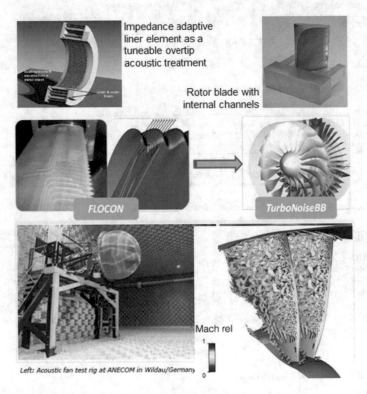

Fig. 12 Examples of FLOCON and TurboNoiseBB key features

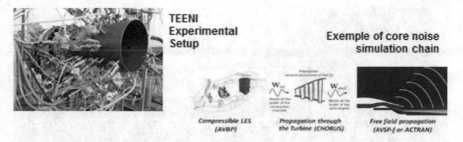

Fig. 13 Examples of TEENI and RECORD key features

Core Noise: The TEENI project dealt with experimental identification of engine core noise emission. Turboshaft exhaust noise is assumed to be a mix between combustion and turbine noise, with very little jet noise. It is representative of what is generally called core noise on aircraft engines. TEENI helped to provide further insight on this complex issue, thanks to a simpler geometry and absence of strong competing noise sources (such as jet and fan noise). Combustion noise was identified at low frequencies and its relative importance shown to be increased downstream of the High Pressure Turbine. In a complementary fashion, RECORD investigated the core noise generation mechanisms caused by the combustor and the combustor-turbine interaction. Through development and validation of core noise prediction methods ranging from low order modelling approaches to high-fidelity compressible Large Eddy Simulation (LES). These methods were validated by different experimental test cases focusing on the different aspects concerning aerodynamics of the flow field, the combustion process and the flow conditions in a real turbine stage (Fig. 13).

Jet Noise: ORINOCO was one of the first projects co-funded by the European Commission and the Ministry of Industry and Trade of Russian Federation, dedicated to advanced jet noise control based on plasma actuators. The use of plasma actuators is a novel concept that required fundamental approaches to understand the interaction mechanisms with the main jet shear layer and the resulting radiated sound. Several plasma techniques were improved and developed.

Addressing another key jet noise related area, the JERONIMO project supported an improved understanding of the physical mechanisms of UHBR Engine s(BPR > 12) installed jet noise with jet-wing interaction, allowing to develop and validate Engine jet noise prediction processes based on anechoic Wind-Tunnel tests and numerical simulation (Fig. 14).

Airframe Noise: the VALIANT projects focused on broadband airframe noise modelling, tackling both landing gears and high lift devices, the two main contributors. Large Eddy Simulation-based approaches provided promising results, but emphasized the critical needs in terms of CPU and available memory for such types of calculations. Recommendations for simplification of design assumptions were made to address these issues considering current capability limitations (Fig. 15).

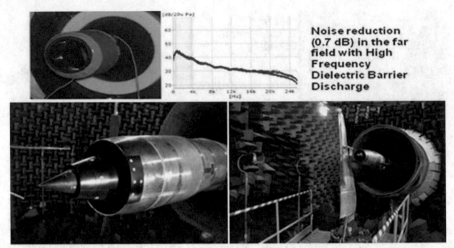

Installed UHBR exhaust system at CeprA19 anechoic wind tunnel

Fig. 14 Examples of ORINOCO and JERONIMO key features

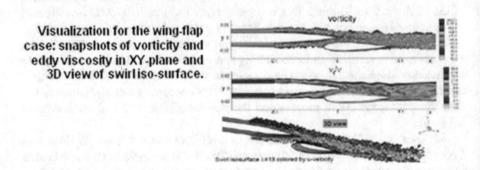

Fig. 15 Examples of ORINOCO and JERONIMO key features

Acoustic Liners: In the footsteps of advances registered in OPENAIR on advanced active liners, ENOVAL further supported the investigation of the electromagnetic concept based on electronically controlled areas of loudspeakers. In light of the new perspectives brought by the fast emergence of additive manufacturing capabilities, passive liners innovative concepts were also investigated by AERIALIST a project specifically addressing the use of meta-materials applied to the reduction of engine and airframe noise emission. This effort has been complemented by dedicated liner activities on-going within the frame of the larger ARTEM project, addressing inno-vative passive and adaptive liner concepts and their implementation issues relevant to the use of UHBR turbofans and distributed propulsion configurations (DEP) on future aircraft concepts (Fig. 16).

To build on such achievements and consolidate the anticipated noise benefits, a further effort is currently going on in two complementary directions:

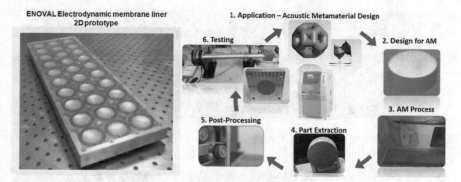

Fig. 16 Examples of ENOVAL and AERIALIST key features

(a) Pursuing the work on future advanced prediction techniques through 3 clustered projects launched in 2020:

- DJINN, which will develop a new generation of reliable computational fluid dynamics (CFD) methods to assess promising noise-reduction technologies for future integrated propulsion aircraft. Improved CFD methods for multi-physics modelling utilising high-performance computing are expected to reduce design times and costs by around 25% compared to large-scale testing.
- INVENTOR which will study the physics of noise generated by landing gears and high-lift devices to achieve further progress as airframe noise remain a major source of noise during landing operations.
- ENODISE which aims at reducing aircraft gaseous and noise emissions by improving the integration of the propulsion system with the airframe. To this end, it will investigate the existence of local/global integration optima via an innovative experimental methodology combined with reduced order modelling and machine learning strategies.

(b) Moving previous collaborative projects findings closer to industrial application, taking advantage of the CLEAN SKY 2 call for projects mechanism. As of the end of 2019, 16 noise related project involving 26 research partners had benefitted from such support. These aimed in particular at:

- Further integration of aeroacoustic design methods into industrial processes: Open Rotor Low Noise design (PROPMAT, SCONE), installed jet noise (INSPIRE) and core noise assessment (CIRRUS).
- Further maturation of advanced noise reduction technologies, taking on the promising active liner concept studied in ENOVAL (SALUTE) as well as several low noise OGV concepts first investigated in OPENAIR, FLOCON and TurboNoiseBB (INNOSTAT). Experimental evaluation on a fan rig will allow to compare these solutions and anticipate on industrial application issues (Figs. 17 and 18).

Fig. 17 Simulation of installed jet noise (INSPIRE)

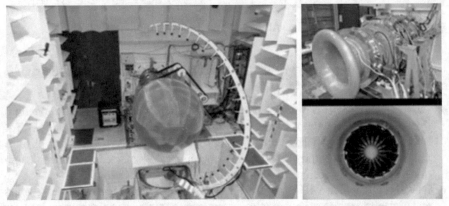

Acoustic liners and OGV design validation at PHARE acoustic test rig in Lyon

Fig. 18 Acoustic liners and OGV design validation at PHARE acoustic test rig in Lyon (SALUTE and INNOSTAT)

Novel Propulsion Concepts and Airframe Configuration—Achievements and Perspectives

Ultra High By-Pass Ratio (UHBR) engine Concept: Further noise reduction at source is expected to be provided by the availability of ultra-high by-pass ratio propulsion concepts (from 12:1 up to 20:1) fitted on aircraft applications with entry into service (EIS) from 2025 onwards. While leading to lower noise by design, the by-pass ration increase is anticipated to lead to additional weight and drag due to the increase of the fan diameter.

The overall target of the ENOVAL project was then to reduce CO_2 emissions by up to 26% compared to the year 2000 reference engine, where the specific contribution of ENOVAL will be 3–5%, while providing an incremental benefit up to −1.3 dB on the way to the 2050 ACARE noise target (Fig. 19). These objectives have been enabled by addressing the following breakthroughs technologies, to be validated at component level up to TRL5:

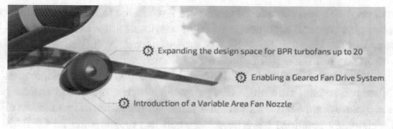

ENOVAL Key Technology Objectives for UHBR engines

Fig. 19 ENOVAL Key Technology Objectives for UHBR engines

- Expanding the design space for Turbofans up to BPR 20: Advanced Fan Blades, Light weight Intermediate Case, Holistic design concepts for shorter and thinner Nacelle with enhanced integration and acoustic liners, Improved LP-Turbine designs.
- Enabling a Geared Fan Drive System for the very large, long range engines
- Introducing a Variable Area Fan Nozzle (VAFN) for optimum stability and design for low pressure ratio fans

Other innovative UHBR options such as the counter rotating fan system were investigated in the COBRA project co-funded by the EU and the Russian Federation (see Fig. 20).

Counter-Rotating Open Rotor (CROR) engine concepts: CROR engine concepts have re-emerged in recent years as a serious option to provide the needed fuel burn benefits implied by the targets set for aviation CO_2 emissions reduction. It is then per se a significant contributor to the ACARE CO_2 reduction target, whilst not a key contributor to the achievements of ACARE noise target.

COBRA: Contrafan mechanical and noise design optimisation

Fig. 20 COBRA: Contrafan mechanical and noise design optimisation

However, noise was considered as a major issue in the initial investigation effort of such engine concept, which culminated in a series of noise evaluation flight tests performed in the US in 1986–1987. As a consequence of this situation, a significant effort has been and still is dedicated to noise reduction as part of European research programmes dedicated to CROR engine concepts. Investigated areas encompass advanced aeroacoustic blade design methodologies, optimised engine geometry and cycle, advanced installation configurations, reduction of installation effects. Such technology solutions have been explored in projects such as DREAM and NACRE, then in CLEAN SKY, Further wind tunnel experiments carried out under CLEAN SKY 2 have helped to consolidate the assessment of noise levels expressed in the previous report while a full scale engine demonstrator ran to confirm the high interest in such propulsion concept. At this stage, based on results from model tests in anechoic wind tunnel (TRL4), CROR-powered aircraft with an EIS around 2030 can be expected to produce noise levels similar to those of turbofan-powered aircraft recently certified. When placed in perspective with the best expectations resulting from the original 1987 post flight-test assessment, this represent a typical 20 dB noise reduction on a cumulative basis, a spectacular achievement for the European research effort initiated in 2008 through the DREAM project (Fig. 21).

The noise benefits provided by novel aircraft configurations are considered as a key factor to meet the ACARE noise targets. In the shorter term, tail-mounted tube and wing concepts are expected to provide a typical 2 dB noise reduction through masking effects. In the longer term, wider options associated blended-wing body concepts such as embedded nacelles or distributed propulsion systems should also significantly contribute to further noise reduction.

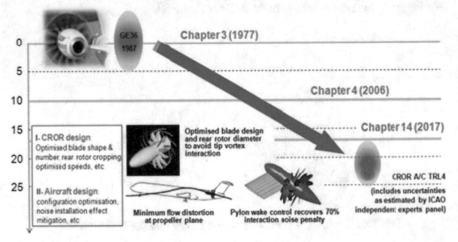

Fig. 21 Evolution of anticipated noise levels from CROR-powered aircraft

Targeting the potential noise benefits expected from novel aircraft configurations, dedicated activity was initiated in ROSAS, then pursued in a multidisciplinary framework through the successful Framework 6 project NACRE which explored several new aircraft concepts tailored to address specific subsets of design drivers, including the Pro Green (PG) aircraft concept, paying major emphasis on the reduction of environmental impact of air travel.

Of interest for noise, key NACRE achievements included:

- Relevant work on Powered Tails and Advanced Wings for the Pro-Green concept, aiming at high environmental performance (noise and CO_2 emissions): Contrafan and Open Rotor propulsion systems were integrated with a noise-shielding empennage and assessed (see Fig. 22).
- Progress on the understanding of the complex Flying Wing configurations, opening a new path for the promising over-body-engine configuration; Detailed aerodynamic and acoustic assessment of selected concepts were performed in particular for the over-the-buried 3-engine aircraft configuration.
- Tough not flight-tested yet, full development of the IEP concept into a demonstrator, establishing the foundations of a new kind of test facility in Europe based on real-atmosphere free flight and available for future projects (see Fig. 23)

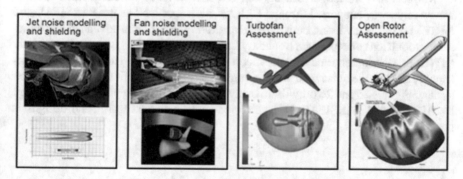

Fig. 22 Scope of NACRE activities on pro-green powered tail configurations

Fig. 23 Development steps of the IEP demonstrator

Fig. 24 Noise shielding
demonstrator

Innovative business jet aircraft architectures with an engine noise shielding after-body were also investigated in CLEAN SKY. Noise assessments was performed using prediction tools and validated and calibrated using dedicated wind tunnel and static tests (see Fig. 24).

With regard to the longer term agenda, an active novel aircraft architectures effort is now underway supported through CLEAN SKY 2 as well as the PARSIFAL and ARTEM projects involving both innovative aircraft designs and advanced low emissions propulsion concepts such Boundary Layer Ingestion (BLI), Blended Wing Body (BWB) or Distributed Electric Propulsion (DEP). Boxwing/Prandtl Plane concepts should also provide noise reduction through optimised installation of the propulsion system. Figures 25 and 26 provide an overview of such installation concepts covered by ARTEM and PARSIFAL. The noise reduction expected from such designs is covered into the next Chapter.

Noise Abatement Procedures (NAP)—Achievements and Perspectives

As an integral element of the solutions to achieve the first 2020 ACARE target (50% noise reduction), work on noise abatement procedures have progressed steadily over the years addressing the successive steps towards maturity:

- Develop flight trajectory concepts minimising noise impact
- Develop robust aircraft flight management capabilities
- Enable optimum implementation into Traffic Management Area.

At research level, noise abatement operational concepts have been involved in the following projects:

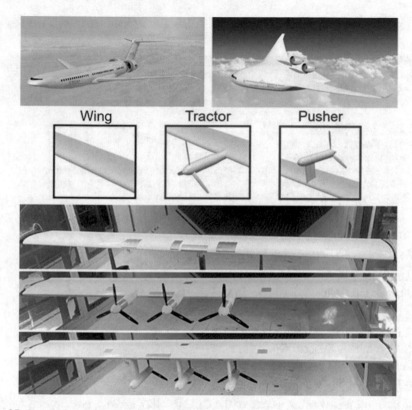

Fig. 25 Installation concepts studied in ARTEM

Considered engine locations: (a) under rear-wing location, (b) rear fuselage location and (c) above front-wing location. The engine acoustic source is represented by the red dot.

Fig. 26 Installation concepts studied in PARSIFAL

Departure.

- Optimised Noise Abatement Departure Procedures Sourdine
- Automated Management of optimized NADPs Sourdine 2
- Multi Criteria Departure Procedures CLEANSKY (SGO) (see Fig. 26)

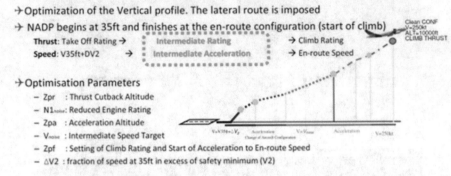

Optimization of a Noise Abatement Departure Procedure (NADP) , with multiple criteria ➜ Multi-Criteria Departure Procedure (MCDP)

➜ Optimization of the Vertical profile. The lateral route is imposed

➜ NADP begins at 35ft and finishes at the en-route configuration (start of climb)

Thrust: Take Off Rating ➜   Intermediate Rating   ➜ Climb Rating

Speed: V35ft+DV2   ➜   Intermediate Acceleration   ➜ En-route Speed

➜ Optimisation Parameters

- Zpr   : Thrust Cutback Altitude
- N1ₙₒᵢₛₑ: Reduced Engine Rating
- Zpa   : Acceleration Altitude
- Vₙₒᵢₛₑ : Intermediate Speed Target
- Zpf   : Setting of Climb Rating and Start of Acceleration to En-route Speed
- ΔV2 : fraction of speed at 35ft in excess of safety minimum (V2)

Fig. 27 Multi criteria departure procedures

Arrival

Steeper approach

- Adaptive Increased Glide Slope (A-IGS)CLEANSKY (SGO) (see Fig. 27)
- Adaptive Runway Aiming PointCLEANSKY (SGO)
    Continuous Descent Operations
- Fully Managed CDO OPTIMAL
- Adaptive Flight Path Angle SESAR.

In the most recent effort, as part of the CLEAN SKY activity, new approaches for the Management of Trajectory and Mission were addressed through the following concepts (Fig. 28):

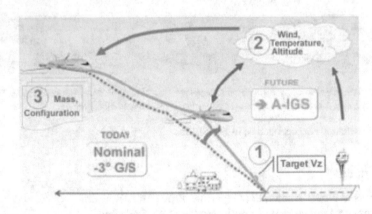

Fig. 28 Arrival adaptive increased glide slope (A-IGS)

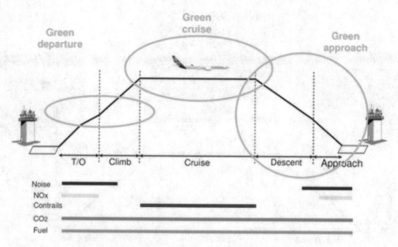

Fig. 29 Trajectory optimisation in 3 phases

- Green Trajectories, based on more precise, reliable and predictable 3 dimensional flightpaths, optimised for minimum noise impact and low emissions, including agile trajectory management, in response to meteorological hazard.
- A Green Mission from start to finish, with management of revised climb, cruise and descent profiles, which permit multiple criteria optimisation (especially of noise, emissions, fuel, time); including management for weather conditions, which could negatively impact the aircraft optimum route and result in additional fuel consumption.

Based on the results from several previous studies (e.g., OPTIMAL, FLYSAFE, etc.) these concepts needed to be enabled by the development of flight management systems (FMS) and guidance algorithms to provide high accuracy flight path prediction and control, encompassing flight phases (see Fig. 29):

- During departure and climb, the developed functions ensures a combined optimisation of CO_2, noise and NO_x emissions, meeting the constraints of each mission (including airport procedures, atmospheric conditions, aircraft state, etc.).
- During climb, the crew is able to optimise the vertical and lateral trajectory according to aircraft performance and weather conditions (mainly in respect of wind).
- In descent and approach phase, enhancements to the already studied CDA concept (from, e.g.,) are available, in order to enable CDA to be achieved even in dense traffic situations.

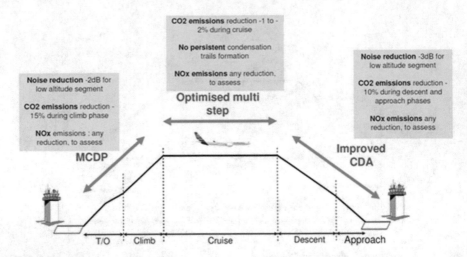

Fig. 30 FMS green functions: green objectives

Mission optimisation systems provide capabilities to build complete flight profiles optimised for multiple criteria including environmental impacts, combining selected climb, cruise and descent trajectories. This also encompasses the accurate real-time prediction of the complete flight profile and of its environmental impacts. The new functions are to be ultimately be integrated into the target product: the green FMS and demonstrated in cockpit simulations (see Fig. 30).

Improvement in the Understanding of Aircraft Noise and Its Impacts

While a significant effort was directed towards the achievement of the -10 dB per operation target, a second noise objective defined by the ACARE SRA aims at ensuring that such benefit from technology and operational solutions effectively leads to there being no impacted people outside airport boundaries, provided the appropriate practices and policies are in place. Pan-European research activities were then initiated in the area related to Management of Noise Impacts. Three projects stand out in direct relation with the possible evolution of aircraft technical characteristics, along the following lines:

- How can the evolution of aircraft noise signature at takeoff and landing can positively affect impacts felt by surrounding communities (COSMA)?
- How to evaluate the noise impact of Open Rotor driven aircraft during cruise operations over ground (NINHA)?
- How to evaluate the impact of supersonic transport sonic boom phase on populations (RUMBLE)?

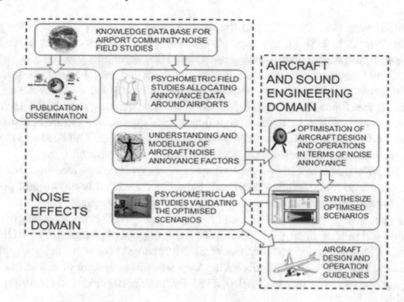

Fig. 31 COSMA multidisciplinary approach to establishing low annoyance criteria for aircraft operations

In this area, international collaboration has been very much in the picture, results from COSMA and NINHA were shared with US counterparts in dedicated workshops in view of involving cooperation with follow-up research efforts while Russia has been directly involved in the RUMBLE project activities from the beginning.

Supported by a well-balanced partnership, COSMA established a unique approach to aviation noise research, targeting significant progress in the understanding of community noise impacts while consolidating the relationship between the technology optimisation process and how the resulting aircraft sound is perceived (see Fig. 31).

COSMA's scientific concept led to innovative ways of combining sound engineering and noise effects analysis to generate low noise impact design recommendations for future aircraft. Extensive field studies around European airports, combined with psychometric laboratory studies formed the basis to establish optimal aircraft noise characteristics regarding lower annoyance. Specific sound synthesis techniques developed within COSMA re-created a realistic simulation of global airport operations and were then applied to the optimisation of flight procedures when associated with anticipated technology benefits.

Beyond its scientific results, COSMA is also leaving a legacy of harmonised test protocols and innovative simulation tools for future projects to use in improving the understanding of aviation noise effects.

NINHA specifically addressed the noise impact of aircraft with novel engine configurations in high altitude operations. The potential introduction of aircraft fitted with advanced counter-rotating open rotor (CROR) engine power plants should contribute significantly to the reduction of fuel burn and gaseous emissions. In the 1980s, prototypes of the first generation of open rotor engines were developed and tested. It was found that the noise generated by these engines, even in the en-route flight phase, could be considered significant, thus questioning public acceptance. Since then significant effort has been dedicated to improving the CROR aero-acoustic design, with promising results, as discussed earlier in this chapter.

The NINHA project re-assessed whether noise issues away from airports (i.e. during mid- to high-altitude operations, hereafter called "en-route") might potentially hinder the introduction of this new generation of power plant. The project was organized around 3 main challenges: (1) Adapt existing models for long-range propagation and validate them, (2) Predict noise levels on ground generated by CROR en-route, (3) Assess ground noise impact of CROR relative to conventional powerplant, including turboprops. A groundbreaking step was the validation of the long-range propagation model by means of a dedicated flight test performed on the A400M (see Fig. 32).

Based on improved long-range propagation models and 2012 CROR technology capabilities, an updated vision statement on en-route noise levels of CROR configurations was put in perspective with respect to initial efforts carried out around 1986–1988 by NASA and the industry. The NINHA project then established that at cruise the noise of CROR engines with today's technology will be significantly reduced from that of the UnDucted Fan of the 1980s, the maximum noise level being equivalent to that of today's Turbo-Propeller driven aircraft in cruise condition.

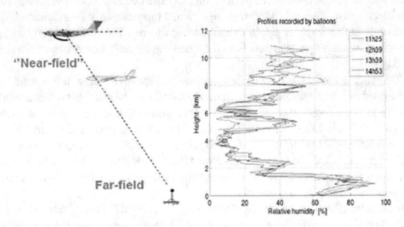

Fig. 32 Overview of NINHA flight test set-up to validate long distance noise propagation models

Fig. 33 Scope of RUMBLE activities

Impacts-oriented activities are also complemented by RUMBLE, which currently addresses the en-route noise produced by supersonic civil aircraft. As such, it aims at producing the scientific evidence requested by national, European and international regulation authorities to determine the acceptable level of overland sonic booms and the appropriate ways to comply with it (see Fig. 33).

Another issue is directly associated with supersonic aircraft is the capability to manage landing and takeoff noise levels compatible with subsonic aircraft standards, and if not the case to identify the level of acceptability in the general public for such operations, even limited in number. The recently launched SENECA project is related to a clear need to clarify such questions marks, as expressed by regulatory bodies.

The need to consider a wider and more global effort on aviation noise impacts was at last addressed through the ANIMA project.

As exposed earlier in the "Phased Strategy" section, it was recommended that, in order to exploit new technology and low noise operations developments and to enable integrated impact mitigation solutions, a dedicated effort would need to:

- improve and continuously update the understanding of how noise from air transport operations implemented through new Air Traffic Management solutions affects people
- provide the technical support to successful implementation of planning policies compatible with traffic growth for the long term benefit of the communities

This also included Enabling Factors such as tools supporting transparent communication policies covering relevant indices, flight path / operations on-line forecast and tracking as well as comprehensive assessment of environmental interdependencies and monetisation of impacts.

Implementing such longer term recommendations further, ANIMA addresses key aspects of noise effects and impacts in:

- carrying out a critical review and assessment of noise impacts and existing management practices to establish best practices' guidelines for an effective management of annoyance beyond ICAO Balanced Approach
- developing a better understanding to address community annoyance, sleep disturbance and improve quality of life
- developing Noise Management Toolset to empower non-specialists with decision support capability
- testing and validating with end-users (airports and community) an "Aviation noise community platform", gathering tools and best practices, facilitating consensus building and engaging communities in the mitigation process, ensuring exploitation of the results.

Achievements from the ANIMA project are presented in detail in several chapters of this book.

Assessment of Progress Relative to ACARE Noise Reduction Targets

At this stage, three formal evaluations relative to the 2020 ACARE targets have been performed in succession (AGAPE-2011,OPTI-2013, X-NOISE/ACARE-2015) while a new assessment is close to completion relative to the final position on 2020 targets as well as a first vision of progress relative to 2050 goals. These efforts are described in the following sections.

Background and Methodology

The initial SRA1 approach presided over the definition of the ACARE 2020 noise targets, as originally based on the Balanced Approach concept developed by ICAO:

- The first noise target aims at reducing noise emission of flying vehicles by half, which was translated in quantitative terms as an average reduction of 10 decibels per operation, taking into account both technology benefits and operational improvements.
- The second noise target aims at ensuring that the 10 dB benefit in noise emission anticipated for fixed-wing aircraft effectively leads to there being no impacted people outside airport boundaries, provided the appropriate management practices are in place. For the sake of evaluation, this was translated in quantitative terms to a 65 Lden target at airport boundaries.

The first two elements of the Balanced Approach concept (noise reduction at source, noise abatement procedures) constitute the identified contributors to the 10 dB reduction aircraft noise target, and can be further described in terms of associated technical and operational solutions as shown below:

- Quiet Aircraft contributor associated solutions: Noise Reduction Technologies (NRT) generation 1 and 2, Novel aircraft and engine/powerplant architectures
- Noise Abatement Procedures contributor associated solutions: Improved Operating Practices with Current Concepts / Optimized Operations with New Technology/ATM-ATC Integration.

In the course of previous formal progress assessment exercises (AGAPE, OPTI), a methodology for the evaluation of progress has been established within the noise community, based on two complementary tools described below.

Firstly, the internationally recognized TRL scale, that enables keeping track of the situation of individual technologies identified in the SRA1 as key elements of the ACARE technology-oriented solutions for noise reduction. This tool is being used to measure the progress in term of strategy implementation considering the initial technology panel promoted in the SRA1.

Secondly, on a dedicated Aircraft Noise Technology Evaluation (ANTE) process, involving a predictive model with the capability to roll up the benefits of individual technologies at solution level for a number of current and advanced aircraft engine configurations. This tool is being used to establish quantitatively the progress achieved at solution level as well as overall versus ACARE targets, including operational aspects, when applied to a typical airport platform in 2020. Initiated via the SILENCE(R) project, it has been implemented since 2001 through the string of major EU projects dealing with aircraft noise reduction.

Thus, in the following pages, the TRL status will be mostly used to support the expert group qualitative analysis in terms of assessing the implementation of the initial SRA1 strategy, evaluating the size of the associated effort and identifying potential gaps emerging from either technical difficulties or insufficient funding support or new priorities and technological avenues. In parallel, the existing Technology Evaluator exercises will support the quantitative analysis of achievement relative to the ACARE targets.

Figures 34 and 35 summarise the selected methodological approach, fully applicable to the primary aircraft noise target described in the SRA1, i.e., 10 dB reduction per aircraft operation.

As hinted at on the noise effort projects roadmap (Fig. 7), the status of research progress concerning technology and operational solutions can be assessed concentrating on the results achieved by larger projects. These full-scale demonstrator projects exploit the upstream research achievements of smaller projects carried out at EU or national level. The resulting solutions at TRL 6 level allow industry efforts to take over further maturation dealing with key industrial application issues.

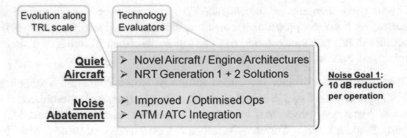

Fig. 34 Tools aimed at noise target evaluation for fixed wing aircraft

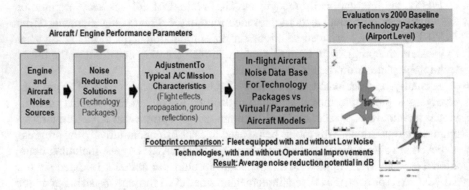

Fig. 35 Aircraft noise technology evaluation (ANTE) process

The dedicated technology evaluation methodology has been implemented in such large key projects to evaluate and provide a measure of anticipated achievements at TRL6 level, consistent with the ACARE framework and objectives established for 2020. The previous AGAPE and OPTI assessments relied then on the results of OPTIMAL, SILENCE(R), and VITAL confirming that TRL6 solutions (NRT Generation 1 + Noise Abatement Procedures (NAPs),) supported the achievement of the mid-term target of −5 dB per operation. In dealing with the further steps towards the −10 dB target (NRT Generation 2, Novel Architectures), the 2015 assessment exercise benefited from the achievements of the OPENAIR project as well as interim results from CLEAN SKY in specific areas related to business jets and regional aircraft in particular.

Changing Boundary Conditions

The initial recommended approach presiding over the definition of the ACARE 2020 noise targets remains valid, as originally based on the general Balanced Approach concept developed by ICAO. The first two elements of the concept (noise reduction at source, noise abatement procedures) constitute, in practice, the identified contributors to the 10 dB reduction aircraft noise target.

However, a number of external factors are worth mentioning, as further developed below. When the SRA1 was established, environmental targets (Noise, NO_x, CO_2) were established in parallel without considering possible interdependencies, either at technology or operational level. Historically, the industry has had to comply with standards set in ICAO for noise and NO_x. It has been the industry's ability to reduce these effects, whilst at the same time for the benefit of operators improving fuel efficiency, that has allowed aviation to grow rapidly in recent years without losing general public support. However, aviation more recently faces two important shifts. First, there is an increasing focus on climate change as a key environmental issue, indeed the dominant one for some stakeholders, leading to a new CO_2 standard soon to be established by ICAO. Second, as a result of the potential introduction of new technologies, there may be a need for stakeholders to make more explicit choices about which environmental effects to prioritise. Broadly, it is likely that in future generations of aircraft, manufacturers will have the option to pursue incremental developments on current technology. It is suggested that this could deliver substantial reductions in noise and improvements in fuel efficiency, which although worthwhile are not in the same order of magnitude. An alternative would be the use of technology such as open rotor engines for narrowbody aircraft, which could deliver large improvements in fuel efficiency, while achieving less progress on noise reduction.

In parallel with this evolution driven by global environmental issues, there is evidence of increased sensitivity to noise in local communities impacted by aviation operations despite significant reduction of aircraft source noise over the years. The air transport growth perspectives in Europe then remain conditioned to improvements in all three elements of the Balanced Approach. As proposed as part of the SRIA 2050 Volume 2 airport noise section, it is essential that a greatly increased effort is launched to address the key aspects of noise effects and impacts and ultimately to support the implementation of successful policies complementing the technological achievements expected to be realised.

These different aspects have been considered in performing the progress assessment reported in the following sections and in formulating associated recommendations for future research.

2015 Progress Assessment

The 2015 progress assessment exercise was performed within the framework of the X-NOISE EV project at ACARE's request. An approach by consensus based on experts judgement, assessment of the TRL situation and results from the technology evaluation exercises was used, coming up with updated progress achievement figures and formulating associated recommendations for future research. The exercise findings are summarised below.

Since the year 2000 a number of civil air transport aircraft have been certified by the European industry. The 2015 situation provided a representative panel of effective implementation of state-of-the-art Generation 1 Noise Reduction Technologies (NRT) delivered to TRL6 by completed research programmes such as SILENCE(R) and VITAL. The observed average achievement, together with 2 dB operational benefits of noise abatement procedures, was totalling 5 dB of the ACARE target, as shown in Fig. 3.

In dealing with the further steps towards the −10 dB target (NRT Generation 2, Novel Architectures), the 2015 assessment exercise benefited from the achievements of the OPENAIR project as well as interim results from CLEAN SKY in specific areas related to business jets and regional aircraft, in particular.

At the end of the OPENAIR project, 15 "Generation 2" Noise Reduction Technologies achieved TRL 4/5 through large scale testing in wind tunnels and/or dedicated engine fan or exhaust rigs. These technologies were aimed primarily at Short-Medium Range and Long-Range aircraft fitted with advanced ducted turbofans. Through CLEAN SKY, additional efforts reached similar TRL achievements on complementary noise reduction solutions aimed at Regional Aircraft (low noise landing gear and highlift devices) and Business Jets (U-Tail). In addition to technology solutions, CLEAN SKY also brought further consolidation of noise abatement procedures benefits. Finally, CLEAN SKY produced a first noise evaluation of the Contra Rotating Open Rotor (CROR) engine concept at mission level on a Short-Medium Range aircraft.

When combining the CLEAN SKY interim analysis (2014) with the OPENAIR final analysis at airport level and considering the relative importance of business and regional operations, it was concluded that a typical 2.5 dB additional benefit, relative to the 5 dB already consolidated at TRL 6, can be expected from Generation 2 Noise Reduction Technologies provided such technologies mature to TRL6 in time for 2020. The combined CLEAN SKY/SESAR effort on low noise abatement procedures was also expected to provide further consolidation towards TRL6 of the 2 dB benefit registered earlier.

These findings have then been implemented into the ACARE general indicator established to provide a measure of overall achievement relative to a given 2020 target.

Five years ahead of 2020, the progress registered since 2000 was considered significant, reaching an excellent level of completion with about 64% of expected benefits secured, due to effective implementation of the research roadmap and associated priorities. In terms of identified contributors, the investigation and development of recommended ACARE solutions have been well supported at European level over the years, complemented by a steady activity at national level.

Relative to the second ACARE 2020 noise target (number of people impacted outside airport boundaries), a pilot study led to the following observations:

- Benefits of each individual element differ significantly (very airport dependent)
- The effect of Land Use Planning may be of the same order of magnitude as that of noise reduction at source
- A combination of the two actions above are required to maintain the future population affected below 2000 levels.

The full assessment process, however, will require a very significant amount of input information and need effective support, if it is to be in place and validated ahead of the next assessment exercise. In the meantime, dedicated research actions should address the development of updated dose–response relationships to allow a translation from exposure (in terms of Day-evening-night equivalent level (Lden)) to annoyance suited to the characteristics of today's and tomorrow's operations.

Considering the evidence of increased sensitivity to noise in local communities impacted by aviation operations, it is, in fact, essential that a greatly increased effort is launched at the earliest opportunity to address the key aspects of noise effects and impacts and ultimately support the implementation of successful policies complementing the technological achievements expected to be realised. A strongly coordinated and integrated approach would definitely provide a significant added value relative to the business as usual (BAU) approach currently generally prevailing in Europe in this area of noise impacts.

Recent Developments in Assessing ACARE 2020 and 2050 Noise Targets and Associated Recommendations

In 2020, ACARE WG3 has undertaken an exercise to determine the progress towards the ACARE goals, looking at both the ACARE 2020 as well as the ACARE 2050 goal. A group of experts from industry, Eurocontrol and airlines have gathered material to support the assessment which has resulted in a draft report within ACARE, expected to be used in a future publication.

In Section "Background and Methodology", the ACARE 2020 goals translation into quantitative terms has already been discussed. Similarly, for the ACARE 2050 goal we can translate the 2050 noise goals of reducing the perceived noise by 65% (relative to the year 2000), as average reduction of 15 decibels per operation, taking into account both technology benefits and operational improvements.

This reduction has been defined as the TRL6 technology readiness level.

For the 2020 assessment, the experts used the XNoise led 2015 assessment as a starting point and evaluated the 2015–2020 timeframe. During this period, the AFLONEXT project was the only project that brought additional noise technologies to TRL6 trough full scale flight testing. However, as there was no evaluation at aircraft/fleet level, the contribution to the ACARE goals could not be quantified for this project.

Several noise research efforts between 2015 and 2020 changed focus towards novel configurations. On aircraft configurations, the ARTEM project made progress and on engine configurations, the Open Rotor engine required significant research. The traditionally noisy unducted engine architecture, forced resources to support several wind tunnel campaigns to bring innovative noise solutions for a society demanding low CO_2 transport.

For the 10 dB goal of ACARE 2020, the latest overall Technology evaluation assessment, combined with engineering judgement on the benefits from Noise reduction at the source and Noise Abatement Procedures were assessed to have progressed by 6,4 dB. This means that, with a gap of 3.6 dB, this ACARE 2020 goal has NOT been achieved.

With respect to the 2nd ACARE 2020 noise goal related to the elimination of nuisance outside airport boundaries, quieter aircraft have entered the fleet since 2000. However, the number of annoyed people within the noise contours has increased from 2005 to 2017 by about 10% (ref EASA Environmental report 2019). This means that this ACARE 2020 goal has NOT been achieved.

To determine whether we are on track for the 2050 goal, the experts took a look into the advancement in time. They also considered a look into the noise technologies that have been incorporated into real products. Figure 36 shows (in dots) certified noise level improvements of new Airbus aircraft since 2000. It shows that most of the new aircraft that entered service around 2015 are slightly noisier than what would have been required to follow the (grey) ACARE 2050 target line. The crosses in the Fig. 36 show intermediate projections made in Clean Sky2 by their Technology Evaluator process[1] and indicate that current noise technology development is not sufficiently progressing to match the trend line towards the 2050 noise objective.

In order to resolve this gap, and in view of comparable research developments at the international level, establishing a more ambitious, more integrated framework is needed. It should support forward looking, longer term noise research to address a common goal, binging together the wide range of expertise needed across the board of solutions, from technology enablers to understanding impacts. The basic scope for such an effort has been provided through the SRIA Volume 2, which is addresses, beyond others, all the options of the ICAO CAEP Balanced Approach and is discussed in more detail in the next paragraph.

A major risk in achieving the 2050 noise goals is the current focus for CO_2 reduction. The trade-off between especially CO_2 and Noise (but also NO_x) can be significant, as is illustrated in Fig. 37[2], which shows the inter-relational trend of the

---

[1] Airbus paper @ AEC 2020 conference Bordeaux.

[2] Prof Dimitri Mavris - Georgia Tech @ AEC 2020 conference Bordeaux.

aircraft/engine environmental between Noise, CO_2 and NO_x. It can be seen that if CO_2/Fuel Burn is prioritised, the Noise (as well as the No_x) performance will be compromised.

It can already be seen in current low CO_2 research for air transport that some of the promising solutions of the future, are expected to have a trade-off between CO_2 and noise. This is especially the case for:

- Advanced UHBR engine, where the short and thin nacelle needed for drag and weight reduction causes distorted flow into the fan, less liner area and –thickness, all increasing the noise.
- BLI Boundary Layer ingestion engines have an even more distorted flow into the fan, again increasing noise.
- Open Rotor architectures, inherently without nacelle to attenuate the noise, require new strategies and an unconventional approach towards annoyance reduction.

In conclusion, relative to the ACARE noise targets, the aircraft noise research effort can be considered as no longer on track to meet its objective, and thereby requiring significant support to get back on track. Actions critical to the ultimate success of the comprehensive overall approach, already summarised in the 2015 assessment, remain valid through the following recommendations:

- Bring the most promising Generation 2 noise reduction technology toTRL6, through an appropriate full scale validation effort across the board (engines, nacelles, landing gears, airframes).

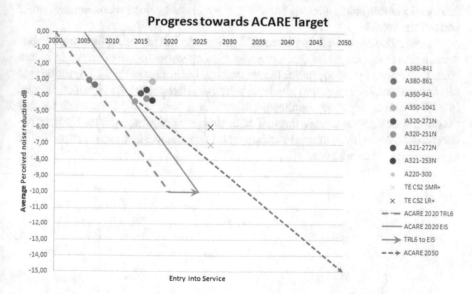

Fig. 36 Airbus recently certified aircraft and future projections relative to ACRE objectives

Fig. 37 Example of
aircraft/engine design space
trade-offs on
Noise/CO$_2$/NO$_x$

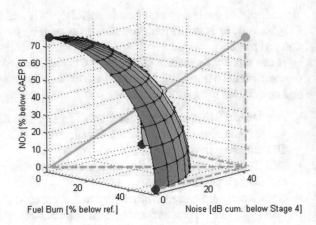

- Very significantly increase the effort dedicated to Low Noise Aircraft configurations noting that whilst programme prospects are better concerning novel engine architectures, the effort on aircraft configurations is significantly lagging behind.
- Take advantage of the sustained effort on low noise operational procedures to consolidate wider implementation capability.

Key uncertainties concerned the capability to support successful OPENAIR technologies to TRL6 through static and flight demonstrations and the needed emergence of ambitious multidisciplinary initiatives dedicated to environmentally friendly advanced aircraft configurations and design, which had gained momentum in other parts of the world.

It should also be pointed out that current progress results are valid for ducted turbofans engine concepts, while Open Rotor engine concepts had re-emerged in recent years as a serious option to provide the needed fuel burn benefits implied by the targets set for aviation CO$_2$ emissions reduction. While the European research effort initiated in 2008 could already show significant achievements (see Fig. 21), it was outlined that the effort should be maintained through dedicated research aimed at rotor blade aeroacoustic design, engine/airframe installation and flow control techniques, in particular to consolidate such advances.

Addressing the Longer Term Objectives—Noise and the ACARE SRIA

Identifying the Conditions for a Successful Strategy Aimed at the 2050 Noise Target while Considering a Wider Noise Research Effort Allowing Full Implementation of the Balanced Approach

Discussing implementation approach beyond the mere list of priority topics, the ACARE Strategic Research and Innovation Agenda (SRIA) has clearly stated that the 65% noise reduction targeted for 2050 "should be achieved through a significant and balanced research program. Such activity should be aimed at developing novel technologies and enhanced low noise operational procedures, complemented by a coordinated effort providing industry, airports and authorities with better knowledge and impact assessment tools. Such package will ensure that the benefits are effectively perceived by the communities exposed to noise from air transport activities.

Another objective of the SRIA is the capability to use the European airspace flexibly to facilitate reduced environmental impact from aircraft operations in the context of an air traffic management system allowing 24-h efficient operation of airports. Investigating and understanding the conditions allowing night-time operations from a community noise standpoint is certainly a pre-requisite in that regard.

Two workshops organised by X-NOISE under the ACARE auspices supported further development of a way forward (Future Trends in Noise Research, Brussels, October 2014 and Managing Aviation Noise Impacts—Mapping Future Research Priorities, Iasi, May 2015).

In the process, a conceptual roadmap to achieve the ACARE 2050 noise objectives within a 24/7 operations context has been defined (see Fig. 38), it does associate technological and operational solutions coupled with a significant effort aimed at noise effects on the communities:

- Capable of achieving the 65% noise reduction target set for 2050 when associated with novel aircraft / engine architectures
- Contributing to the «24/7 operations» goal through significant reduction of community impacts.

To address the key issues put forward by the workshops participants, the added value of a coordinated and integrated approach at European level was strongly emphasized. Effective engagement of stakeholders in research, joining efforts, sharing data and strategic advice was strongly advocated during the workshops. The basic features of a coordinating structure embedded in the effort discussed above has then been developed. It should among other activities:

- Coordinate research activities related to aviation noise through common national / EU projects roadmap, progress assessments versus the ACARE targets, stimulate appropriate international cooperation and dissemination.

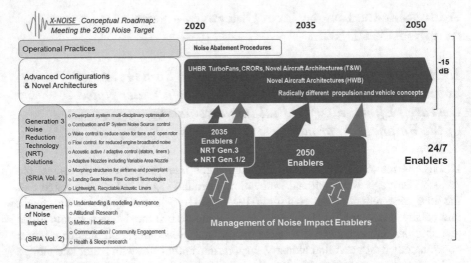

Fig. 38 Conceptual Roadmap to 2050 noise targets vs ACARE SRIA Vol.2 noise solutions

- Through shared strategic vision with on-going and planned national and self funded efforts, ensures gaps and priorities are being addressed altogether to achieve the programme technical and scientific objectives
- Establish a Noise Effects Knowledge Base Covering Long Term / Short Term Annoyance and Sleep Disturbance, Capable to bring in data from other compatible studies (national and international)
- Ensure that the most recent knowledge on noise impacts and annoyance in particular, is reflected in the criteria used for steering the technology effort, with the aim of capitalising on reduced noise at source to provide further effective improvement for the communities
- Rely on an advisory board involving stakeholders and authorities, thus closely associated with the proposed research initiative.

At this occasion, an overview of community noise effects research carried out in European countries was presented. It was noted that half of the overall national funding was dedicated to Health alone and that other aspects such as annoyance and sleep would strongly benefit for increased support at European level. It was also recommended that wider international cooperation be sought, in such areas where knowledge development is not concerned with competitive and industrial property issues.

In fact, several European experts and scientists participated in the ICAO Environmental Impacts workshop held February 2015 in Washington DC which ultimately led to a white paper on noise impacts presented at the CAEP/10 meeting (February 2016). Conclusions related to sleep disturbance and annoyance read as follows: "Undisturbed sleep is a prerequisite for high daytime performance, well-being and health. Aircraft noise can disturb sleep and impair sleep recuperation. Further research is needed to (a) derive reliable exposure–response relationships between aircraft noise exposure and sleep disturbance, (b) explore the link between noise-induced sleep disturbance and long-term health consequences, (c) investigate vulnerable populations, and (d) demonstrate the effectiveness of noise mitigation strategies."

"There is substantial evidence that aircraft noise exposure is associated with annoyance indicators, and exposure–response relationships have been derived to estimate the expected percentage of highly annoyed persons at a community level. Still, several personal and situational factors importantly affect the annoyance of individuals. Recent evidence for an increase in the annoyance response at a given exposure level indicates the need for updating exposure–response curves based on recent studies using harmonized methods, as well as verifying the circumstances leading to a heightened community response." It should also be noted that the World Health Organisation WHO has issued new recommendations in 2018 for environmental noise in the European region, including quite low exposure thresholds for avoiding health impacts resulting from Aircraft Noise. This confirms the findings made earlier on the higher annoyance perceptions from air traffic than from other transport modes like road or rail.

Land-Use Planning (LUP) was also discussed. The CAEP/5 meeting (2001) endorsed a balanced approach to aircraft noise mitigation (known as ICAO Balanced Approach), one pillar being improved land use planning and control. Years later, LUP is the only pillar of the Balanced Approach that is perceived as not delivering according to expectations.

One reason for such a situation lies in the lack of an explicit indicator to measure progress and then LUP contribution to the evolution of the noise situation around airports. Not delivering according to expectations also creates significant barriers to airport expansion, generating situation of conflict between several aviation stakeholders, including opportunities for complaints. Few and isolated research projects have tackled LUP related issues during the last decade, while the pressure and airports constraints to expansion has grown exponentially.

Another reason to improve LUP is to assist the ACARE noise goal on "no impacted people outside airport boundaries", or "no impacted people inside the 65 Lden contour, an objective also recently adopted by the USA CLEEN III project. Proper LUP is one of the means to achieve this goal, but delegated and scattered LUP responsibilities in many countries make it difficult to consolidate benefits from these measures. On the contrary, there is a severe risk at this moment that the number of impacted people increase due to conflicting policies in the field.

Research shall definitely help **establish a process for assessing the effectiveness of land use planning** as a function of time and evolution of other aspects of the Balanced Approach. A current challenge is to develop guidance material for policy makers and communities to encourage optimally compatible land-use around airports, as a critical part of any successful noise control strategy. Ensuring at existing airports that further residential developments do not endanger reduction of noise exposure already achieved (prevention method) constitute a key topic for investigation, while conversion of incompatible land-use shall remain a permanent concern. Useful research aimed at policy makers and non-specialists should then include a compact, practical, easy to understand and easy to use toolbox and a consistent set of indicators to evaluate the success of integrated land use in a sustainable airport context.

In summary, such research shall propose innovative and flexible land use management schemes for future airport developments. Associated tools will address support to decision making, information and community awareness. Interdependencies between noise and emissions may be also considered, as LUP is equally important for noise management as it is for controlling airport local air quality.

The concept behind the ANIMA project is clearly borne out of these conclusions, addressing the wider research agenda associated airports and communities issues while providing an opportunity for coordination with other efforts aimed at source noise reduction and moving the definition of a detailed global research strategy forward.

Remaining on the strategic aspects, as the focus of this chapter, the ANIMA work programme clearly answered the expressed request to establish a common strategic research roadmap for aviation noise reduction, addressing the development of new technologies and methodologies.

As such, the ANIMA project has put in place a collaborative process to establish and update a common strategic research roadmap through a scenario-based approach associated with a dedicated simulation process. Keeping in perspective the 2035 timeframe for the sake of the exercise, efforts have aimed at developing a roadmap covering all concerned areas of research and assess its elements based on an enhanced set of criteria including impacts driven targets. The role of this activity is shown on Fig. 39 and detailed in the next section.

Establishing the Common Strategic Research Roadmap for Aviation Noise Reduction

Under the support of the ANIMA project, the European aviation noise research community known as XNOISE has expanded the roadmaps towards the reduction of aviation noise annoyance. Besides the traditional "source noise reduction" roadmaps, a way forward has been defined on a range of associated elements like noise impact understanding, airport management and enablers to support the proposed solution contributors. Starting from a set of available roadmaps, a harmonisation has been conducted on the format and layout of the roadmaps. Taken into account the technology status of 2020 brought by recent projects and activities, a vision for the way forward on each element of the roadmap has been developed and consulted with the network. The final version will be published at the end of the ANIMA project (Fig. 40).

In order to verify if the roadmaps are sufficiently challenging, a scenario based approach is applied. Figure 41 provides an overview of the technology contributions

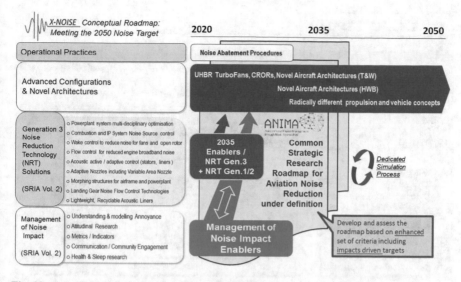

Fig. 39 Role of ANIMA in developing the noise research strategy for 2035

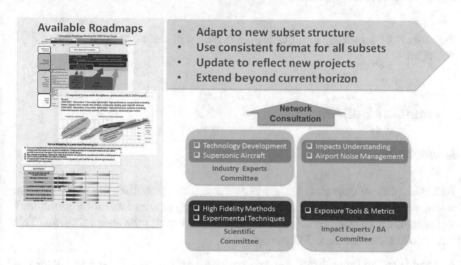

Fig. 40 Roadmap definition and harmonisation

that are considered for implementation into the scenarios. Besides these source noise technologies, the scenarios will also include Noise abatement procedures and Land use planning. The results will be evaluated for a set of 8–10 airports, representing small, mid-size and large European airports.

Figure 42 shows the list of roadmaps developed and the major topics within each roadmap. For each topic, a list of sub-topics describes the current TRL status and a consolidated expert view on how each activity could be further matured to support

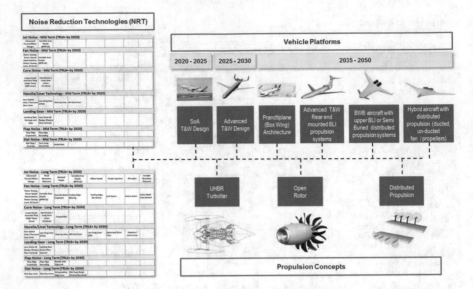

Fig. 41 ANIMA Technology matrix

the mission of achieving an air transport system that can sustainably achieve 24/7 operation.

Community Building

General Overview

As a main pillar of their global coordination activities, X-NOISE, then ANIMA have steadily brought together over the last 20 years a technical and scientific community with capability to advance aviation noise research beyond the state of the art. This has been achieved through 3 key objectives:

- Identifying and mobilising the best expertise all over the European Union and Associated States, addressing multi-disciplinary needs to support individual projects.
- Growing the base of SMEs involved in upstream research
- Supporting the global environmental research agenda of the EU and assessing progress made towards the ACARE targets, through wide-ranging experts committees.

A network of national focal points (NFPs) has been established covering the EU and a number of associated states. Working around a common set of well disseminated priorities and objectives, it has favored a better exploitation of innovative upstream

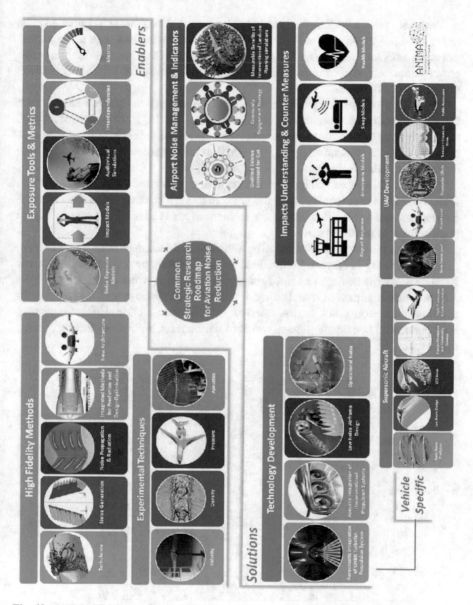

Fig. 42 ANIMA/XNoise roadmaps

research developed at national level into larger European projects aimed at down-stream research. An overview of national efforts has also made been available based on successive reporting from all NFPs. This has also allowed a better dissemination at national level of EU projects research findings and acquired knowledge.

With help from a bottom-up process seeking novel ideas through dedicated calls or brainstorming seminars as well as key dissemination actions such as an annual scientific workshop and web-based information features, this has made for an active scientific and technical community at European level, focused on improving aviation noise mitigation options.

Including all the parties that have been involved, up until the end of FP8 (Horizon 2020) in aviation noise research proposals submitted for EC funding, the grand total of participating organisations is now well over 200 since the early steps of the noise effort in FP4. As shown in Fig. 43, the SME trend in particular has kept growing, representing today more than 25% of that total figure.

A total of 26 European Union member states and associated states as well 4 "Third Countries" have been involved over the years in this effort, covering as shown on the projects roadmap (Fig. 7) source noise reduction, noise abatement procedures and management of noise impact.

Following the footsteps of successive X-NOISE CSAs, a dedicated "Global Coordination" activity is currently embedded in the ANIMA project. As such, former X-NOISE experts committees and National Focal Points network contribute to the ANIMA efforts aimed at establishing a "common strategic research roadmap for aviation noise reduction", as discussed in the previous section. Efforts have also been successful in extending the network of national focal point as discussed in the next section.

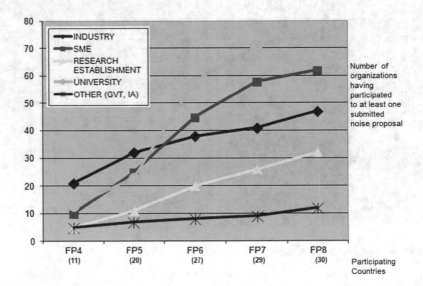

Fig. 43 Evolution of participation to noise project proposals

Integration of Research Community

Initial efforts were developed in line with the EU policy addressing the development of the European Research Area (ERA). In the process, the concept of network National Focal Points (NFPs) emerged as a way to the reach a maximum of research teams potentially interested in aviation related noise research, while keeping the network deliberations to a reasonable size.

As an essential element of the 3-pillar approach described earlier, the network of National Focal Points (NFP) covers the European Union as well as neighboring countries involved in the EU research framework programmes. As such, it aims at achieving a better integration of the aircraft noise research community, looking forward to bringing in innovative ideas, fostering new collaborations and disseminating solutions. Ultimately, it support the development of an up-to-date picture of the aviation noise research effort including EU-funded and national projects, providing organisations interested in performing or exploiting aircraft noise research with better visibility. The expected output from this process is a research taking advantage of local expertise, also addressing local priorities as part of the whole effort.

As can be see seen on Fig. 44, network Focal Points contributing to this day to the network activity, ensure a wide coverage of EU states and neighbouring countries.

Fig. 44 European Aviation noise research network—Current Geographic coverage

The role of NFP is expected to fit each country's specificity relative to the wider issues of mitigating aviation noise and providing solutions to reduce its impact. As an indication, it can typically involve the following activities:

- Summarise national aviation noise issues and associated research priorities
- Provide advice on the Common Research Roadmap for Aviation Noise Reduction developed by the network for the benefit of the EC.
- Identify relevant national programmes as well as potential input of interest for future European efforts and possible areas of collaboration within future projects.
- Contribute information to develop an overall roadmap of national programmes at European level
- Map and promote expertise of local researchers towards the European network as well as national authorities
- Inform local researchers on aircraft noise research strategic priorities at EU level.

The network membership is complemented by several experts committees gathering academia, research establishment and industry experts to contribute and review priorities for future research in the areas of technology as well as impact and balanced approach related issues. Figure 45 describes the process by which network participants support the definition and update of a common research roadmap.

Bringing participants together, a one day "Full Network " meeting has been organized every year since 2002 in co-location with a Scientific Workshop and annual Committee meetings. This has been an opportunity to involve and inform all network members. Typically, the meeting would proceed with the following programme of sessions:

- Update on the Common Research Roadmap for Aviation Noise Reduction
- Status on European Research Projects: reports from project coordinators
- Presentation of national research activities through a rotating scheme.

The Annual Scientific Workshop itself, jointly organized with the CEAS Aeroacoustic Specialist Committee, aims to address a priority topic relative to the research agenda. It is the occasion for the best experts in the field to present the most

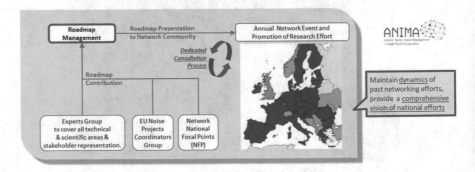

Fig. 45 Common research roadmap definition and update process

recent findings as well as for the whole network community to gather and exchange information.

At last, to support a steady stream of participation from local researchers and feed into the strategic vision process, "Calls for Novel Ideas" have been featured, seeking innovative proposals and project ideas. The OPENAIR, ANIMA and ARTEM projects typically took advantage of such exercises in the early stage of proposal development.

Furthermore, to consolidate the vision developed through individual focal point reports, a study on national efforts was carried out across the countries involved in X-NOISE activities, related to noise reduction at source on the one hand and management of noise impacts on the other hand. This survey was conducted in view of preparing the October 2014 ACARE workshop "Future Trends in Aviation Noise Research" discussed earlier. An example of its findings is presented in Fig. 46.

A follow up exercise is on-going under the ANIMA project that will produce high level roadmaps of national activities consistent with the Common Strategic Roadmap developed in parallel.

At last, it should be mentioned that Egypt have been associated to X-NOISE activity up its end, with the objective of developing a regional Mediterranean network.

Countries with a national aeronautics / transport research programme, which includes noise aspects (industry oriented)

ACARE 2035 Solutions	Summary	
	Research Efforts TRL 2 to 5 only (excluding large scale demos) 5-Year estimate: Recent & On-Going National Efforts (2011 - 2015)	
	Completed Research Projects	Completed PhDs
Powerplant system multi-disciplinary optimisation (incl. efforts in Computational AeroAcoustics and Source Modelling, except combustion)	35%	46%
Development of noise prediction and integrated system capability on UHBR powerplants (incl. efforts in Nacelle Acoustic Design and Installation Effects Modelling)	16%	20%
Combustion and IP System Noise Source Control	8%	5%
Lightweight, Recyclable Passive Liners	3%	6%
Wake control to reduce interaction noise for fans (e.g. trailing edge passive) and open rotor (e.g. pylon blowing adaptive)	5%	3%
Flow control for reduced turbomachinery broadband noise	8%	5%
Acoustic active control : Active stators, Active / Adaptive Liners	5%	5%
Adaptive Nozzles including Variable Area Nozzle	3%	5%
Morphing structures for airframes, nacelles and engines	0%	0%
Landing Gear Noise Flow Control Technologies	8%	5%
Novel Aircraft Configurations maximising noise benefits from masking effects: Airframe / nacelle / engine multi-disciplinary optimisation	8%	2%

Focus on PhDs (industry oriented & generic)

	Industry Oriented	Generic
	Completed PhDs (supported by national funding)	Completed PhDs (supported by national funding)
Computational AeroAcoustics and Source Modelling - Fan / Turbine Noise	46%	33%
Computational AeroAcoustics and Source Modelling - Jet Noise	46%	33%
Computational AeroAcoustics and Source Modelling - Airframe Noise	2%	14%
Computational AeroAcoustics and Source Modelling - Combustion Noise	5%	14%
Acoustic Liners / Propagation models	26%	14%
Active Noise Control Systems	9%	14%
Noise Reduction through Flow Control	12%	11%

- More than 100 PhDs in total
- 50% of national efforts on source modelling (core and propeller noise excluded)
- Activity on active / flow control remains low key, particularly in terms of systems aspects
- New absorbing concepts and materials to be pursued more agressively

Fig. 46 Vision of national activities aimed at Source Noise Reduction

Collaboration Efforts and Outreach

As lasting organisations beyond the limited timeframe of individual projects, research networks ensure a much needed structural continuity aimed at longer term strategies. In linking together, the research networks also play a key role in addressing wider issues such as the global environmental research agenda of the EU. Outside the European aviation noise community, the network activity have aimed over the years at exchanges and external collaboration in four main directions:

- Interdependencies issues with its European counterparts focused on aviation gaseous emissions (AERONET and FORUM-AE CSAs)
- Synergies with the noise research priorities expressed by ground transportation European experts bodies (CALM CSAs)
- International collaboration involving joint projects with Russia and regular exchanges with transatlantic counterparts (USA, Canada, Brazil) as well as support to EU initiatives (CooperateUS, Cannape, SUNJET and ICARe CSAs).
- Information and expertise contribution to regulatory bodies (ECAC, ICAO CAEP).

Noise—Emissions Interdependencies

X-NOISE had established over the years a close working relationship with the aviation emissions community (AERONET, ECATS, FORUM-AE), resulting in several co-organised seminars (2002, 2008) and in a proposed roadmap for European Aviation Environmental Modelling, including the definition of a Durable Organisation Structure presented to ANCAT in 2013 (see Fig. 47).

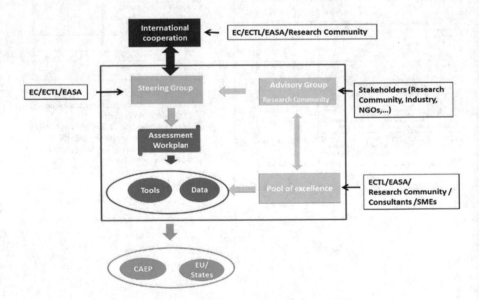

Fig. 47 Durable and cost efficient European Environmental Assessment Structure (EA2S)

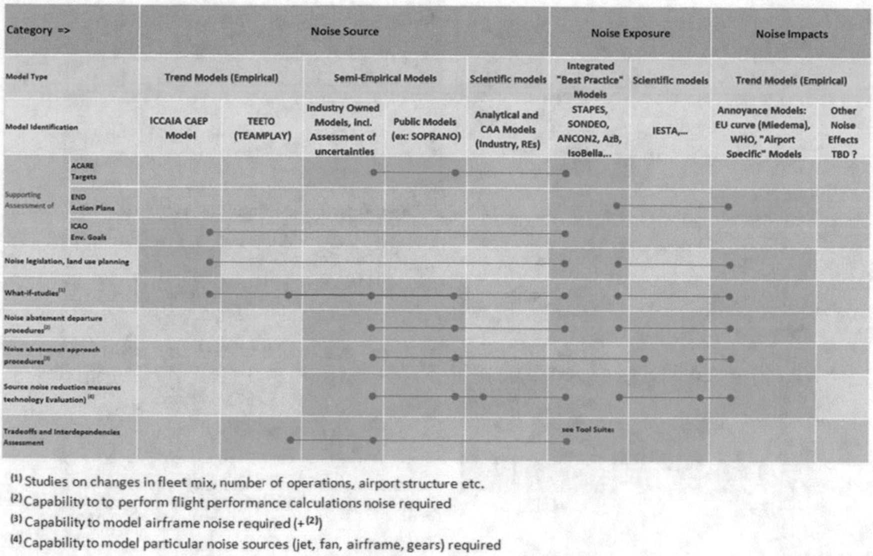

Category =>	Noise Source					Noise Exposure		Noise Impacts	
Model Type	Trend Models (Empirical)		Semi-Empirical Models		Scientific models	Integrated "Best Practice" Models	Scientific models	Trend Models (Empirical)	
Model Identification	ICCAIA CAEP Model	TEETO (TEAMPLAY)	Industry Owned Models, incl. Assessment of uncertainties	Public Models (ex: SOPRANO)	Analytical and CAA Models (Industry, REs)	STAPES, SONDEO, ANCON2, AzB, IsoBella…	IESTA,…	Annoyance Models: EU curve (Miedema), WHO, "Airport Specific" Models	Other Noise Effects TBD ?
Supporting Assessment of — ACARE Targets			•	•		•			
Supporting Assessment of — END Action Plans						•			
Supporting Assessment of — ICAO Env. Goals	•					•			
Noise legislation, land use planning	•					•		•	
What-if-studies[1]	•		•			•		•	
Noise abatement departure procedures[2]				•	•	•			
Noise abatement approach procedures[3]				•	•		•	•	
Source noise reduction measures technology Evaluation) [4]			•	•	•	•	•	•	
Tradeoffs and Interdependencies Assessment			•	•		see Tool Suites •			

[1] Studies on changes in fleet mix, number of operations, airport structure etc.
[2] Capability to to perform flight performance calculations noise required
[3] Capability to model airframe noise required (+ [2])
[4] Capability to model particular noise sources (jet, fan, airframe, gears) required

Fig. 48 Aviation Noise Models—Characteristics/Field of Application

As a follow-up action, a full picture of the situation on noise prediction models was developed for the 2016 Workshop on European Aviation Modelling Capabilities (Fig. 48).

Synergies with European Research Policies on Ground Transportation

In connection with the establishment of the Directive 2002/49/EC " Assessment & Management of Environmental Noise, the CALM initiative was the result of a close collaboration between the European Commission services responsible for developing the new European noise policy, DG Environment, and the DG Research. It aimed at ensuring that initiatives concerning research on noise reduction were in line with the requirements of the directive and vice versa.

The main CALM deliverable has been a strategy paper titled « Research for a Quieter Europe in 2020» first published in 2004 then updated in 2007. The CALM network expertise focusing primarily on ground transportation, it was advocated that a collaboration with X-NOISE was established in order for the noise research policy strategy to benefit from the work done for air transport within the ACARE framework. The strategy is expressed along two axes, perception and emission. While on the emission aspects (noise reduction at source), the proposed roadmaps are specific of transportation modes (i.e. X-NOISE/ACARE for air transport), the perception aspects priorities are identified across the board as shown on (Fig. 49).

Overall, an excellent level of collaboration was developed between ground and air transportation during these strategic exercises.

However, given the absence of a follow-up of CALM after 2007, it remains puzzling that no post-2020 overall European noise research strategy has been expressed so far in a similar way.

Fig. 49 Key aspects of European noise research strategy concerning perception

International Collaboration

Efforts aimed at international collaboration have focused in the first place on the countries featured in Fig. 1, with the understanding that counterparts pre-existed thanks to running research programmes in a similar area.

Over the years, with X-NOISE, then ANIMA support, a series of joint seminars organised through these counterparts have been held with US, Canada, Russia and Brazil. Links were also established with several CSAs involved in exploring international cooperation such CooperateUS, CANNAPE, SUNJET and recently ICARe.

As of FP6, this has led to participation in noise project proposals of US, Canadian, Japanese, Russian and Brazilian entities. Advantage was taken in particular of "Joint Calls" issued with Russia, Japan and Canada.

The ANIMA effort have recently focused on updating the topics of common interest and collaborated with ICARe to this end, completing in effect the three-point programme laid out for its "global coordination" activity (Fig. 50).

Fig. 50 International collaboration activity in ANIMA

Along with the definition of quite varied interests depending on the countries involved (from technology to impacts), it was clear from all discussions held with counterparts in this context that bi-lateral focused research was favored over larger multi-lateral efforts.

The existence of well-identified counterparts is key in successfully investigating topics and conditions for international collaboration. In the case of noise, these fruitful discussions were facilitated by the occurrence of formal networks (US, Canada) or informal ones (integrated national research projects in Japan, Russia and Brazil). Networks collaboration have been in fact an important factor in identifying or aligning projects on shared priorities.

Organised since 2005, the 3AF / AIAA Aircraft Noise and Emissions Reduction Symposium (ANERS) has provided such opportunity for network exchanges at regular intervals (every two year). To figure out the interest in going a step further into maintaining some form of constant dialog, a "Network of Networks Seminar" was organised by X-NOISE, taking advantage of the ANERS event held in September 2015. Identified networks involved in aviation environmental research included:

- US FAA Center of Excellence ASCENT (formerly PARTNER): Noise Emissions, Alternative Fuels
- Canada Network of Excellence GARDN: Noise, Emissions
- EU X-NOISE CSA: Noise
- EU FORUM-AE CSA: Emissions
- EU CORE-JetFuel CSA: Alternative Fuels
- ECATS International Association: Emissions
- ANNA (Aircraft Noise Non Acoustic effects): Noise.

The seminar session gathered representatives from the networks mentioned above as well as participants from Brazil, Russia, and Japan research programmes.

Following the debate, there was definite support for a yearly event dedicated to exchanges on international programmes, with a preference for the March to May slot. It was recommended that, while addressing the whole aviation environmental agenda, there should be a stronger focus on a specific topic, different at each event. The ANERS Symposium came close to the appropriate formula, but being organised by scientific societies (3AF, AIAA) with overheads and set conference organisation practices, the event programme somewhat lacked flexibility and remained conditioned by financial considerations.

In order to seek wider international involvement and participation, the capability to provide a Webinar option for abroad participants was put forward. At last, on years where a face to face event could not be organised, exchanging on international programmes by way of a webinar was recommended.

Unfortunately, the end of activity for the 3 European CSAs involved (X-NOISE, FORUM-AE, CORE-JetFuel) created a situation in which only ANIMA could further explore international collaboration, albeit in a very focused approach related to noise.

However, as could be verified through discussions during ANERS 2017 and ANERS 2020, the international interest is there for the network of networks formula provided appropriate support from the European side.

Information and support to regulatory bodies

A discussed above, dedicated seminars supported the work of ECAC ANCAT aimed at Interdependencies modeling. In another area connected both to ANCAT and the Environmental Noise Directive (END), several seminars (2003, 2005, 2010, 2013) were organized to discuss and debate noise mapping techniques in an air transport context.

Since 2006, the network has coordinated the European input to the report on noise research update provided at each plenary ICAO CAEP meeting (every three year). This report covers altogether the research programmes shown in Fig. 1 and has also been featured in a summary form in the ICAO Environmental Report published those same years.

Moreover, three international technology seminars were organised to kick-off the preparation of important ICAO CAEP events such as the first ICAO Noise Technology Workshop and the first and second ICAO Independent Experts Noise Technology Reviews (2001, 2008, 2011). These involved specialists and regulators from US, Canada, Japan, Russia and Brazil in particular. Similarly, the network coordinated European noise projects input to the new Integrated Independent Experts Review (noise and emissions combined) performed in 2017.

Lessons Learned

Along the strategy implementation effort described in the previous section, a numbers of lessons could be acquired from successful initiatives as well as occasional setbacks. The regular oversight provided by the global picture on worldwide research and ways of dealing with such efforts in various countries also provided interesting perspectives.

Reflecting on the Technology Focused Effort and the Extended Research Scope

- Strategy wise, the EU competitive funding mechanism, while quite different from the top-down US approach to similar efforts, allowed the successful implementation of the phased approach established around the ACARE SRA and SRIA. The coordination layer provided by a noise dedicated CSA has allowed regular exchanges with ACARE in terms of future priorities as well as assessment of progress made. This layer will remain essential as the research effort widens its scope towards achieving the 2050 target.
- Large noise dedicated, industry-led projects (Ref SILENCER / AFLONEXT) have proven their capability to bring new technologies to TRL6 in a very effective and cost efficient approach. These large noise projects have also been successful in

bringing on board a critical mass of multidisciplinary expertise in order to ensure a viable exploitation into an aircraft/engine, as production technologies needed to be matured in parallel to the noise technologies.

- However, a key observation has been that for these technologies to reduce noise further as a feature of the next engine/aircraft commercial programmes, a full scale demonstration such as engine static test or aircraft flight test was required to achieve this level of maturation. When dedicated funding for such demonstrations were not available, this created a situation where a number of promising technology solutions such as put forward by the OPENAIR project sat at TRL5 for some time, seriously restricting the panel of technologies able to support the satisfaction of the ACARE 2020 noise targets. As such, progress towards the ACARE Noise goals is not fully in line with expectations at this stage, which increases the risk that public acceptance will force authorities to take action against growth of air transport.

On a more global note:

- The whole effort involving some 35 projects in 25 years has associated industry, research establishments, academia and SMEs participants in a very collaborative manner, which led to a large number of medium size projects coordinated by research establishments.
- Thanks to networking / global coordination efforts:
  - contributions to the ACARE SRA and SRIA were developed in similar collaborative fashion.
  - the progressive extension of the research scope has been the opportunity for great collaboration between technology and impacts experts. The expected widening of the research calls for maintaining this approach in the future
  - dedicated activities have been instrumental in establishing successful international cooperation channels.
  - support to ICAO information activities on research and technology reviews have been timely and well perceived.

Noise vs Emissions

- Interdependencies between Noise and emissions (CO_2/NO_x/Particles) need to be quantified to provide visibility on how the trades could influence the achievement of environmental objectives. Such quantification is recommended from both an aircraft/engine design point of view as well as for the operational aspects of air transport.
- On the technical aspects of trades and interdependencies, an active dialog has been maintained with the emissions/modelling community. However, the coordination structures (CSA) handling these aspects have been absent the last few years. This is

a problem to consider when time comes for the next environmental goals progress assessment.

- Moreover, the European priority for aviation noise annoyance reduction appears strongly reduced given the changing focus on climate change (Green Deal). Experience tells that when funding is provided for a wider "environmental" scope, stakeholders tend to apply a limited part of it towards noise reduction.

EU Transport Noise Policy

- An excellent level of collaboration was developed between ground and air transportation at the occasion of the successive strategic exercises handled by the CALM network in preparation of the strategy paper « Research for a Quieter Europe in 2020». However, given the absence of a follow-up of CALM after 2007, it remains puzzling that no post-2020 overall European noise research strategy has been expressed so far in a similar way.
- In the absence of such a long term shared vision, dilution of funding through clustering of efforts with all transport modes remains a clear concern.

Future Aircraft and the Future of Aircraft Noise

Karsten Knobloch, Eric Manoha, Olivier Atinault, Raphaël Barrier,
Cyril Polacsek, Mathieu Lorteau, Damiano Casalino, Daniele Ragni,
Gianluca Romani, Francesco Centracchio, Monica Rossetti, Ilaria Cioffi,
Umberto Iemma, Vittorio Cipolla, Aldo Frediani, Robert Jaron,
and Lars Enghardt

Abstract In order to cope with increasing air traffic and the requirement to decrease
the overall footprint of the aviation sector—making it more sustainably and acceptable for the whole society—drastic technology improvements are required beside
all other measures. This includes also the development of novel aircraft configurations and associated technologies which are anticipated to bring significant improvements for fuel burn, gaseous and noise emissions compared to the current state and
the current evolutionary development. Several research projects all over the world
have been investigating specific technologies to address these goals individually,
or novel—sometimes also called "disruptive" —aircraft concepts as a whole. The
chapter provides a small glimpse on these activities—mainly from a point of view
of recent European funded research activities like Horizon2020 projects ARTEM,
PARSIFAL, and SENECA being by no-way complete or exhaustive. The focus of
this collection is on noise implications of exemplary novel concepts as this is one of
the most complicated and least addressed topics in the assessment of aircraft configurations in an early design stage. Beside the boundary layer ingestion concept, the
design process for a blended wing body aircraft is described, a box-wing concept is
presented and an outlook on emerging supersonic air transport is given.

Giving some views on possible future civil aircraft architectures—including supersonic ones—and
the possible consequences on noise at source

K. Knobloch (✉) · R. Jaron · L. Enghardt
Department Engine Acoustics, German Aerospace Center (DLR), 10623 Berlin, Germany
e-mail: Karsten.Knobloch@dlr.de

E. Manoha · O. Atinault · R. Barrier · C. Polacsek · M. Lorteau
ONERA—The French Aerospace Lab—Centre de Châtillon, Department of d'Aérodynamique,
Aéroélasticité et Acoustique (DAAA), 92322 Châtillon Cedex, France

D. Casalino · D. Ragni · G. Romani
TU Delft, Faculty of Aerospace Engineering, 2629 HS Delft, The Netherlands

F. Centracchio · M. Rossetti · I. Cioffi · U. Iemma
Department of Engineering, Università Degli Studi Roma Tre, 00146 Roma, Italy

V. Cipolla · A. Frediani
Department of Civil and Industrial Engineering, University of Pisa, 56126 Pisa, Italy

© The Author(s) 2022
L. Leylekian et al. (eds.), *Aviation Noise Impact Management*,
https://doi.org/10.1007/978-3-030-91194-2_5

Keywords Future aircraft configurations · Aircraft noise · Noise characteristics ·
Aircraft design for civil airliners · Noise reduction · ARTEM

The Need for Novel Aircraft Configurations

Karsten Knobloch

Air travel is undoubtedly one of the great achievements of scientists, engineers,
and many different professions which make the whole system of today's air transport
running smoothly. Business and leisure travel connecting almost all parts of the
world and affordable prices for a large part of the population – at least in developed
countries – brings literally "the world together".

However, the constant increase in passengers traveling, number of aircrafts and
number of flights is directly connected to some drawbacks, which are inherent to
all traffic systems: increase in resource consumption (e.g. fuel), absolute increase of
emissions, and increased annoyance from air traffic related noise. Although aviation
was accountable for only 3.6% of the total greenhouse gas (GHG) emissions in EU28
area [1], it is ranked second for transport related GHG emissions after road traffic,
and of increasing relevance with non-transport sources of GHG emissions declining
[2]. The European Aviation Environmental Report 2019 [3] stated an increase of
total passenger kilometers (departures from EU28 + EFTA states) of 60% between
2005 and 2017, an increase of people within 55 dB (L_{DEN}) noise contours around
airports of 12%, despite a decrease of average noise per flight of 14% in the same
time frame. The overall fuel consumption increased between 2005 and 2017 by 16%
while the average fuel consumption (per passenger kilometer) decreased by 24%.

This strongly underlines the constant technology development towards more efficient, less fuel-consuming and less noisy aircrafts, introduction of improved flight
procedures, efficient management practices, and the effect of fleet renewal. However,
the pace of evolutionary improvements – in the past based to a large extent on
improved aircraft engine technologies – is not sufficient to counterbalance current
and expected future growth of air traffic.

It is worth underlining that the period in which this book is written is characterised
by the outbreak of the COVID-19 virus, which in addition to the worldwide health
emergency, has caused an economic crisis that has curtailed the air traffic. According
to observers from international organisations such as IATA [4] or ICAO [5], although
the shock introduced by the pandemic is changing the global air traffic market in
the short-term, air transport will recover its positive growth rate, hence, soon the
environmental impact of aviation will be again a priority.

With this short general examination of the recent past and the current situation of the air transport sector it becomes clear that significant improvements in
all connected disciplines are required to counterbalance the expected growth and
beyond: to decrease the footprint of aviation in terms of use of resources, emissions,
and noise exposure.

This chapter focuses on the impact of aircraft technology, specifically novel aircraft configurations which differ from the tube-and-wing design of almost all current commercial aircraft. While some of these "novel" configurations have been discussed already for quite some time, usually the focus of past assessment was on specific benefits (e.g. fuel consumption) of the respective configuration—neglecting other important aspects which need to be addressed as well for a successful aircraft development process. A detailed assessment of expected aerodynamic performance, flight mechanics, fuel burn, and emitted noise already in an early stage of the design process is of utmost importance to initiate further activities in the long-lasting, expensive, and complicated process of the design and introduction of a completely new aircraft. The focus here is on noise implications of exemplary novel concepts as this is one of the most complicated and often least addressed topics in the assessment of aircrafts at concept stage. The selected activities—being by no-way complete or exhaustive—are based on recent and on-going EC funded research activities.

In particular, following four topics are addressed in this chapter:

- A new propulsion concept using embedded engines (BLI) on an otherwise nearly conventional tube-and-wing aircraft design is described and its implications on noise is assessed.
- A Multidisciplinary Conceptual Robust Design Optimisation (MCRDO) framework is described which gives a good impression of the complexity and interactions of the individual disciplines in an aircraft design process. An application of this framework for the design of two novel aircraft blended wing aircraft configurations is described as well.
- The box-wing concept – aiming at an improved aerodynamic performance – as explored within the recent H2020 project PARSIFAL is presented briefly and major project results are described.
- Finally, a further spotlight is shed on developments for super-sonic civil air transport – which is expected to resume in the near future. The latter does not imply the expectation for positive contributions of supersonic transport to the desired reduction in noise and resource consumption, but shall provide a short overview of current worldwide activities which are expected to influence the air transport sector and aircraft noise of the future.

Activities on the first two topics have been carried out within ARTEM, an EC funded Horizon2020 project started in 2017 and running until 2022 (grant number 769350). In ARTEM, partners from research centers, industry, and academia joint in order to help closing the gap between noise reductions obtained by current technologies – as already applied or being matured in large EU technology projects such as OpenAir and CleanSky – and the long-term goals of ACARE, i.e. a noise reduction of 65% for each aircraft operation in year 2050 compared to the reference year 2000 value.

The main topics of ARTEM are novel liner concepts and metamaterials capabilities for reduction of noise propagation, the reduction of noise generation itself by understanding, reducing, or tailoring component interactions, and the prediction and assessment of the effect of these noise reduction measures on aircraft level. The aim is to develop these "Generation 3" noise reduction technologies (NRTs) to a technology readiness level (TRL) of 3 (experimental proof of concept) to 4 (technology validated in lab).

The application of noise reduction technologies depends strongly on the aircraft configurations itself which requires the detailed consideration of potential future configurations. This was the reason to implement a specific work package within ARTEM dealing with the assessment of technologies on novel aircraft configurations, starting with a detailed definition of these configurations. The noise signature of the anticipated configurations will be strongly influenced by the interaction of several aircraft components: the interaction of airframe, high-lift-system, and propulsive jet of the engine(s), the interaction of airframe and engine inlet, the interaction of the landing gear with the airframe. These effects – which directly involve the noise generation – are investigated in the ARTEM framework by dedicated experiments and high-fidelity numerical calculations. The development of tools, their validation, and their application to investigate noise reduction potential of certain technologies is the major output of ARTEM.

The following subsections thereby highlight selected topics which are typical for many potential future aircraft concepts – while others had to be neglected for the sake of conciseness.

Boundary Layer Ingestion and the NOVA Concept: Implications on Noise

Eric Manoha, Olivier Atinault, Raphaël Barrier, Mathieu Lorteau, Cyril Polacsek

Damiano Casalino, Daniel Ragni, Gianluca Romani

In the last few decades, the constant strive for lower noise and fuel consumption of transport aircraft has led to a significant increase of the bypass ratio of coaxial turbofan engines, resulting in an increased fan and nacelle diameter. This tendency now reaches its limit for turbofan engines conventionally installed under the wing. An alternative is to partly bury the engine, in the rear fuselage for conventional tube-and-wing aircraft or above the airframe for more innovative blended wing body aircraft. The theoretical propulsive benefit of this "boundary layer ingestion" relies on the reduction of the exhaust jet wasted kinetic energy and filling-in the airframe wake velocity defect. Another positive consequence, beyond the reduction of the exposed frontal area of the aircraft, is the reduction of the overall aircraft mass and drag, due to the nacelle pylon removal and the lower wetted surface area. On the acoustic point of view, a potential noise reduction is expected from the partial or total shielding of the fan/OGV (outlet guide vane) noise sources by the airframe, as long as the amplitude of these noise sources does not significantly increase with respect to a conventional podded implementation, due to the space–time distortions of the flow

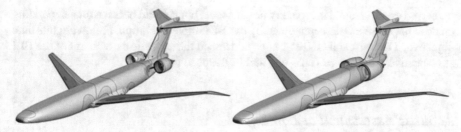

Fig. 1 NOVA aircraft configurations: conventional w.rear fuselage engines and BLI configuration

ingested by the fan. Up to now, the acoustic balance between these opposite effects of the BLI concept had not been assessed.

The NOVA Aircraft

In this general context, ONERA has designed the NOVA (Next generation ONERA Versatile Aircraft), which integrates the best available technologies for optimal propulsive performances. NOVA's architecture includes a wide lifting fuselage and a wing with high aspect ratio and downward oriented winglets [6, 7]. A modern ultra-high bypass ratio (UHBR) engine with bypass ratio of 16 has been specifically designed and implemented in several NOVA versions, either conventional (in isolated nacelles implemented under-wing or at rear fuselage) or semi-buried in the rear fuselage side, ingesting the boundary-layer developed along the whole fuselage length (Fig. 1).

Through several projects, these NOVA versions have been subjected to intensive aerodynamic numerical simulations by ONERA to compare their global propulsive performances.

ARTEM Objectives and Work Sharing

In ARTEM, the objective pursued by ONERA and the Technical University of Delft (TU-Delft) was to tentatively assess the acoustic performances of the NOVA aircraft versions, and especially to evaluate the impact of the BLI configuration on the overall noise through comparisons to the other more conventional configurations where the engine is considered as isolated. A secondary aim was to evaluate up-to-date numerical simulation methods, including the generation of aeroacoustic noise sources in the local aerodynamic flow field and the noise propagation to the far field, accounting for the shielding effects.

The global issue of aircraft noise is especially critical in the airport areas, at take-off and landing. At high altitude/speed, the noise from conventional aircraft perceived on the ground is typically not an issue, but it could become one, depending on possible noise increase generated by the BLI implementation. Following this dual objective, TU-Delft and ONERA have addressed the acoustic assessment of the BLI at respectively low speed (take-off) and high speed (cruise).

Acoustic Assessment at Low Speed

TUDelft's investigated the broadband and tonal noise generated by the fan embedded in BLI configuration, accounting for the turbulent flow developing over the fuselage at take-off with power cut-back. Due to confidentiality constraints on the BLI implementation designed by ONERA, the analysis used a generic benchmarked engine, the Low-Noise NASA Source Diagnostic Test (SDT), both isolated and integrated in BLI configuration. For this comparison, a global rescaling of the engine diameter and the nacelle length was required. Moreover, for the BLI version, the difficult design of an S-duct generating reasonable fan inflow was also needed (Fig. 2).

The numerical flow solution was obtained with the LBM (Lattice-Boltzmann method) solver 3DS Simulia PowerFLOW®. Then, the acoustic far-field was computed by using a FW-H (Ffowcs-Williams and Hawkings) integral solution from a permeable integration surface which encompassed the engine and a portion of the fuselage to partially take into account the acoustic scattering on the aircraft. The other installation effects due to the non-uniform BLI were implicitly taken into account by the installed engine simulation, as in Fig. 2. Compared to the isolated implementation, the BLI fan-stage is characterised by strong in-plane azimuthal velocity and blade loading fluctuations, and non-axisymmetric and incoherent rotor wakes. This results in far-field noise spectra with no distinct tonal components and high broadband levels. In Effective Perceived Noise levels, the BLI case turned out to be noisier than the isolated one, by 4 EPNdB at front side and 18 EPNdB at aft side (Fig. 3, see also [8]). A detailed analysis of the local aerodynamic flow field shows that the increase in noise results from a strong separation induced by the S-duct in the fan-stage rotor. This confirmed the necessity of a coupled optimisation of the blade loading with the duct shape to expect reasonable performances of the ducted configuration compared to the isolated one. This work in ARTEM sets one of the first

Fig. 2 NASA-SDT generic engine: isolated nacelle and BLI implementation in NOVA aircraft

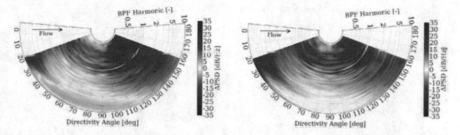

Fig. 3 Far-field noise directivity on ground arc (left) and sideline arc (right) centered around the fan, difference between BLI and isolated engine

Fig. 4 Mean axial velocity (m/s) over an iso-radial cut at 50% height for BLI case

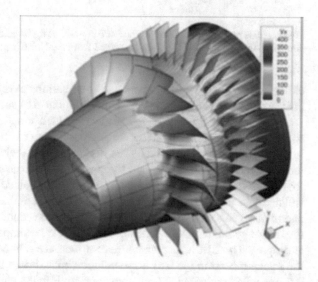

demonstrations of such a necessity to match combined aerodynamic and aeroacoustic constraints for the BLI propulsion concept.

ONERA investigated the tonal noise generated in cruise conditions in the intake and bypass duct of the turbofan engine specifically designed for the NOVA program, either implemented in BLI or isolated [9]. The harmonic loadings are inputted to an in-house FWH solver (ONERA FanNoise tool) adapted to annular duct propagation (Goldstein's formulation) [10].

Acoustic Assessment at High Speed

ONERA investigated the tonal noise generated in cruise conditions in the intake and bypass duct of the turbofan engine specifically designed for the NOVA program,

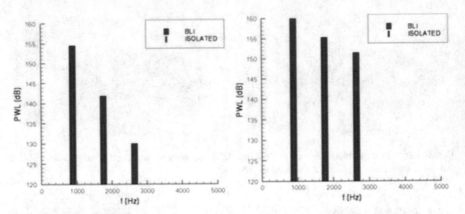

Fig. 5 Noise power spectra in the intake (BPF1 to BPF3) for baseline and BLI cases obtained from direct URANS (shockwaves, left) and assessed from FanNoise (rotor blade sources, right)

either implemented in BLI or isolated. fluctuations over blades/vanes provided by unsteady-RANS computations achieved with the *elsA* solver (Fig. 4).

At such transonic speeds, the blade tip Mach number is greater than one and shock waves propagate in the intake and contribute to the sound power, in addition to the loading noise generated by the flow interacting with the fan blades. Although the CFD mesh was not designed to ensure direct acoustic predictions, it has been found that the shocks are a dominant source at the blade passing frequency for the isolated case (sound power level PWL is +6 dB higher than the one due to RSI noise) and might balance the BLI effects on the harmonic sound power levels generated in the intake by the fan/OGV loadings (see the PWL comparisons between Fig. 5 left and right). The BLI is responsible for an increase of +6 dB of the overall PWL in the intake related to the RSI noise sources, that should be probably lower (around +5 dB) when including shocks contribution. Finally, harmonic loading noise from the OGVs in the bypass duct (expected to be a dominant source contribution in this region) gives rise to an increase of the OAPWL with BLI equal to +5.5 dB.

Conclusions and Perspectives on the BLI Study

This study presents original high-fidelity CFD/CAA simulations of a full-scale innovative aircraft concept comprehensive of a fan/OGV implemented in a BLI configuration. Both, at low speed (take-off) and high speed (cruise), a clear understanding is provided of the increase of the noise levels in BLI integrated configurations compared to more conventional podded turbofan installation. However all these results should be considered with the following reservations:

- At low speed, the study relies on the adaptation of a generic benchmarked engine, including the design "from scratch" of a new S-duct through a small number of iterations (due to limited project resources), leading to a strong flow separation ingested by the fan. It shows that such an optimised design is critical to minimise the impact on the BLI noise "penalty".

- At high speed, the method provides in-duct sound levels which reflect the source mechanism of fan/OGV interactions but do not account for any shielding effects by the S-duct and the airframe. Additionally, the completeness of this study should include an acoustic assessment at low speed of this BLI implementation designed by ONERA.

- Both studies at low and high speed do not include any passive or active flow/noise control technology that could significantly improve the acoustic characteristics of the BLI installation. Most promising passive devices include distortion-tolerant fan blades and acoustic liners in the S-duct. Beyond this, active technologies like smart adaptive blowing in the S-duct could counter inflow inhomogeneities and prevent flow separations and associated excess noise.

FRIDA, a Framework for Innovative Design in Aeronautics

Francesco Centracchio and Umberto Iemma

The tool FRIDA (FRamework for Innovative Design in Aeronautics) is the Multi-disciplinary Conceptual Robust Design Optimisation (MCRDO) framework developed by the Aerospace Structures and Design group of Roma Tre University. FRIDA is a conceptual optimal design environment capable of addressing innovative, unconventional configurations taking into account objectives and constraints related to all the aspects typical of a long-term design, including the environmental sustainability, community acceptance and life-cycle costs. The framework can guarantee the robustness and reliability of the design in presence of uncertainties using different approaches to the quantification of the statistical properties of the solutions. The project was conceived at the end of the nineties in response to the growing interest of the aeronautical community in disruptive, environment-friendly configurations for civil aviation. The need for breakthrough solutions to cope with the increasing severity of the environmental constraints imposed a rethinking of the conceptual phase of the design, introducing prime-principle-based models to deal with the lack of empirical or analytical methods. FRIDA's ancestor academic tool MAGIC (Multidisciplinary Aircraft desiGn of Innovative Configurations [11]), although limited to the optimisation of the lifting system using local line-search schemes, was one of the first aeronautical MDO frameworks including a community noise module to define objectives and constraints related to the acoustic impact. MAGIC was conceived to handle the conceptual design of unconventional concepts, like the Prandtl-Plane (PP) or the Blended-Wing-Body (BWB) configurations [12]. The FRIDA project was initiated during the FP6 project SEFA (Sound Engineering For Aircraft, 2004–2007) as a breakthrough enhancement of MAGIC. The most important improvements introduced were the extension to global optimisation using heuristic methods, the enrichment of the noise module with source-related models as a complement to the

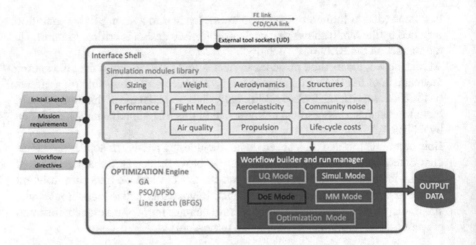

Fig. 6 Conceptual map of the FRIDA framework

classic noise-power-distance maps, and the development of an efficient parsing structure to make the framework easily expandable and linkable to external tools. These modifications made possible the first attempt of integration of sound-quality-based metrics in the design of an aircraft, opening the road towards the systematic use of perception-related objectives and constraints since the early conceptual phase ([13–14a]). Over the following decade, FRIDA has been extensively used and improved within the context of FP7 and H2020 projects COSMA (FP7, 2009–2013), OPENAIR (FP7, 2009–2014), ANIMA (H2020, ongoing), and ARTEM (H2020, ongoing) in order to address the increasing complexity of the requirements with a state-of-the-art, multi-objective, robust and reliable approach suited to the most advanced Simulation-Based Design Optimisation (SBDO) frameworks. The conceptual layout of FRIDA is depicted in Fig. 6. A parsing shell acts as an interface with the external world and manages the I/O structure, as well as the link with external simulation tools (under development). The *Simulation modules library* includes all the disciplines relevant for the design of a sustainable, innovative concept that can be activated through an appropriate description of the workflow. The *Interface shell* can be used to define at run time variables and parameters sets, allowing for a dynamic management of the design/analysis space. The optimisation core currently implements local line-search methods (BFGS, CG), genetic algorithm and deterministic and stochastic particle swarm methods (PSO/DPSO). The accuracy level of the physical models implemented in FRIDA can capture the relevant physics of the phenomena involved along the entire aircraft flight envelope. The implemented algorithms are, whenever possible, prime-principle-based under specific assumptions to reduce the order of complexity. The overall computational cost turns out to be compatible with the high number of evaluations typical of an optimisation process. The framework is completed by a *workflow builder* and *run manager* in charge of the translation of the user directives in a well-defined workflow. An example of workflow used in the

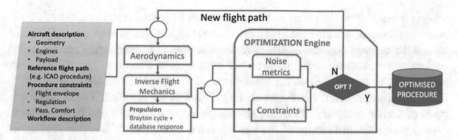

Fig. 7 FRIDA workflow used in ANIMA for optimisation of procedures

H2020 project ANIMA for the optimisation of Noise Abatement Procedures (NAP) is presented in the block diagram of Fig. 7. The modules activated are *Aerodynamics*, *Inverse Flight Mechanics* (calculates the aircraft settings for a given flightpath), and *Propulsion* (defines the engines rpm for a given thrust requirement using an in-house implementation of Brayton cycle coupled with a database of engines performance). The optimal flightpath is obtained using an evolutionary global optimiser (GA or PSO).

The *Workflow Builder* can combine any module to address a large variety of problems like quantification of uncertainties (*UQ mode*), design space exploration and design of experiment (*DoE mode*), or surrogate model definition (*MetaModelling mode*). Among the most recently introduced features, it is worth mentioning the link of the FRIDA flight simulation environment with the *Flight-Gear* (https://www.flightgear.org/) simulation tool and the open flight-dynamics library *JSBSIm* (http://jsbsim.sourceforge.net/). The interface is currently limited to the basic aerodynamic performance and the longitudinal dynamics of the aircraft. The full implementation of the *JSBSim* input format in the FRIDA I/O structure is under development. A preliminary example of a piloting session is available on a publication dedicated to "Flight simulation session of the HEP REBEL from ARTEM project" (https://doi.org/10.5281/zenodo.4650343, credits Marco Stefanini).

Aircraft Design and Noise Prediction for Novel Concepts

Francesco Centracchio, Monica Rossetti, Ilaria Cioffi and Umberto Iemma

The achievement of the ambitious noise reduction targets indicated in the ACARE Flightpath 2050 is subordinate to the identification of breakthrough solutions. Indeed, the simple progressive improvement of the current technologies wouldn't reach the level of noise abatement required to guarantee the quality of life of the community surrounding the airports in a scenario of a constantly growing market demand. One of the novel designs analysed and assessed in the project ARTEM is the Blended Wing Body (BWB) concept. It can be considered as a hybridisation of the flying wing concept, where the entire airframe generates lift, and a classic wide-body tube-and-wing aircraft, with the payload area located in a large central structure. The idea behind the concept consists in blending the wings and fuselage into a unified lifting surface, with a section of the large center body shaped like an airfoil. This configuration can be considered as unconventional in the context of civil aviation, even if the earliest designs of BWB airliners date back to the nineties. The main advantages of such a configuration are a high aerodynamic efficiency and the capacity to deploy a significant amount of lift for the same flying conditions. This results in a low consumption of fuel and the possibility to take off and land with a reduced use of the high-lift devices, which are responsible for a large portion of the aerodynamic drag and noise. An additional benefit of this configuration is obtained by the possibility to install the propulsion system on the top the center body and exploit the effect of shielding provided by the large surface of the payload-carrying structure. In the ARTEM project, three BWB configurations have been developed on the basis of two mission profiles: one regional/short range mission (<900 nautical miles) and one long-haul profile of 5500 *nautical miles*. The long-range configuration denominated *BOLT* (Blended *wing body with Optimised Low-noise Technologies)* has been equipped with two last-generation UHBR turbofan engines, with an expected payload of 400 passengers in a two-classes cabin layout. The short-range *REBEL (REgional Blended-wing-body Electric-propelled),* although originally conceived to be equipped with hybrid electric propulsion, has been actually designed in two versions: a baseline configuration equipped with conventional technologies and turbofan engines (REBEL-C) and the REBEL-HEP propelled by a distributed hybrid electric system [15–17]. A pictorial rendering of the ARTEM BWB fleet is presented in Fig. 8.

Fig. 8 Pictorial rendering of the ARTEM BWB fleet in flight

The research activity performed within the framework of the ARTEM and ANIMA projects was devoted to the development and assessment of efficient and accurate models for the aeroacoustic assessment of this class of aircraft. A campaign of numerical simulations is foreseen in ARTEM using the analysis tools available at the Italian Center of Aerospace Research (*CIRA*, [18, 19]), whereas in ANIMA the Aerospace Structures and Design group of the *Roma Tre* University is in charge of the development of suitable surrogate models for the estimate of the shielding effect to be integrated in the design toolset [20–22].

Other ARTEM partners have used the BWB design for the assessment of noise shielding capabilities and improvement of models for the prediction of this effect. During the final project phase, the noise reduction technologies will be applied to the BWB configurations BOLT and REBEL, and by the subsequent simulation of flight trajectories the noise signature will be predicted and auralised for comparative listening tests.

Benefits of the Box-Wing Architecture for Passenger Aircraft—The H2020 Project "PARSIFAL"

Dr. Vittorio Cipolla and Prof. Aldo Frediani

PARSIFAL stands for "PrandtlPlane Architecture for the Sustainable Improvement of Future Airplanes". Funded by the EU under the Horizon 2020 Programme between 2017 and 2020, the PARSIFAL project has investigated the feasibility and performed a comprehensive impact assessment of the application of the *box-wing* architecture to short-to-medium range aircraft.

The box-wing configuration derives from the studies carried out in 1920s by Ludwig Prandtl, who described the box-wing, i.e. two horizontal wings connected at their tips by properly designed vertical wings (Fig. 9, left), as the "best wing system". According to Prandtl studies published in 1924 [23], in fact, among all the possible lifting systems the box-wing is the one capable to minimise the induced drag, once lift and wingspan are given. In order to pay tribute to Prandtl, the research team from University of Pisa, that in the 1990s started to study the engineering application of

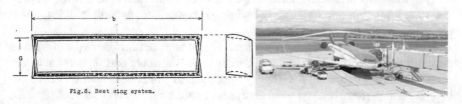

Fig.8. Best wing system.

Fig. 9 Prandtl's "best wing system" (left) and the PrandtlPlane object of study of the project PARSIFAL (right)

the "best wing system" concept to aircraft design, gave the name "PrandtlPlane" to the resulting aircraft Fig. 9, right).

One of the practical consequences of the improved aerodynamics is that the box-wing allows for an increase of the so-called *span efficiency*, one of the most significant parameters defining the lift-to-drag ratio of an aircraft, without increasing the wingspan.

Thinking at todays need of reducing the fuel consumption, hence the environmental impact, of aircraft, one of the possible approaches is acting on the aircraft architecture in order to improve the lift-to-drag ratio, also called aerodynamic efficiency. One of the ways to achieve such result is to improve the span efficiency by increasing the wingspan, which can be applied without any change to the architecture, hence by replacing in-service aircraft with larger ones, or introducing unconventional architectures such as truss-braced wings or folding wings, today also objects of research and development programs.

Aircraft with larger wingspan are good solutions as far as constraints from available apron space and other airport infrastructures are not taken into account, since they allow the increase of passenger numbers with the same amount of aircraft movements.

The PARSIFAL project has been carried out by the University of Pisa (Italy), in the role of coordinator, ONERA (France), the Delft University of Technology (Netherlands), ENSAM – Arts et Métiers (France), the DLR (Germany) and SkyBox Engineering, a SME from Italy. Although the PrandtlPlane configuration is suitable for different aircraft categories, the PARSIFAL project has been focused on short-to-medium haul aircraft, which typically feature a wingspan within 36 m as they belong to ICAO "C" category, such as the Airbus 320 or Boeing 737 aircraft. The reason for this choice is the fact that about 75% of passenger aircraft belong to this category, hence the compliance with such a standard implies a stronger impact.

The aerodynamic advantages of the box-wing have been utilised solely for the increase of passenger capacity (i.e. the payload) and not for an increase of cruise speed, the mission range, or other key performance parameters.

This requirement has been met by designing a double aisle fuselage specifically conceived for the box-wing system, with the result of increasing the number of maximum passengers from below 200 for the conventional aircraft to more than 300 in the PrandtlPlane case.

The possibility to increase the fuselage dimensions and, at the same time, keeping the 36 m wingspan constraint and obtaining a higher aerodynamic efficiency is a peculiarity of the box-wing system.

Starting from these two main choices, the consortium has carried out several design loops, adopting first low and medium fidelity tools and then refining the results by means of a multidisciplinary approach, in which high-fidelity tools have been used to perform aerodynamic analyses at both transonic and subsonic speeds, structural weight estimations, mass and balance characteristics evaluations, aeroelastic analyses, flight dynamic simulations, sizing of control surfaces, engine sizing and integration, landing gear sizing, etc.

The same approach and same tools have been adopted to study the CeRAS CSR-01 [24], a short range conventional reference aircraft which the PrandtlPlane has been compared to.

All the output obtained from these analyses have been then used to perform comparative impact assessment between the PrandtlPlane and the reference aircraft. In detail, these analyses have been focused on evaluating the atmospheric emissions during the whole mission and their influence on global warming, noise footprint during take-off and climb and associated psychoacoustic metrics, impact on turnaround time and airport logistics and, finally, the influence on direct operating costs and profitability from an airline point of view.

The most significant result is the following: increasing the number of passengers of about 50% allows for a reduction of fuel per passenger, hence emitted CO_2, up to 22%. As detailed in [24a], the associated Global Warming Potential, calculated on a 20 years horizon, would be reduced by about 17%. In addition, the day-evening-night average level of noise is decreased for a given airport with assigned daily passengers traffic. On airport operations side, the turnaround of 300 passengers PrandtlPlane will be take only 5–10 min (depending on the considered operation scenario as defined in aircraft manufacturers' manuals, more than a 200 passengers conventional aircraft, e.g. outstation, full service, etc.) without any additional burden in terms of apron space or slots needed [24b]. Finally, direct costs per available seat-kilometer could be cut by about 12%. Although the aircraft purchase cost would go up of about 60%, the reduced direct operating cost would give the airlines the possibility to reach the same break-even point of conventional competitors with an average ticket price reduction of 13%.

As said, these numbers come from the assumption of a growing market with some saturation constraints. For a more comprehensive analysis, the same comparisons have been performed considering a "less disruptive" 240 passengers PrandtlPlane, i.e. provided with a conventional single aisle fuselage (similar to the A321). By comparing the results with those of the reference conventional aircraft, similar margins of improvement have been estimated for both CO_2 (about -20%) and direct operating costs (about −10%). More details can be found in [24b–24d] and the official reports of the project [25, 26].

As a final remark, it is worth to underline that the achieved results do not take alternative fuels or alternative propulsion systems into account, but are only due to the box-wing architecture. Further investigations on the synergies between the box-wing and hydrogen-based or hybrid-electric propulsion systems are ongoing and will be a significant part of the future research on the PrandtlPlane.

Sustainable SuperSonic Transport: Technical Challenges and Noise Certification

Robert Jaron and Lars Enghardt

Fig. 10 Aerion Supersonic AS2 business jet (left ©Aerion Supersonic) and Boom Supersonic Overture airliner (right ©Boom Supersonic)

About 25 years after Concord's last flight, commercial civilian supersonic aircraft may possibly enter the market once again as early as in the mid-2020s. Among others, two American manufacturers have announced an approaching market entry. Aerion Supersonic is designing a supersonic business jet cruising at Mach 1.4 with a cabin large enough for 12 passengers (Fig. 10). Boom Supersonic announced an airliner cruising at twice as fast as today's commercial airplanes and a passenger capacity up to 88.[1] It is expected that public acceptance for supersonic transport can only be achieved by concepts that are both sustainable and indistinguishable from subsonic aircraft in terms of their noise annoyance. Combining the high engine thrust required for supersonic cruise with sustainability poses a major challenge for manufacturers and researchers. Moreover, at the time being, there are no applicable noise certification standards for supersonic aircraft. Thus, the International Civil Aviation Organisation (ICAO, or national certification authorities) needs to define standards for the advent of this new generation of supersonic aircraft to ensure their public acceptance. With regard to noise emissions, two major challenges have to be tackled by establishing new certification standards in parallel to the technology development: Firstly, the landing and take-off (LTO) noise and secondly, the sonic boom and its impact on human beings.

The next generation of supersonic aircraft is expected to be restricted to flying subsonically over land, as technologies for so-called low-boom designs have not yet reached market readiness. The impact of supersonic booms caused by conventional designs on humans, animals, and building structures is still unacceptable. For that reason, flying supersonically over land is prohibited in several countries and even over water supersonic speed is only allowed with a sufficient buffer distance to the coastline, as secondary booms can propagate long distances depending on the prevailing atmospheric conditions. Thus, the major challenge of the next generation of supersonic aircraft will be to meet LTO noise levels comparable to subsonic aircraft. Supersonic aircraft are facing major tradeoffs between cruise performance and LTO noise. With regard to cruise performance, small engine diameters with high

---

[1] Aerion announced the termination of this AS2 business jet programme just when the present book was about to be edited. However, some similar concepts may appear again in the near future.

jet speeds are required. This is because the pressure drag induced by the frontal cross-sectional area increases considerably at transonic and supersonic cruise speeds. As a consequence, the volume distribution along the longitudinal axis of the aircraft is decisive for the wave drag. Using a design of three engines, the volumes can be distributed much more homogeneously with three small engines, one of which is placed in the vertical stabiliser. However, since installed jet noise scales approximately to the sixth power of jet speed, small engines with high jet speeds lead to very high LTO noise levels. Regarding the take-off certification procedure, there are substantial differences between subsonic and supersonic aircraft: Subsonic aircraft engines are dimensioned for takeoff conditions. In contrast, supersonic aircraft have the highest thrust requirements at top of climb leading to massive excess thrust at takeoff conditions. Given the tradeoff between LTO noise and cruise efficiency for supersonic aircraft, meeting the LTO noise targets will likely require not only technical developments, but also changes to the takeoff procedures specified in the noise regulation rules.

Already in 1979, Grantham and Smith [27] investigated LTO procedures of a supersonic airliner with a payload of 273 passengers cruising at Mach 2.62 for community noise abatement purposes. To reduce the flyover noise levels, they suggested a fast climb out at lower climb rate with higher speed. As supersonic aircraft have a significantly higher minimum drag speed, higher speeds allow a higher thrust cutback and in consequence a reduction of flyover EPNdB levels. To reduce sideline certification levels as well, they suggested a thrust reduction before the cutback point. The reduction level was limited in order to maintain the flight height or a four percent climb gradient for the emergency case of one engine being inoperative. Since then, this procedure has become known as programmed thrust lapse rate (PLR) takeoff. To ensure the same thrust reduction in daily operations and to relieve the pilot from too many actions during takeoff, only a fully automated FADEC (Full Authority Digital Engine Control) controlled thrust reduction should be considered for new certification procedures in addition to the allowance of a higher climb out speed V_2. Recently, NASA and JAXA studied the acoustic advantage of the adjusted takeoff trajectory on a supersonic business jet with three engines, a cruise speed of Mach 1.4 and a MTOW of 45 [28, 29]. With 10% PLR thrust reduction and a 15 kn higher climb out speed ($V_2 + 35$ kn), they found a cumulative EPNdB reduction in the order of 3 dB. Nevertheless, the margin to the Chap. 14 noise regulation rules achieved was so small that other noise reduction features were investigated, as well. By enhancing the bypass ratio from 2.9 to 3.6, the cumulative noise benefit was estimated to be 5.3 EPNdB, albeit accompanied with a 4.1% penalty regarding the possible flight range. Using chevrons on engine exhausts, an estimated noise reduction potential of 2.7 EPNdB was identified, this time accompanied with a penalty of 2.8% in flight range. Another promising technical option to reduce jet noise is the so-called mixer ejector nozzle, which enables the virtual increase of the takeoff bypass ratio by injecting air from outside the engine into the jet through a variable opening to efficiently reduce the jet velocity. Variable mixer ejectors were investigated by NASA [30] and in the EU FP6 project HISAC with an expected benefit of approximately 7 EPNdB. Here, the identified drawback is an increased drag coefficient and extra weight. In view of these results, there is an urgent need for extensive research and development to adapt takeoff procedures and engine technologies in order to achieve the targeted noise levels at reasonable costs in terms of the anticipated flight range.

The second major challenge to establish supersonic aircraft on the market in the long term is the sonic boom. So far, there are two approaches of lowering the annoying pressure fluctuations impinging on the ground. Firstly, the so-called Mach cut-off procedure, where the airplane is flying just over Mach 1.0 but below Mach 1.15 For a steady standard atmosphere, flying in this very speed range will theoretically result in the sonic boom waveform not reaching the ground. Since this approach is strongly dependent on the atmospheric conditions as well as the uncertainties predicting these conditions, for the time being, it will quite likely not be allowed. The second possibility is a so-called low-boom design. After propagating through the atmosphere, the pressure signature of a conventional sonic boom has the very noisy and detrimental shape of a N-wave. Due to a controlled nose shock with gradual pressure increase to the wing, the ground signature can be changed to a sine wave with lower gradients compared to shocks of conventional designs. Low-boom airframes are optimised to substantially lower the annoyance of the sonic boom. The objective of such optimisations is to minimise the human response to indoor and outdoor sonic boom predicted by accurate methods. Precise sonic boom prediction relies on the CFD simulation of the near field pressure field, the analytic propagation through the atmosphere considering atmospheric turbulence [31] and the modeling of the ground topology. In the H2020 project RUMBLE (2017–2020), different sonic boom prediction tools were developed and assessed [32]. The need for reliable sonic boom prediction has already been addressed in three workshops at the AIAA SciTech conference in 2014, 2017, and 2020 [33]. Regarding the human response one has to find an appropriate metric [34] but also conduct listening tests and sleep studies.

In 2021, two H2020 EU projects started with the aim to further improve the detailed assessment of supersonic aircraft on the one hand regarding LTO noise and on the other to investigate sonic boom. The projects will support the establishment of new certification standards ensuring sustainability and public acceptance of novel supersonic aircraft models. In the project SENECA (LTO noise and emissions of supersonic aircraft, 2021–2024), supersonic aircraft concepts will be designed and multidisciplinary optimised regarding airframe and engine architecture as well as the LTO trajectories. The aim is to reduce LTO noise and emissions as well as the global climate impact. In total, four different aircraft concepts will be examined: two business jets with cruise Mach numbers of 1.4 and 1.6 and two airliners cruising at Mach 2.0 and 2.2. The certification authorities will be given the scientific evidence to change the LTO procedure by handing over a comprehensive and reliable database of virtually flown landings and takeoffs with varying climb out speeds and thrust reductions before cutback and also with different aircraft and engine architectures, seeking for the lowest community noise. In particular various engine bypass ratios, the number of engines, the position of the engines as well as the nozzle and inlet geometry will be examined in terms of their tradeoffs between flight range and their noise reduction potential. The project MOREandLESS (MDO and Regulations for Low-boom and Environmentally Sustainable Supersonic Aviation, 2021–2024) will also investigate the sustainability of upcoming supersonic aircraft focusing on higher cruise Mach numbers starting with airliners cruising with Mach 2.2 going up to the hypersonic regime with Mach numbers of 5. In addition to the multidisciplinary optimisation of the aircraft concepts, experimental and numerical investigations of jet noise are envisaged. The impact of the aircraft shape on the sonic boom will be investigated by shooting projectiles in the shape of aircraft out of a gun on an outdoor testing track. The resulting sonic boom will be measured by microphones. Furthermore, to support the modeling of sound boom propagation and variability due to meteorology and turbulence in different environments, in rural areas and in urban areas, various measurements will be performed.

For the near future, a first low-boom flight demonstrator, the NASA x-59 is scheduled to make its first flights in 2022 in the USA. With regard to the global nature of air transport and with the difference between reactions of local communities, some flight tests may also be planned in Europe, in coordination with the European and EU national authorities for civil aviation. The NASA X-59 program based on the Lockeed-Martin QuietSST aircraft aims at proving low-boom technologies at large scale and gathering data on boom propagation and human response to the low-boom flight (Fig. 11).

In the long term, low-boom concepts will have to be scaled for higher payloads. Furthermore, one has to handle the integration of more than one engine to comply with civil aircraft safety regulations. Merging low-boom volume distributions with more than one engine and meeting LTO noise regulations will be the biggest challenge to finally achieve publicly accepted supersonic flight over land.

Fig. 11 NASA/Lockeed-Martin low-boom demonstrator X-59 (©Lockeed-Martin)

Concluding Remarks

The current chapter has presented only a small number of aircraft concepts which are currently under consideration for the improvement of efficiency and sustainability of future air transport.

Generally speaking, the research and development costs for future improvements are expected to be rather high. With current engine technology having reached a high level of maturity and complexity, the further increase in bypass ratio will be somehow limited by detrimental effects like drag, weight, (under wing) installation space etc. Geared turbofans have made a significant contribution by reducing the rotational speed of the fan at the cost of increased weight and cost for the gear itself. Future engine installations with bypass ratios of 16 or beyond will face integration issues and interaction effects as can be seen from inflow distortion effects for the BLI concept.

Open rotors instead of nacelle-mounted turbofan engines – which have not been discussed here – are another option for increased efficiency, but inhibit again the different noise characteristics demanding for adapted noise reduction technologies. The distributed propulsion – being driven by small turbo-prop engines or electric motors – are currently being considered at least for short-range aircraft (like the ARTEM-REBEL configuration). Here, the interaction and phasing effects are one of the major topics – beside the generation and distribution of electric energy for electrified versions. Electric driven propulsors are not necessarily "quiet" a priori, as the well-known fan noise sources and interactions are present as well. Moreover there are unknown effects of mutual interactions in the case of multiple propulsors.

For the aircraft fuselage, a clear trend towards lift-providing structures is visible – which is consistent with the airframes of NOVA, BOLT, and REBEL presented here in the framework of the ARTEM project. TU Delft has pursued its "Flying V" – a variation of the blended wing concept [35]. AGILE, CENTRELINE, IMOTHEP and NACOR [36, 37] are recent or on-going EC funded projects which also deal with future aircraft configurations – mainly from performance and aerodynamic point of view.

For the introduction of disruptive configurations like the blended wing body, rather drastic changes are likely being required for current airport facilities, maintenance procedures, but also for design and manufacturing routines. So far, there is a certain lack of data and therefore in reliability of all predictions with respect to performance, aerodynamics, but especially also noise emissions of these configurations.

NASA and other institutions have made considerable efforts in the detailed assessment of current and future aircrafts. Khorrami and Fares [38] demonstrate these activities – including tools and simulation validation from sub-structures and wind tunnel models up to full scale-flight test with a Gulfstream research aircraft from aerodynamic and aero-acoustic perspective – the latter highlighted as being the more demanding. Spakovszky [39] provides a summary of activities ranging from quiet aircraft demonstrator (QAT), over the Silent Aircraft Initiative, to environmentally responsible aircraft (ERA) programs including also the well-known MIT-driven D8 "double bubble" concept. The challenges are the same as briefly addressed in this chapter here: integration effects (like inflow distortion), shielding effects, and the modelling of noise reducing technologies. The interested reader can find a good list of current references on these programs' outcomes in [39].

Examining all these activities it becomes clear that, beside academic research by university and research centers, significant contributions from aircraft and engine manufacturers are needed in order to obtain higher TRL levels for the novel configurations. Airbus has recently revealed MAVERIC, a blended wing body demonstrator [40], which is likely to collect valuable validation data for future studies – thereby giving hope for the realisation of novel – more sustainable aircraft configurations in the coming decades. It must be also reminded that all those technological improvements have to be ultimately accepted by the market, i.e. the development and introduction are strongly dependent on regulations, and on the overall competitive advantage they may provide (where also noise reduction translates back to earned or saved money).

Acknowledgements The main authors wish to thank all contributors to this chapter – most co-authors are mentioned directly in conjunction with their respective contributions. The funding provided by the EC for the project ARTEM under grant number 769350 is gratefully acknowledged. Also, many other projects receive funding by the EC (grant numbers are given above). Finally, we would like to thank the ANIMA team for initiating the current book project and giving us the opportunity to share our views and findings on the future of air transport.

References

1. EEA (2018) Greenhouse gas—data viewer. https://www.eea.europa.eu/data-and-maps/data/data-viewers/greenhouse-gases-viewer
2. EEA (2017) Trends and projections in Europe 2017, EEA Report No 17/2017. https://www.eea.europa.eu/publications/trends-and-projections-in-europe-2017
3. European Aviation Environmental Report 2019, Published by EEA, EASA, and EurControl. https://doi.org/10.2822/309946
4. Pierce B (2020) COVID-19 outlook for air transport and the airline industry. IATA, November 2020
5. Effects of novel coronavirus (COVID-19) on civil aviation: economic impact analysis. ICAO, February 2021
6. Wiart L, Atinault O, Boniface J-C, Barrier R (2016) Aeropropulsive performance analysis of the NOVA configurations. In: 30th congress of the international council of the aeronautical sciences, Daejeon, South Korea, September 2016. https://www.icas.org/ICAS_ARCHIVE/ICAS2016/data/papers/2016_0092_paper.pdf
7. Wiart L, Atinault O, Grenon R, Paluch B, Hue D (2015) Development of NOVA aircraft configurations for large engine integration studies. In: 33rd AIAA applied aerodynamics conference, Dallas, TX, June 2015. ISBN 9781624103636; https://doi.org/10.2514/6.2015-2254, https://doi.org/10.2514/6.2015-2254.
8. Romani G (2020) Numerical analysis of fan noise for the NOVA boundary-layer ingestion configuration. Aerosp Sci Technol 96:105532. https://doi.org/10.1016/j.ast.2019.105532
9. Godard B, De Jaeghere E, Ben Nasr N, Marty J, Barrier R, Gourdain N (2017) A review of inlet-fan coupling methodologies. In: Proceedings of ASME Turbo Expo 2017, Paper GT2017-63577

10. Lewy S, Polacsek C, Barrier R (2014) Analytical and numerical prediction of harmonic sound power in the inlet of aero-engines with emphasis on transonic rotation speeds. J Sound Vibr 333(26)

11. Morino L, Bernardini G, Mastroddi F (2006) Multi-disciplinary optimisation for the conceptual design of innovative aircraft configurations. CMES 13(1):1–18. Tech Science Press

12. Iemma U, Diez M (2005) Optimal life-cycle-costs design of new large aircraft including the cost of community noise. In: ICCES 2005, IIT Madras, 1–6 December 2005, Chennai, India

13. Iemma U, Diez M, Marchese V (2006) Matching the aircraft noise to a target sound: a novel approach for optimal design under community noise constraints. In: 13th international congress on sound and vibration

14. Diez M, Iemma U (2012) Multidisciplinary conceptual design optimisation of aircraft using a sound-matching-based objective function. Eng Optimisation 591–612

14a. Centracchio F, Burghignoli L, Iemma U (2021) Multiobjective optimisation of flight paths for noise level mitigation and sound quality improvement. Noise Mapp 8(1):268–280. https://doi.org/10.1515/noise-2021-0022

15. Centracchio F, Rossetti M, Iemma U (2018) Approach to the weight estimation in the conceptual design of hybrid-electric-powered unconventional regional aircraft. J Adv Transp

16. Burghignoli L, Centracchio F, Iemma U, Rossetti M (2018) Multi-objective optimisation of a BWB aircraft for noise impact abatement. In: 25th international congress on sound and vibration, ICSV25, Hiroshima, Japan

17. Bernardini G et al (2020) Numerical characterisation of the aeroacoustic signature of propellers array for distributed electric propulsion. Special issue "Airframe noise and airframe/propulsion integration" of applied science (MDPI). Appl. Sci. 10(8):2643. https://doi.org/10.3390/app10082643

18. Visingardi A, D'Alascio A, Pagano A, Renzoni P (1996) Validation of CIRA's rotorcraft aerodynamic modelling system with DNW experimental data. In: 22nd European rotorcraft forum, Brighton, UK

19. Barbarino M, Bianco D (2018) A BEM–FMM approach applied to the combined convected Helmholtz integral formulation for the solution of aeroacoustics problems. Comput Methods Appl Mech Eng. https://doi.org/10.1016/j.cma.2018.07.034

20. Centracchio F, Burghignoli L, Rossetti M, Iemma U (2018) Noise shielding models for the conceptual design of unconventional aircraft. In: 47th international congress and exposition on noise control engineering, Inter-Noise 2018, Chicago, Illinois, USA

21. Burghignoli L, Rossetti M, Centracchio F, Iemma U (2019) Noise shielding metamodels based on stochastic radial basis functions. In: Proceedings of the 26th international congress on sound and vibration, ICSV 2019

22. Centracchio F, Burghignoli L, Rossetti M, Iemma U (2018) Noise shielding models for the conceptual design of unconventional aircraft. In: INTER-NOISE 2018—47th international congress and exposition on noise control engineering: impact of noise control engineering

23. Prandtl L (1924) Induced drag of multiplanes. Technical Report TN 182, NACA 1924

24. CERAS CeRAS-CSR01: short range reference aircraft. https://ceras.ilr.rwth-aachen.de

24a. Tasca AL, Cipolla V, Abu Salem K, Puccini M (2021) Innovative box-wing aircraft: emissions and climate change. Sustainability 13:3282. https://doi.org/10.3390/su13063282

24b. Picchi Scardaoni M, Magnacca G, Massai A, Cipolla V (2021) Aircraft turnaround time estimation in early design phases: simulation tools development and application to the case of box-wing architecture. J Air Transp Manage 96. https://doi.org/10.1016/j.jairtraman.2021.102122

24c. Abu Salem K, Cipolla V, Palaia G, Binante V, Zanetti D (2021) A physics-based multidisciplinary approach for the preliminary design and performance analysis of a medium range aircraft with box-wing architecture. Aerospace 8(10):292. https://doi.org/10.3390/aerospace8100292

24d. Abu Salem K, Palaia G, Cipolla V, Binante V, Zanetti D, Chiarelli M (2021) Tools and methodologies for box-wing aircraft conceptual aerodynamic design and aeromechanic analysis. Mech Ind 22:39. https://doi.org/10.1051/meca/2021037

25. PARSIFAL Project Consortium. Report on operational and economic assessment. PARSIFAL Project Deliverable D1.2, 2020 https://www.parsifalproject.eu/PARSIFAL_DOWNLOAD/PARSIFAL_D12.pdf

26. PARSIFAL Project Consortium. PrandtlPlane performance analysis and scaling procedures. PARSIFAL Project Deliverable D3.4, 2020. https://www.parsifalproject.eu/PARSIFAL_DOWNLOAD/PARSIFAL_D34.pdf

27. Grantham WD, Smith PM (1979) Development of SCR aircraft takeoff and landing procedures for community noise abatement and their impact on flight safety. Supersonic Cruise Research, pp 299–333

28. Akatsuka J, Ishii T (2020) System noise assessment of NASA supersonic technology concept aeroplane using JAXA's noise prediction tool. AIAA Scitech 2020 Forum

29. Berton JJ, Huff DL, Geiselhart K, Seidel J (2020) Supersonic technology concept aeroplanes for environmental studies. AIAA Scitech 2020 Forum

30. Hendricks E, Seidel J (2012) A multidisciplinary approach to mixer-ejector analysis and design. In: 48th AIAA/ASME/SAE/ASEE joint propulsion conference

31. Lazzara DS, Magee T, Shen H, Mabe JH (2020) Sonic boom performance of low-boom aircraft in non-standard atmospheres. AIAA Scitech 2020 Forum

32. Carrier G, Normand P, Malbequi P (2021) Analysis and comparison of the results of two ray tracing-based sonic boom propagation codes applied to the SBPW3 test cases. AIAA Scitech 2021 Forum

33. Rallabhandi SK, Loubeau A (2021) Summary of propagation cases of the third AIAA sonic boom prediction workshop. AIAA Scitech 2021 Forum

34. DeGolia J, Loubeau A (2017) A multiple-criteria decision analysis to evaluate sonic boom noise metrics. J Acoust Soc Am 141(5):3624

35. Flying V concept of TU Delft. https://www.airlineratings.com/news/future-travel-flying-v-takes-flight/. Last accessed on 9.3.2021

36. Méheut M et al (2020) Conceptual design studies of boundary layer engine integration concepts. In: Aerospace Europe conference 2020, 25–28 February 2020, Bordeaux, France

37. Iwanizki M et al (2020) Conceptual design studies of unconventional configurations. In: Aerospace Europe conference 2020, 25–28 February 2020, Bordeaux, France

38. Khorrami MR, Fares E (2019) Toward noise certification during design: airframe noise simulations for full-scale, complete aircraft. CEAS Aeronaut J 10:31–67. https://doi.org/10.1007/s13272-019-00378-1

39. Spakovszky ZS (2019) Advanced low-noise aircraft configurations and their assessment: past, present, and future. CEAS Aeronaut J 10:137–157. https://doi.org/10.1007/s13272-019-00371-8

40. https://www.airbus.com/newsroom/press-releases/en/2020/02/airbus-reveals-its-blended-wing-aircraft-demonstrator.html. Last accessed on 9.3.2021

Competing Agendas for Land-Use Around Airports

Fiona Rajé⊙, Delia Dimitriu, Dan Radulescu, Narcisa Burtea, and Paul Hooper

Abstract This chapter describes the core aspects of the land-use planning (LUP) element of the Balanced Approach (BA) by acknowledging the potential of effective LUP as one of the few anticipatory tools available to manage noise. It explores the planning shortcomings that fail to stop encroachment and, thus, the need for remedial mitigation actions such as sound insulation, compensation and buy-out. It goes on to outline core future challenges and steps to develop a better spatial understanding of noise through improved understanding of people's soundscapes (e.g. via the ANIMA app). To illustrate how LUP challenges can be addressed, the chapter also presents case studies from Iasi Airport and on insulation campaigns, in Marseille and Heathrow respectively. It concludes with an exploration of the lessons that can be taken from LUP experience and examines how more comprehensive communication and engagement with key stakeholders underpins more effective application of planning tools.

Keywords Land use planning · Encroachment · Balanced approach · Mitigation · Planning tools · Preventive controls

Introduction

Regulatory responses to aircraft noise are influenced at the global level by the UN International Civil Aviation Organisation (ICAO), and specifically its 'Balanced Approach' to noise management, adopted at the ICAO 33rd Assembly on Aircraft noise in 2001 [17]. The rationale for the Balanced Approach was built on the concept that airports face their own specific circumstances in terms of levels of traffic, the volume of nighttime flights, proximity of the airport to residential areas, and attitudes

F. Rajé (✉) · D. Dimitriu · P. Hooper
Department of Natural Sciences, Manchester Metropolitan University, Chester Street, Manchester M1 5GD, UK
e-mail: F.Raje@mmu.ac.uk

D. Radulescu · N. Burtea
National Research and Development Institute for Gas Turbines COMOTI, 220D Iuliu Maniu Blvd., District 6, OP76, CP174, 061126 Bucharest, Romania

© The Author(s) 2022
L. Leylekian et al. (eds.), *Aviation Noise Impact Management*,
https://doi.org/10.1007/978-3-030-91194-2_6

of local residents to noise. By providing a simple framework, focusing on the core aspects of noise management, airports would therefore be able to have the flexibility to adopt their own approaches as appropriate to their own situation. This also recognises that Member States may already have their own noise regulations and policies in place.

The Balanced Approach provides a flexible way to identify and transparently address specific noise problems. It comprises four principal elements:

1. Reduction of noise at source—by encouraging the development and use of quieter aircraft.
2. Land-use planning and management—to prevent noise sensitive developments close to airports and flight paths, and to mitigate noise impacts (i.e. through sound insulation).
3. Noise abatement via alternative operational procedures that separate aircraft from noise sensitive areas or reduce sound generated by aircraft by following low noise procedures such as reduced use of thrust.
4. Operating restrictions on aircraft at sensitive times (e.g. at night) or in terms of absolute numbers of movements.

As well as these guiding principles, an accompanying document 'Guidance on the Balanced Approach to Aircraft Noise Management' was produced to support airports in implementing interventions within these core elements. It is important to note that this guidance states that operating restrictions should only be applied as a last resort, after the other elements have been considered and applied, where appropriate. This acknowledges the key role played by aviation in the global socio-economic system, and that reductions in noise can be achieved at a lower economic cost when a stronger focus is placed on the other Balanced Approach elements.

The ICAO Balanced Approach was adopted into European Law through EU Directive 2002/30/EC, which was later replaced by Regulation (EU) No 598/2014. In the EU, legislation is set centrally, while implementation into local law occurs at the Member State level. This ensures that the exact implementation of the four Balanced Approach elements is at the behest of the contracting states, which can also choose to delegate their powers to a competent authority. Below this level, airports are generally encouraged to implement their own specific interventions designed to reduce impact, although this is commonly carried out in collaboration with external stakeholders, particularly National Air Navigation Service Providers and Civil Aviation Authorities. This approach ensures that aircraft noise problems at individual airports can be managed in both an environmentally and economically responsible way—achieving maximum environmental benefit in a cost-effective manner.

A snapshot review of Balanced Approach implementation across EU Member States undertaken at the start of the ANIMA project revealed considerable inconsistency in the implementation of the provisions of Regulation (EU) No598/2014. It concluded with the following core messages which were corroborated at a mixed stakeholder meeting (the Impact and Balanced Approach Expert Community supporting ANIMA—see Heyes et al. [13]):

- The ICAO Balanced Approach is a good basis for action to reduce noise exposure, but guidance is required on the appropriate use and efficacy of different elements.
- Given that it is never possible to reduce noise exposure to zero, it is necessary to engage with affected communities, and to consider this issue in the context of the costs and benefits that accrue to them from living near to the airport, and of aviation in general.
- It is important that such engagement is a two-way process: of dissemination from the airport to communities and listening by the airport to community concerns, insight and priorities.
- All airports, of any size, need to consider aircraft noise and anticipate the consequences of growth. The 50,000-movement/ year figure for the application of the Environmental Noise Directive (END) is too simplistic and needs to be reconsidered. One solution could be to have a pre-qualification criterion that requires airports to begin the process of building noise management capacity and engagement with stakeholders, particularly on the issue of land-use planning.
- Management of noise impacts needs to be informed by quality data. Existing reliance upon noise modelling outputs or complaints analysis to inform Balanced Approach implementation can lead to sub-optimal outcomes. Appropriate engagement and dialogue between airports and their surrounding communities is an important prerequisite to assessing the nature and extent of noise problems and appropriate responses. Further policy and good practice guidance is considered to be helpful to facilitate this.
- It is clear that the industry is committed to reducing noise impact, but doing so requires collaboration across the board, between aviation stakeholders, and between different airports.

Specifically, in respect of Land Use Planning (LUP), the review highlighted the use of a range of anticipatory and mitigation tools. It also underlined that—in the desire to tailor to local conditions and only apply controls where necessary to avoid/minimise noise impact—there is considerable inconsistency in the utilisation of LUP provisions between Member States and airports therein [14]. A key explanation for the range of LUP outcomes is that at the heart of the decision-making process is the need to reconcile many, at times competing, demands; such as those of conservation, agricultural, highways and railways, recreation, municipal utilities, commercial, industrial, residential and institutional developments. The challenge for responsible authorities is to ensure a balance of uses that optimises social, environmental and economic benefits.

Land Use Planning, or land use management controls for an airport, attempts to achieve optimal utilisation of land through the use of zoning linked to noise exposure. This can be an effective method for limiting the increase in the number of residents located near airports, people who could become affected by aircraft noise in the future. Unfortunately, there has been very limited systematic evaluation of the use of land use planning tools to minimise noise impact over the last decade since the initiation of the ambitious ICAO/CAEP 5 work programme on Airport Planning and Land Use Planning. During this period, however, many airports have suffered from encroachment by noise-sensitive developments and, thus, the constraints to infrastructure growth have increased significantly. There remains a need for the assessment of land-use planning for noise impact prevention and mitigation if tools are to be developed that can help policy makers and communities (ICAO resolution A37-18).

The key challenge in attempting such assessments is recognition of the range of planning interventions available and how best to tailor their selection and implementation to particular airport contexts. The range of instruments available includes those which are anticipatory (such as noise zoning, transfer of development rights and comprehensive planning), those which are reactive (such as noise insulation programmes, real estate disclosure and building codes) and those which are financial (such as tax incentives, capital investment planning and noise-related airport charges).

The implications from other ANIMA deliverables (e.g. D2.4 [12] and D2.5 [16]), and in keeping with the priorities for communication and engagement, are that such tailoring is best achieved through consultation with local decision makers, urban planners, local communities, and other parties affected by noise impact. This should allow for the most effective utilisation of the land use planning tools available in the design of prevention and mitigation solutions.

This stakeholder consultation and engagement needs to explore the use of land use planning instruments both individually and in combination to assess their potential to address challenges such as:

- Changes to population distribution around airports (density and location).
- Provision of effective protection against night noise.
- How best to optimise the consequences of operational changes (e.g. optimising synergies between operational changes and land use instruments).
- How best to define and track the effectiveness of land use planning.

The development of guidance material and the adoption of proactive approaches to the use of LUP powers have led to some examples of good practice, for example, at Kiev and Catania airports (see [15]). Both have shown how national legislation helps the land use planning process and ensures that zones surrounding the airports are subjected to as little as possible uncontrolled or business driven development. Both airports recognise the key role of collaboration and communication between airport and relevant local authorities. This approach enables the needs of each party to be shared and understood, and the long-term implications arising from the potential development of noise sensitive buildings close to airports considered and controlled by regional decision makers. Thus, the long-term health and economic future of the region can be safeguarded—the airport is better able to grow, whilst the health impacts of living near to an airport can be mitigated.

Another example of good practice comes from Australia. A recent report by TO70 ([37], p. 25) for the City of Canning examined the case for a new runway at Perth Airport. It highlighted the importance of the land use planning near airports, stating that

...land use planning is the process in which noise sensitive areas, such as residences, hospitals and schools are not placed on or near the area surrounding airports. Land use planning is usually conducted by local and state councils and should follow state planning policies to avoid development in high noise areas surrounding airports. Australian standards (AS2021-2015) and NASF, Guideline A provide guidelines for appropriate areas for building and development.

Many land use management strategies utilise passive sound mitigation measures, which consist of the use of noise-isolating materials and various forms of noise insulation. Homes and noise sensitive buildings situated near airports are usually insulated with assistance from the government or airports themselves via noise insulation schemes. Active noise management involves reducing noise through operational procedures or reducing noise from the source.

State and local government are required to use ANEF contours as guidance during land use planning. Australian Standards AS2021—2015 use ANEF contours to guide land use planning for local councils.

To70 ([37], p. 26) continued by setting out the requirement for noise contours to be produced every five years at Perth Airport to assist land use planning around the airport. The report states further that

It is important to point out that this example of good practice also comes with a caveat: while integrated land use planning is desirable, the frequency and intensity of noise need to be considered too, "as these factors play a major role in annoyance" (p. 26). In addition, there is a need to recognise "that ANEFs are based on a forecast of aircraft movement and therefore actual noise experienced will vary" (p.16).

Other examples of good practice can be found: Box 1.

Box 1 Existing best practices on land use planning (a)

Effective Land Use Management to avoid the encroachment of incompatible developments near airports tends to be found where a more *integrated approach is taken to the development of planning strategies and systems of planning control use appropriate methods/tools to understand the extent of the spatial impact of airport activities.* This often takes the form of a wider vision of the airport and the city/region that minimises future constraints on air traffic development.

In Australia and the US, LUP is considered an environmental protection action policy tool which is integrated into the overall planning system. However, the responsibility and capacity for planning and implementation is different in the selected case-studies illustrated below.

The Australian Experience

The approach to land-use planning around airports became an important public policy issue following the privatisation of airports in 1990 [11].

State plans and strategies cover four relevant approaches: *land use planning directives*, *regional planning aspirations* and *structures, aviation-related statements,* and some *airport-specific instruments.*

Regional strategies provide a higher degree of spatial resolution. In metropolitan areas, the airports are considered 'specialised activity centres' in recognition of their function as gateways for economic growth. For example, the Metropolitan Strategy for Sydney specifies Sydney Airport and its environs as a specialised centre generating 'metropolitan-wide benefits' with over 36,000 jobs (about a third at the airport itself). This shows a clear approach to a wider planning system which includes the airport and its requirements in the municipality strategic development.

The Aviation-related and airport-specific directives are more targeted and connect to ensuring adequate noise and safety buffers around airports. Central to these provisions is the use of Australian Noise Exposure Forecasts based on summing the energy from individual aircraft effective perceived noise levels. These are required from airports every 5 years and must forecast the consequences of airport development and associated air traffic changes to a minimum of 20 years. These contours can then inform appropriate zoning of development types around the airport on the basis of future changes rather than existing noise foot printing, thereby in theory future-proofing the airport against the encroachment of incompatible land uses [36]. Protecting the environment of nearby communities through noise mitigation is included in this provision, as should residential areas be predicted to fall into unacceptable levels of exposure in future years (i.e. > 20 ANEF), there is a mandate for the provision of sound insulation.

The recognition of airports as 'special use' land use zones, is designed to ensure, through the ANEF, the imposition of noise protection buffers. Of course, there are challenges with the production of ANEF as they are only as accurate as the forecasts for infrastructure and fleet changes, but nevertheless the forecasting out to a minimum of 20 years is intended to provide an effective protection against encroachment.

The US Approach

The US efforts present a vivid picture of the importance of land use compatibility planning, linking the development approach to methods and tools that illustrate and assess the air traffic growth. There are some good examples of integrated planning approaches at State level, but not yet at the Federal level. The Airport Cooperative Research Program (ACRP) includes a land use survey and case studies that explore issues relating to LUP around airports. The planning system has an integrated approach, a two-way planning system that links land use planning vertically and horizontally to other planning processes. The planning system also has an iterative character that allows *continual adaptation* and avoids the one-time establishment of a plan that could soon become outdated. The plan is created jointly by all institutions involved in the development of *an airport and the city or region it serves*, and it is implemented separately by sectors, being coordinated by a lead agency. Coordination of all stakeholders' requirements is well managed throughout the process.

While it dates back a few decades, the example of land-use management around *Washington Dulles International Airport* illustrates some of the US best practices in tackling the challenging aspect of LUP (Dulles Int'l, n.d.). This airport opened for service in November 1962, but, from the initial planning and development phase, aircraft noise, and its impact on regional communities, was a primary consideration. Thus, the airfield design limits close-in residential development by integrating: "11,000 + acres within the airport perimeter, centrally-positioned runways, and a large area (8,000 feet) buffers from runway endpoints to the estate perimeter". This illustrates that the predicted development of the airport was considered from the start. In addition to the planning process, the FAA (the initial airport operator until 1987) and the Airports Authority have worked closely with the two neighbouring Counties (Fairfax and Loudoun) to deliver residents *compatible land-use protection* through an efficient county planning and zoning strategy. Consequently, in 1993, Loudoun County was identified as a *national frontrunner* in land-use planning associated with a growing international airport.

Box 1 Existing best practices on land use planning (b)

The Singapore Story

This case study is another illustrative example of the integrated approach of an airport and a city, in which a harmonised planning system serves both the citizens and the airport.

The comprehensive planning that went into the development of Changi Airport, and the integrated manner in which it was carried out is considered best practice by many. Singapore's approach involved unique *urban-planning constraints and trade-offs brought about by both civilian and military airports*, and took account of ways to exploit airport developments to catalyse urban and economic development. Further information about the land use planning around Changi Airport can be found at.

https://www.clc.gov.sg/docs/default-source/urban-systems-studies/uss-int egrating-the-planning-of-airports-and-the-city.pdf

Despite evidence of good practice at specific airports, the broader challenge is for the sector as a whole to harness the full range of planning powers to ideally prevent noise problems in an optimal manner. And, where this has not been possible, airports would mitigate impacts by the use of the likes of insulation and buy-out schemes or when other options have been exhausted, compensation programmes. A key achievement of effective land use planning would be the avoidance of further residential developments in areas that would endanger the reduction in noise impact previously achieved and the conversion of existing incompatible land-uses to ones more in keeping with the prevailing noise environment. Achieving this end is by no means straightforward and takes concerted action involving a range of stakeholders if planning priorities are to be harmonised and airport development protected from future constraint. Our case from Iasi Airport in Romania demonstrates why it is essential to start the process of stakeholder engagement at an early stage in airport development if encroachment of incompatible land uses is to be avoided.

Iasi Airport Case Study

Introduction

Iasi International Airport, known officially as "Aeroportul Internaţional Iaşi, România" (ICAO: LRIA, IATA: IAS), is located in the North-Eastern part of Romania. Situated at a distance of 3.48 km East from the Iasi city, the airport has an elevation of 411 FT, with a reference temperature of 30 °C [5] (see Fig. 1). The airport offers three domestic routes (Bucharest, Cluj-Napoca and Timisoara) and multiple international flights to 15 countries (Israel, France, Great Britain, Italy, Spain, Belgium, Germany, Cyprus, Netherlands, Denmark, Austria—regular; Egypt, Tunisia, Turkey, Greece—seasonal). Eight airlines operate at Iasi Airport (TAROM, BLUE AIR, WIZZ AIR, AUSTRIAN AIRLINES, AMC AIRLINES, AIR BUCHAREST, AEGEAN and

Fig. 1 Location of Iasi Airport, Romania. Google Maps, Iasi International [18]

ONUR) [1]. In terms of infrastructure, the airport has 1 heliport, 3 terminals and one runway (RWY 14/32) of 2400 m [2].

Iasi Airport is the regional airport for the North-East of Romania, serving a population of around 4 million and a catchment area of approximately 37 000 km² that includes the County of Iasi, along with the neighbouring counties (Bacau, Botosani, Neamt, Suceava and Vaslui) and the Republic of Moldova (see Fig. 2).

In terms of connectivity, the airport is linked to both road and railway infrastructures, facilitating access from nearby communities, as well as from different cities across Romania.

Before COVID-19, Iasi Airport was considered to be a fast-growing airport, and this can easily be seen in the increasing number of aircraft movements from Fig. 3. Even so, Iasi Airport has remained throughout the years among the top five Romanian airports with the highest air traffic density [19–24].

Fig. 2 Catchment Area, Iasi Airport. AEROPORTUL IASI—Date demografice, [3]

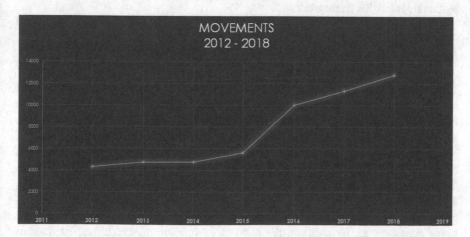

Fig. 3 Evolution of aircraft movements on Iasi Airport (2012–2018). Iasi Airport [18]

In the immediate vicinity of the airport there is an important actor from the Romanian aerospace industry sector (aircraft maintenance), "Aerostar SA—MRO Iasi". In addition, other strategic facilities can be found very near to the airport, such as the military base "Batalionul 151 Infanterie" and the aerodrome for general aviation "Aeroclubul "Alexandru Matei" Iasi".

Experience with Aircraft Noise Management Prior to ANIMA

In 2005, the Environmental Noise Directive (2002/49/EC) was adopted into National legislation through the Governmental Decision "H.G. nr. 321/2005" [34]. Its alignment to the Environmental Noise Directive was further addressed through various updated versions of H.G. no. 321/2005, through: H.G. no. 674/2007 [33], H.G. no. 1.260/2012 [32] and H.G. no. 944/2016 [31]. According to these provisions, Iasi Airport (classified as an 'urban airport') had to prepare Strategic Noise Maps and Action Plans, in 2012–2013 and 2017–2018, although the number of aircraft movements was below 50,000 movements/ year threshold for their mandated production. Based on the findings from the first and second round of Strategic Noise Maps, the airport management concluded that encroachment was the most important concern, especially in the case of a fast-growing airport. The flight paths and the process of encroachment over time (700 m distance from the runway threshold to the closest fence surrounding a residential building—2020) is shown in Figs. 4 and 5. Therefore, the airport's long-term strategy was defined such that maximum effort would be made in support of legislative changes for land-use planning implementation in conflict areas, while, at the same time, focus would also be on raising awareness about the need to consider residential developments and airport operations in a coordinated manner in order to reduce the number of people exposed to aircraft noise. However, land-use planning with the aim of managing aircraft noise was absent within the National legislative framework.

Annoyance Case Study (2015)

Within a joint initiative (Romanian Social Survey on Noise Annoyance), Iasi Airport was involved in an annoyance study, in partnership with INCDT-COMOTI, the Faculty of Psychology (University of Bucharest, Romania) and SINTEF (Norway). The initial research driver was to understand and investigate the real situation behind noise complaints. The study used a survey based on the psychometric characteristics of WHOQOL ("The World Health Organisation Quality of Life") [39] and on the standard questions for noise annoyance from ISO 15666 (ISO/TS 15,666:2003, 2003). The outcomes from this study highlighted the need for collaboration between stakeholders to reach a common understanding of the context and the issues in aircraft noise management. In addition, the absence of funding opportunities, the lack of expertise and trained experts in Romania on aircraft noise and annoyance, together

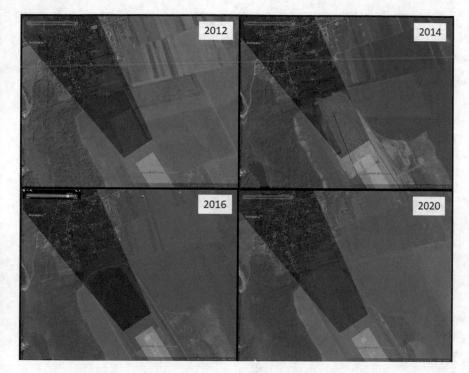

Fig. 4 Flight paths over Aroneanu Village at the end of RWY14. Iasi Airport [18]

with the lack of available research at a National level were identified as the main barriers to further addressing annoyance.

Progress Within the ANIMA Project

Legislative Framework in Romania

At the beginning of the ANIMA Project (October 2017), Iasi Airport started to be proactively engaged within the task "Pan-European Review of Existing Regulations and Mitigation Strategies" [14], in direct collaboration with INCDT-COMOTI and researchers from Manchester Metropolitan University. At this time, the transposition of the Environmental Noise Directive (H.G. 321/2005) was the most important legislation related to managing aircraft noise and was initiated by the Ministry of

Fig. 5 Residential buildings Aroneanu Village at the end of RWY14. Iasi Airport, [18]

Environment. A second legislative instrument was also available, related to the transposition of the 2002/30/EC Directive[1] regarding operating restrictions, which had been initiated by the Civil Aviation Authority. The concept of land-use planning was still only theoretical, despite being a widely discussed topic. The findings from this ANIMA task revealed that land-use planning was not entirely absent from the national legislative framework, but spread between different legislation and with no specific provisions for reducing exposure to aircraft noise.

Visit to Heathrow (May 2018)—Learning from an Experienced Airport

In order to foster a better understanding of practices related to land-use planning, a meeting was organised between a noise expert from Heathrow Airport and a representative from Iasi Airport in 2018. The aim of this initiative was to facilitate the transfer of 'best practice' and 'lessons learnt' knowledge from an airport with a longer history of experience in managing aircraft noise through land-use planning, towards

---

[1] Repealed by Regulation (EU) No 598/2014 of the European Parliament and of the Council of 16 April 2014 on the establishment of rules and procedures with regard to the introduction of noise-related operating restrictions at Union airports within a Balanced Approach and repealing Directive 2002/30/EC.

an airport that was at the beginning of this journey. One key conclusion was the fact that various measures applied successfully by some airports may not be feasible or reach the same level of effectiveness, at other airports, due to differences in context (e.g. being a private airport or state-owned), available legislative frameworks and available resources (e.g. funding opportunities, experts and expertise in the region).

Iasi Workshop (July 2018)—Raising Awareness About the LUP Issue (Common to All Airports)

In the same year, an ANIMA workshop was organised in Iasi to raise awareness about the importance of addressing land-use planning with the aim of reducing the number of people exposed to noise, in the context of fast-growing airports and increased encroachment issues.

Various stakeholders (representatives from communities living in the proximity of Iasi Airport, from the Ministry of Environment and the Civil Aviation Authority, an airline, representatives from five Romanian airports—under the scope of the 'Romanian Airports Association', independent experts and ANIMA partners) were involved in these discussions. They presented their views with respect to managing aviation noise in Romania and described challenges that could appear in the absence of effective land-use planning practices. In spite of competing agendas, it was commonly concluded that land-use planning for reducing the number of people exposed to noise had become an urgent matter and joint efforts were needed to address this. Supplementary to this workshop, Iasi Airport and its local ANIMA partner (INCDT-COMOTI) initiated discussions with representatives from the Ministry of Transportation, the Ministry of Health, local and regional authorities, different airlines, the Romanian Air Navigation Service Provider (ANSP)—ROMATSA and research experts (including ANIMA partners), to increase the efforts in raising awareness about the existing challenges and the importance of collaboration to ensure effectiveness. At this point, the most important issue was finding the balance between reducing the number of people exposed to aviation noise, preventing further encroachment and ensuring the capacity of aircraft operations in the future, in line with air traffic forecasts.

All these efforts were furthered within the ANIMA Project during the subtask "Balanced Approach to Noise Management" [15] and the task "Airport Exemplification Case Studies" to be reported in ANIMA deliverable D2.11.

An initial result (2018) from these discussions was the introduction of the requirement to implement Noise Abatement Departure Procedures (NADPs), which was included within the Romanian Aeronautical Information Publication in 2021. For some airports, this requirement became mandatory, while for other airports (Iasi Airport included), it is in the form of a recommendation.

Another important legislative change in 2019 (Noise Law),[2] related to the transposition of the Environmental Noise Directive. The new proposal was initiated and

---

[2] LEGE nr. 121 din 3 iulie 2019 privind evaluarea și gestionarea zgomotului ambiant.

disseminated for discussion by the Ministry of Environment and took into account many issues raised at the workshop in Iasi, mostly related to the clarification of which authorities are responsible for managing noise produced by air traffic, while reinforcing the importance of collaboration between many stakeholders. This legislative change empowered the application of END provisions, changing the status of the legislation from a Governmental Decision to a Law. Through this change, various other stakeholders (the Air Navigation Service Provider, the Civil Aviation Authority, and the Ministry of Health) became responsible for airport noise management in an official capacity, which is a notable difference from the previous situation where the airport was the sole responsible entity (apart from Governmental bodies), but without any decision-making power in most cases (especially on LUP).

Additional legislative changes [27, 30] took place, related to the application of provisions from the Regulation (EU) no. 598/2014, which further enforced the need to have a collaborative approach for the implementation of noise operating restrictions, as well as the need to address land-use planning before considering operating restrictions as a solution. One important change is that the Civil Aviation Authority has to support the environmental protection authority with assistance in the evaluation of aircraft noise impacts at airports (balancing safety and environmental protection requirements) and the ANSP has to provide the information that helps in the process of evaluating compliance with noise operating restrictions.

An updated version of the "Air Code" [28] was also proposed for adoption (2019) and publicly disseminated for feedback. Among its provisions, it requires that Strategic Noise Maps have to be taken into consideration by the airports within their Airport Development Plans, as well as by local and regional authorities for land-use planning within the Urban Development Plans. The same authorities have to implement noise zoning strategies around airports, yet a specific timeframe for implementation and a clear methodology for noise zoning are still missing (methodology expected to be published by the Ministry of Environment). It is also important to note, in this context, that the Ministry for Regional Development and Tourism is currently the only stakeholder responsible for the approval of construction in the vicinity of airports. In addition, avoiding residential developments around powerful sources of noise (airports included) is only at the level of recommendation, with only few provisions and details for practice suggested [35]. Another important provision is that noise exposure has to be one of the criteria used in the design of new operating procedures or modification of existing ones.

Conclusions

Although many steps have been taken towards the implementation of land-use planning provisions, it was initially identified that it would first be necessary to ensure that the existing legislative framework on managing noise from air traffic was complete and harmonised between the environmental protection provisions and the aviation provisions, thus avoiding inconsistencies and barriers in implementation. In spite of the fact that all these efforts were in support of the implementation of ICAO Balanced Approach, especially land-use planning, this initiative is still at an early stage and needs further legislative harmonisation with the urban development legislation, in order to ensure its effectiveness in implementation. However, changes in legislative frameworks require a long period of time and, thus, increase the risk of missing opportunities to use land-use planning as a preventive measure. Therefore land-use measures would be limited to using planning tools to mitigate existing impacts (e.g. buy-out/relocation of residents, demolition of buildings from conflict areas, noise insulation schemes, and closing the airport), which is considered to imply higher costs and require more effort and time to obtain effective solutions. These challenges emphasise the necessity for having a common European strategy on land-use planning (potentially complemented by a common noise metric and approaches to planning more generally), together with available funding opportunities to support implementation at national and local levels.

In the case of small but fast-growing Iasi Airport, there is no budget or other necessary resources allocated for managing airport noise. The absence of a Governmental funding scheme for reactive noise impact measures, dramatically limits the options of any Romanian airport to act upon this issue. The remaining solution is to cooperate with the National legislative bodies, to develop an appropriate framework to prevent further encroachment.

As the overall air traffic evolution and forecast scenarios show a constant increase in aircraft operations, land-use planning and management was considered by Iasi Airport as the best option to ensure the necessary means to maintain or reduce the number of people exposed to aviation noise. From the position of a state-owned airport, Iasi Airport, as well as most Romanian airports, has little decision-making power in land-use and, therefore, communication and engagement with the relevant stakeholders and affected communities has been of utmost importance in opening up the dialogue about developing land-use planning provisions within the National legislative framework.

Insulation Case Study

Introduction

As suggested previously, where anticipatory planning powers do not prevent encroachment, there is a role for mitigation. One example of this approach is the adoption of sound insulation schemes by airports. The cases of insulation at Heathrow and Marseille Airports are examined here to explore this type of reactive intervention.

Background

ANIMA undertook to examine, through qualitative research, whether interventions implemented by airports or other stakeholders in airport regions could have an impact on residents' quality of life [26]. In this study, sound insulation was studied at Marseille and Heathrow Airports. Concerns about aircraft noise impact date back to the 1950s and 1960s when jet engines started to be introduced, and international aviation became more popular [38]. Thus, a key aim of the insulation schemes was to reduce noise complaints and general community dissatisfaction by reducing noise disturbance attributable to aircraft overflights.

Sound Insulation Schemes Studied

Marseille

In 1997, the French state implemented a specific sound-proofing assistance system for large airports. This meant that residents affected by aircraft noise could then receive a grant for sound insulation for their homes. The system was originally managed by the National Environment and Energy Management Agency and financed by a general tax on polluting activities. Now, the grant is exclusively financed by airlines via a tax on aircraft noise pollution (TNSA), levied by the DGAC (Directorate General of Civil Aviation) according to the "polluter pays" principle. The criteria for eligibility around Marseille Airport are that the accommodation is located inside the annoyance map contours[3] and was built before the noise annoyance plan had been created.

In order to explore quality of life and, in particular, the concept of scheme fairness, focus groups were carried out in three areas around Marseille Airport—two in the annoyance noise map and one outside the annoyance noise map contours, following these criteria:

- Eligible to the grant/insulated: City of L'Estaque (Marseille airport).
- Eligible to the grant / non-insulated yet: city of Marignane (Marseille airport).
- Non-eligible / non-insulated: City of Vitrolles (Marseille airport).

The assumption was that people who were situated in the grant area and had already been in receipt of insulation would be more likely to appreciate the intervention than the other participants. Moreover, it was important for us to investigate the perception of those people who could be insulated but had ignored the process of the insulation program. Indeed, it was hypothesised that the insulation scheme is not well known by people, even those who are eligible for it. This could also have an impact on their perception, because it deals with issues of fairness. Finally, the intention was to investigate this kind of intervention in an area with a mild climate,

---

[3] At Marseille, an annoyance map reflects a small section of a larger noise map (called an exposure noise map) which includes most noise affected areas.

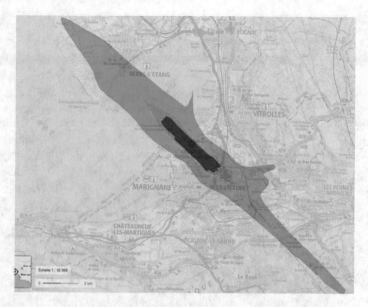

Fig. 6 Annoyance noise map contours valuable for insulation scheme in Marseille. Green areas: 55 dB < Lden < 65 dB;orange areas:65 dB < Lden < 70 dB;red areas:Lden > 70 dB. Kuhlmann et al *Source* [26]

because it was assumed that it would not be as well perceived in comparison to colder areas. It was anticipated that the results could help to better frame the intervention according to the location, that is, that there should be knowledge about the location and potentially a decentralisation of the decision-making bodies.

The annoyance map in Fig. 6 illustrates the range of noise affected areas by different colours.

Another focus group was also conducted in order to consult the people involved in a noise pressure group.

Heathrow

Sound insulation as an intervention to help mitigate aircraft noise impacts around Heathrow began being discussed in the, resulting in a range of schemes being developed over the ensuing 60 years. In the mid-1990s, a voluntary daytime noise insulation scheme was introduced by Heathrow Airport, followed by a voluntary night noise insulation scheme early in the following decade. By 2014, Heathrow had started to offer the Quieter Homes Scheme (QHS) for those residents living closest to the airport within the 69 dB LAeq,16 h aircraft noise contour. An overview of these schemes is provided: Box 2.

Box 2 Brief details of sound insulation schemes at Heathrow Airport

The (Residential) **Day Noise Insulation Scheme** (or Day Scheme) is based on the 1994 69 dB LAeq,18 h contour and is designed to protect those homes exposed to the aircraft noise in the day, including in the early morning arrival period before 06:00. These properties are eligible to receive **50% of the cost** of replacement windows and external doors, or free secondary-glazing, and free loft insulation and ventilation. 9300 homes fall into this scheme's boundary.

The **Night Noise Insulation Scheme** (or Night Scheme) is designed to address the impact of night flights on local residents. The scheme boundary is based on the footprint of the noisiest aircraft regularly operating between 23:30 and 06:00. Eligible properties are entitled to receive **50% of the cost** of replacement bedroom or bed-sitting room windows, or free secondary glazing of bedroom or bed-sitting room windows, and free loft insulation and ventilation. Approximately 37,000 homes fall within this scheme's boundary.

The **Quieter Homes Scheme (QHS)** applies to homes based on the 2011 69 dB LAeq,16 h contour. It covers the **full cost** of carrying out the work which can include loft and ceiling insulation, double-glazing or external door replacements and loft and ceiling over-boarding. Around 1200 homes located close to the airport are entitled to this scheme (Fig. 7).

Unlike the insulation scheme funding model in France, Heathrow has introduced its range of noise control and mitigation measures voluntarily, since legal instruments

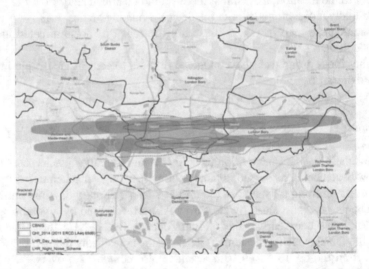

Fig. 7 Boundaries of heathrow noise mitigation schemes. *Source* Heathrow Noise Action Plan 2019–23 @ https://www.heathrow.com/content/dam/heathrow/web/common/documents/company/local-community/noise/making-heathrow-quiter/noise-action-plan/Noise_Action_Plan_2019-2023_Supporting_Annexes.pdf

related to sound insulation at Heathrow have expired. However, the prospect of statutory action is usually highlighted by the government if appropriate 'voluntary' actions are not undertaken at UK airports. For further information about the guidance around voluntary action, please see Box 3.

Box 3 UK Guidance on sound insulation

In the UK, under Sect. 79 of the Civil Aviation Act (as subsequently amended) the government has powers to direct airport operators to implement noise insulation schemes. Although the prospect of statutory action is usually highlighted by government if appropriate 'voluntary' actions are not undertaken, Heathrow, Gatwick and Stansted as designated airports, along with many of the larger non-designated airports in the country have introduced their own noise insulation schemes on a voluntary basis or in response to planning conditions/agreements; schemes operated at other UK airports tend to be derived from or closely resemble the designated airport schemes. Such has been the effectiveness of these initiatives that the Government chose not to amend Sect. 79 in light of conclusions to the consultation on the control of noise from aircraft published in 2003.

The UK government remains committed to the idea that aircraft noise problems are best resolved locally and that airport operators should be expected to take all practical steps to ensure that disturbance to those living in the surrounding area is kept to a minimum ([7]: 7). Indications as to what constitutes 'all practical steps' can be found in the Aerodromes (Noise Restrictions) (Rules and Procedures) Regulations 2003 that implemented ICAO's balanced approach outlined above, and more generally in the White Paper of the same year. The 'balanced approach' was adopted as EU policy in March of 2002 when the European Parliament and Council approved on the Directive 2002/30/EC on the establishment of rules and procedures with regard to the introduction of noise-related operating restrictions at Community airports.

The White Paper "The Future of Air Transport" [7], set out a strategy for the future of the industry in the UK that 'balances' the desire for growth with the need to 'reduce and mitigate the environmental impacts of air transport and of airport development' (p.29). It identified the measures that the government expects airport operators to adopt in order to help those affected by noise when new airport development takes place, these include:

- A continuation of the voluntary noise insulation grant schemes which take as their guideline threshold the 69dBA Leq 16 h contour for 2002.
- The adoption at larger airports (those with more than 50,000 movements a year), of mitigation measures that:
  - Offer households who are subject to high levels of noise (69dBA Leq or more) assistance with the costs of relocating; and

– Offer acoustic/sound insulation (applied to residential properties) to other noise sensitive buildings, such as schools and hospitals, exposed to medium to high levels of noise (63dBA Leq or more).

Thus, in the UK, the extent and generosity of sound insulation schemes is largely determined by voluntary action. The value of these actions in maintaining/improving relationships with local communities is emphasised in the UK Airport Operators Association's Environmental Guidance Manual for Airports.

In order to understand peoples' experience of living in the vicinity of/under en-route paths to/from Heathrow and their views on sound insulation, telephone interviews were carried out in September 2020. While focus groups had been planned for this aspect of the work also, interviews were adopted due to the need for social distancing during the pandemic. Participants were recruited through a local civic group, HACAN (Heathrow Association for the Control of Aircraft Noise) and included ten respondents. This group was purposively selected as their individual membership of HACAN, whose role is to be a voice for those under Heathrow flight paths, indicated that they would have some willingness to discuss issues related to aircraft noise. It should also be noted that there was a likelihood that some of the group may have had a willingness to oppose the airport and its activities too. This is something that the research team were aware of but it was agreed that the group's views would still provide insight into individual views amongst a small self-selected population. The interviews covered residents' satisfaction with their area and issues affecting their quality of life, their views about the airport and about the sound insulation offer, and an exploration of the value they placed on the intervention.

Since this was not a randomly selected group of interviewees but a group for whom noise was clearly already a factor, there needs to be a caveat about the representativeness of the results. Nevertheless, this was a motivated group of individuals who were willing to give their time to discuss quality of life in relation to aircraft noise—something that was of immense value to the researchers during continued restrictions due to the COVID-19 pandemic which prevented the initially planned questionnaire and focus group approach.

All ten interviewees were located to the East of the airport and variously affected by westerly arrivals (close in at Hounslow and further out along the arrival path) or easterly departures (one under the flightpath taking 40% of easterly departure traffic). All had been in their properties for long periods, except for one participant who had moved from an area near the airport to one which was even closer and had been surprised by the apparent increase in noise intrusion, feeling that the move had been a mistake (Fig. 8).

Fig. 8 Location of Heathrow Airport. *Source* Google Maps, 2020

Sound Insulation at Marseille and Heathrow—Research Results

The research undertaken at Marseille, involving four focus groups, suggested that insulation.

- was useful for lessening the effects of noise in wintertime when windows are closed.
- does not have any effect on air pollution caused by aircraft.
- seems to be very effective and can reduce stress and fear of crashes when people are inside their home.
- improves thermal comfort and contributes to a reduction of household energy bills.

The Marseille results revealed that a sound insulation intervention should take account of not only the indoor noise but also the outdoor noise exposure. In addition, they indicated that it is necessary for attention to be paid to the capacity of the intervention to improve social interactions in the respondents' residential area and, in particular, at home. The insulation scheme was seen as a good way to avoid annoyance from indoor noise exposure, but it had to be complemented by other interventions, especially when noise impacted areas are situated in a warm climate area (thus decreasing the time spent with closed windows). Despite these findings, the insulation scheme was well regarded by participants who intended to continue to avail themselves of the intervention.

With specific reference to ineligible participants at Marseille, there was varied knowledge of sound insulation. Nevertheless, participants had a favourable attitude to the insulation scheme procedure itself, even though it was considered unnecessary and ineffective for noise outdoors and during the summer period when windows remain open all day. However, they criticised the delineation of the outline of the noise annoyance map. In addition, they suggested that the annoyance map be scalable to reflect the increase in traffic and be reviewed more regularly. This group expressed concern that enough attention was not drawn to the intervention and that its availability and details had not been sufficiently communicated to the general public and potentially eligible people.

At Heathrow, drawing on the qualitative interviews, the research suggested that there was generally a low level of awareness of what the airport does to minimise noise exposure. Unsurprisingly, then, there was a low level of awareness of insulation provision. Participants drawn from the airport amenity group generally agreed with the principle of addressing the experience of the most noise affected, although the means for determining this was criticised: Some either suggesting that conventional noise measures such as Leq did not adequately reflect lived experience of a series of aircraft noise events of greater intensity than the average noise level, or simply that insulation should extend further out geographically and take account of the increase in numbers of aircraft movements over the years.

Only one participant living in Hounslow (beside Heathrow on the dominant westerlies arrival path) was in an area covered by an insulation scheme (night noise scheme). The sound insulation work had been done before the person moved in, and when they tried to have further work done following the conversion of an attic, this was seen by the airport to be outside scheme provision as it was a new alternation. Ultimately, the participant paid for sound-insulated windows which have improved the situation but not fully remedied it.

All Heathrow participants understood the various sound insulation schemes once they were explained (they had been sent in advance an information sheet on the schemes for use during the interview) and the use of Lmax footprint for the night scheme was considered to be sensible. Overall progression of schemes was not readily evident, especially as the more recent QHS only covers a small number of properties. However, the 100% funding available under the QHS was seen as an improvement; although the 50% offer to pay towards insulation in other schemes was seen to be unfair—with participants querying why residents should have to pay to rectify a problem of the airport's making. Generally, interviewers had to work hard for any evaluative comments about sound insulation as an intervention, with participants feeling it was impossible to provide a view without speaking to those who had been in receipt. Nevertheless, some relevant comments were:

- Future airspace plans are more important.
- Respite is more of a contribution than insulation.
- Description feels technical.
- What's the performance of the insulation provision? (in terms of indoor sound level reduction).
- Offer needs to go further for different scenarios (i.e. consider each operational mode as you are exposed throughout the time when on a particular mode).
- Full costs coverage is a clear improvement.

- Good use of money but other things are important.
- Would be concerned about contractors and quality of installation.
- Offer makes sense from a business perspective, it 'looks good'.
- Looks good on paper but what's the real impact?
- Can vulnerability be factored into the qualification for insulation?

In terms of land use planning around Heathrow, it is important to highlight that participants were generally happy to acknowledge the economic benefits from living near to the airport, although personal accessibility was less of a perceived benefit. The interviews also raised the negative issues around people who are frequent fliers and wider environmental problems (carbon and emissions). There was universal agreement that noise disbenefits outweighed any positive contribution from the airport to local communities. Much of this conversation was overlaid with concerns about the airport's expansion through a third runway: the government decision in favour of which was seen to be misplaced, leading to much criticism of named politicians and processes of decision-making, with communities being 'treated with total contempt'.

The participants described very little direct information from the airport and what little there may have been as tokenistic, leaving people with a feeling of no control. Some had participated in consultations which they felt had had some influence (e.g. over departures after 11.30 pm) but momentum seemed to have waned.

There was a desire to be consulted but there were also fears that the airport would control the agenda and, thus, outcomes. There was clearly room for improvement in communication over how engagement processes can be enhanced to allow for influence over things that currently feel out of control.

Discussion

Despite being leading airports, current sound insulation schemes at Marseille and Heathrow are not directly designed to target and improve residents' quality of life. Instead, they would appear to be part of a suite of noise management tools whose effectiveness in deployment is generally unchallenged by the airport and not sufficiently finessed to meet the needs of local people, when asked about what they would value. In attempting to evaluate the impact of these interventions after the fact, it was clear that this is near impossible as perceptions of appropriateness and impact are overlaid by wider perceptions of the operation/performance of the airport and indeed its development plans.

In addition, it is important to highlight that those insulation schemes that have been implemented have not been systematically evaluated. This can lead to repeated implementation of the same intervention in different contexts and/or continuation of interventions that may not be successful and may not result in the desired outcomes. This is an important consideration in respect to land use planning: without effective evaluation, it is impossible to ascertain whether an insulation scheme is of value to the individuals who are in receipt of the intervention. It is, then, likely to be equally unfeasible to establish whether a community vulnerable to encroachment would find value from airport provision of such sound insulation schemes.

Examining the results for the two airports further highlights that there is a high level of variation in available funds for insulation schemes across nations. Thus, conflicting policies and funding models can make comparison of schemes difficult and confusing.

It is also notable that the results from Marseille and Heathrow Airports show different impacts of sound insulation schemes on residents' quality of life. For example, depending on climate conditions of a region, sound insulation schemes can greatly differ across airports and national boundaries with respect to their impact on people's lives.

Conclusions

Sound insulation interventions have received substantial coverage in the academic and grey literature. However, with no evaluation of the types of schemes, it is challenging to determine best practice or potential for national experience to be globally applicable. Nevertheless, the results show better management approaches may help to more directly address the needs of local communities. Within this context, evaluation of a sound insulation scheme is essential, especially as such an intervention may not lead to the airport's desired outcome or may have potential unintended side effects. By evaluating an intervention, such undesirable impacts can be identified on a timely basis, addressed and the intervention improved accordingly. This form of evaluation can lead to the development of best practice for use of sound insulation in the context of land use planning. To contribute to a more holistic offer, which includes effective evaluation, it is also important to foster effective communication and open dialogue between an airport and its surrounding communities. Such steps can help towards successful sound interventions that are fair and of value to residents.

Overall Discussion

The LUP cases presented above highlight the central importance of enhanced engagement with stakeholders to inform specific interventions. In the case of preventative measures such improvement are needed to ensure coordination between competing land uses and sufficient consideration of future change in planning provisions by all authorities with planning responsibilities. In the case of mitigation measures, engaging with noise affected communities allows a better understanding of'what success looks like' such that measures can be nuance to more directly address issues of concern to local communities and thereby providing more optimised social outcomes valued by those same communities.

With respect to mitigation measures, evaluation of interventions seems to be an important, and as yet, over-looked, contributor to learning about land use planning solutions. A brief review of regulatory and policy guidance on aircraft noise [6] revealed considerable variability in the provision of mitigation measures across the globe. Consequently, there is little standardisation in these areas, which makes tasks such as benchmarking very difficult as quantitative measures of performance have yet to be agreed upon across the airport sector. Further, the range of possible actions and the need to tailor mitigation provisions to local needs means that actions that are perceived to be generous and effective in one location may not receive the same response at another airport. Indeed, any ultimate indicator of the effectiveness of these actions (e.g. responses to community outreach, number of noise complaints, etc.) will be the result of a number of other inputs such as the success of communication strategies more generally and the effectiveness of attempts to manage aircraft noise at source.

Experience suggests that if genuine evaluation of specific interventions is to be attempted going forward, it needs to be built into the process of intervention design, decision-making and implementation. In other words, if the criteria for judging the success of an intervention are agreed from the outset along with the means (e.g. metrics) to monitor and assess achievement against these success criteria, then evaluation processes can at least determine whether the original agreed outcomes have been achieved and, indeed, contribute to any amendment if changes are needed to better address agreed outcomes. Demonstrating such progress with a series of interventions overtime could contribute to more positive airport-community relationships and thus potentially improve some of those non-acoustic factors (e.g. attitude, trust, perception of control) known to exacerbate annoyance.

When utilising planning powers to prevent noise problems around airports, experience points to the importance of coordinated engagement and action by those authorities with planning responsibilities if the future development of airports is to be acknowledged in spatial development plans and thereby constraints, arising from environmental impacts such as noise intrusion, minimised. The EU is championing such approaches through the advocacy of SUMPs[4] (Sustainable Urban Mobility Plans) which provide a welcome addition to the land use toolkit. This approach focuses on the involvement of *citizens and stakeholders*, the coordination of policies between sectors (especially *transport, land use, environment,* development, energy, safety, social and health) and wide-ranging cooperation across different layers of government. Involvement of private actors is also considered to be relevant.

SUMPs highlight the *need to cover all aspects of mobility* (both people and goods), transport modes and associated services in an integrated manner. A plan is designed for the entire "functional urban area", as opposed to a single municipality within its administrative limits. Linking an airport to the neighbouring city or region will involve an integrated land use plan in *sustainable urban/regional transport planning,* combining measures from different sectors and underlying gaps, conflicts and priorities in a harmonised way.

---

[4] Mobility Strategy | Mobility and Transport (europa.eu).

The sustainable development concept provides a prescriptive framework for self-governing parties to *negotiate and settle differences* concerning economic, social and ecological interests over the use of land in a spirit of partnership. The inclusion of airports in this framework addresses the existing 'lack of interest in dialogue' gap between parties concerned in LUP around airports.

Thus, SUMPs are valid, available instruments that provide a framework to help diminish/eliminate the challenges around LUP and airport development. Such an approach to an *airport-city* concept will solve several existing conflicts, as the planning process to develop a metropolitan area will have to consider the sustainability aspects of wider urban mobility, including connectivity to the airport and aspects of noise impact and air quality. Aviation impact would be considered with other integrated transport impacts, that is, the noise from aircraft would be integrated with road and rail noise and the aspects of community wellbeing would be reflected in a holistic manner.

Conclusions

The research previously described suggests that the biggest challenge to land use planning is a history of planning of land use around airports characterised by existing gaps and barriers. It is clear that there is a need to reverse previous poor practice. The examples provided highlight how some airports, local and National governments and other stakeholders have worked together on land use planning, which is more holistic and sensitive to economic, social and environmental needs. Nevertheless, there is still a need for wider stakeholder engagement if LUP options are to be aligned with community interests and thus optimise the social return on investment in LUP interventions.

The research suggested that there was a need for common strategy and sensitive local implementation. While the US can adjust planning processes at a state level, EU planning systems do not afford the flexibility to accommodate for such modification. However, the call for proactive, preventative approaches to systematic land use planning appears to be being answered in the EU through SUMPs.[5] Better integration and more strategic approaches within the context of SUMPs hold value for prevention of encroachment and modifying noise impacts downwards.

Experience suggests that harmonised planning is preventative and, thus, preferable as airports develop. However, there remains a role for reactive approaches which, while they do offer some opportunity for mitigation clearly, have not been complemented by evaluation. Such evaluation would allow learning to be taken forward to the planning and development of more nuanced and tailored interventions. And, if implemented properly, provide an evidence base of the delivery on agreed outcomes

---

[5] *More information available* @ https://ec.europa.eu/transport/themes/urban/urban-mobility/urban-mobility-actions/sustainable-urban_en).

valued by, and agreed with, communities. Thereby, helping to build better airport-community relationships through actions that have demonstratively addressed local needs and experience.

Finally, looking forwards, new ways of understanding personal mobility can contribute to greater knowledge of how space is used in communities and how the experience of noise changes with movements around airport areas. The ANIMA project has looked at extending knowledge of personal mobility and aircraft noise distribution through two studies:

- one using dynamic mapping of noise around airports that uses people's daily travel patterns to determine where they are at a particular time and how their noise exposure changes over a day and,
- the other, looking at the usability of a specially developed online application to gain a greater understanding of the influence of the sound and visual environment on quality of life in airport regions. Using such techniques holds promise for a better understanding of the noise impacts which land use planning seeks to address.

References

1. Aeroport-iasi.ro. (2021) Aeroportul International Iaşi Romania. Available at: https://www.aer oport-iasi.ro/informatii-pasageri ("Info pasageri"/ "Destinatii"). Accessed 30 March 2021
2. Aeroport-iasi.ro. n.d. *IASI AIRPORT—Detalii tehnice*. Available at: https://www.aeroport-iasi.ro:5000/docs/download/Iasi-Airport-Date-tehnice.pdf. Accessed 30 March 2021
3. Aeroport-iasi.ro. n.d. AEROPORTUL IASI—Date demografice. Available at: https://www.aeroport-iasi.ro:5000/docs/download/Iasi-Airport-Date-Demografice.pdf. Accessed 30 March 2021
4. Aisro.ro. (2020) AIS ROMANIA. Available at: https://www.aisro.ro/ ("Publications"/ "AIP"/ "AIP"/ "AD"/ "AD 1 Aerodromes/ Heliports Introduction"/ "AD 1.1 Aerodrome availability"/ "AD 1.1–3 08 NOV 2018"/ "6.1. Noise abatement departure procedures"). Accessed 30 March 2021
5. Aisro.ro (2020) AIS ROMANIA. Available at: https://www.aisro.ro/ ("Publications"/ "AIP"/ "AIP"/ "AD"/ "AD 2 Aerodromes"/ "AD 2.10 LRIA IASI / Iasi"/ "AD 2.02 LRIA Aerodrome geographical and administrative data"). Accessed 30 March 2021
6. CATE (2009) Benchmarking BAA's Aircraft Noise Mitigation and Compensation Provisions. Internal Report. Manchester, CATE.
7. DfT (2003) The future of air transport. Department for Transport, London
8. Eur-lex.europa.eu (2002) Directive 2002/49/Ec of the European Parliament and of the CounciL of 25 June 2002 relating to the assessment and management of environmental noise. Available at: https://eur-lex.europa.eu/legal-content/EN/TXT/PDF/?uri=CELEX:320 02L0049&qid=1617070253212&from=EN. Accessed 30 March 2021
9. Eur-lex.europa.eu (2002) DIRECTIVE 2002/30/EC OF THE EUROPEAN PARLIAMENT AND OF THE COUNCIL of 26 March 2002 on the establishment of rules and procedures with regard to the introduction of noise-related operating restrictions at Community airports. Available at: https://eur-lex.europa.eu/legal-content/EN/TXT/PDF/?uri=CELEX:32002L0030&fro m=EN. Accessed 30 March 2021
10. Facebook.com (2018) IASI Airport. Available at: https://www.facebook.com/IASIairport/pho tos/a.2005486159536463/2005498689535210/. Accessed 30 March 2021

11. Freestone R, Baker D (2010) Challenges in land use planning around Australian airports. J Air Transp Manag 16:264–271
12. Haubrich J, Burtea N, Flindell I, Hooper P, Hudson R, Rajé F, Radulescu D, Schreckenberg D (2018) Recommendations on annoyance mitigation and implications for communication and end engagement. ANIMA Project Report D2.4. https://zenodo.org/record/2616668#.YGRvai 1Q3s0. Accessed 31 Mar 2021
13. Heyes G, Dimitriu D, Hooper PD (2018a) Noise impact mitigation priorities report, ANIMA Project report D2.2. https://explore.openaire.eu/search/publication?pid=10.5281% 252Fzenodo.2578805. Accessed 23 Mar 2021]
14. Heyes G, Dimitriu D, Hooper PD (2018b) Pan-European overview of existing knowledge and implementation of noise reduction strategies, ANIMA Project Report D2.1. https://explore.ope naire.eu/search/publication?pid=10.5281%252Fzenodo.2578793. Accessed 23 Mar 2021
15. Heyes G, Galatioto F (2019) ANIMA D2.5 - Critical review of Balanced Approach Implementation across EU Member States. Zenodo. https://doi.org/10.5281/zenodo.3146128
16. Heyes G, Burtea N, Dimitriu D, Hooper P (2020) Recommendations for the implementation of the exemplification case studies, ANIMA Project Report D2.9. https://zenodo.org/record/463 9650#.YGWUXS1Q2qA. Accessed 31 Mar 2021
17. ICAO (n.d.) live web page—balanced Approach to Aircraft Noise Management. Available at: https://www.icao.int/environmental-protection/Pages/noise.aspx. Accessed 23 Mar 2021
18. Iasi Airport. 2021. *Encroachment, flight paths and evolution of aircraft movements.* [email].
19. Institutul National de Statistica (2016) Transportul Aeroportuar De Pasageri Si Marfuri, Anul 2015. Available at: https://insse.ro/cms/sites/default/files/field/publicatii/transportul_aer oportuar_de_pasageri_si_marfuri_2015.pdf. Accessed 30 March 2021
20. Institutul national de statistica (2017) Transportul aeroportuar de pasageri si marfuri, anul *2016*. Available at: https://insse.ro/cms/sites/default/files/field/publicatii/transportul_aeropo rtuar_de_pasageri_si_marfuri_in_anul_2016.pdf. Accessed 30 March 2021
21. Institutul National de Statistica (2018) Transportul aeroportuar de pasageri si marfuri, anul 2017. Available at: https://insse.ro/cms/sites/default/files/field/publicatii/transportul_aeropo rtuar_de_pasageri_si_marfuri_in_anul_2017.pdf. Accessed 30 March 2021
22. Institutul National De Statistica. 2019. Transportul Aeroportuar De Pasageri Si Marfuri, Anul 2018. Available at: https://insse.ro/cms/sites/default/files/field/publicatii/transportul_aer oportuar_de_pasageri_si_marfuri_in_anul_2018.pdf. Accessed 30 March 2021
23. Institutul National de Statistica (2020) Transportul Aeroportuar De Pasageri Si Marfuri, Anul 2019. Available at: https://insse.ro/cms/sites/default/files/field/publicatii/transportul_aer oportuar_de_pasageri_si_marfuri_in_anul_2019.pdf. Accessed 30 March 2021
24. Institutul National De Statistica (2021) Transportul Aeroportuar De Pasageri Si Marfuri, Anul 2020. Available at: https://insse.ro/cms/sites/default/files/field/publicatii/transportul_aer oportuar_de_pasageri_si_marfuri_in_anul_2020_0.pdf. Accessed 30 March 2021
25. iso.org (2003) ISO/TS 15666:2003. Available at: https://www.iso.org/standard/28630.html. Accessed 30 March 2021
26. Kuhlmann J, Rajé F, Richard I, Ohlenforst B (2020) Evaluations of previous interventions in improving quality of life, ANIMA Project Report D3.6. Accessed 23 Mar 2021
27. Legislatie.just.ro (2020) ORDIN nr. 318 din 25 februarie 2020. Available at: http://legislatie. just.ro/Public/DetaliiDocument/223521. Accessed 30 March 2021
28. Legislatie.just.ro (2020) CODUL AERIAN din 18 martie 2020. Available at: http://legislatie. just.ro/Public/DetaliiDocument/223897. Accessed 30 March 2021
29. Legislatie.just.ro (2019) LEGE nr. 121 din 3 iulie 2019. Available at: http://legislatie.just.ro/ Public/DetaliiDocument/216510. Accessed 30 March 2021
30. Legislatie.just.ro (2018) HOTARARE nr. 639 din 23 august 2018. Available at: http://legisl atie.just.ro/Public/DetaliiDocument/204347. Accessed 30 March 2021
31. Legislatie.just.ro (2016) HOTARARE nr. 944 din 15 decembrie 2016. Available at: http://leg islatie.just.ro/Public/DetaliiDocument/185147. Accessed 30 March 2021
32. Legislatie.just.ro (2013) HOTARARE nr. 1.260 din 12 decembrie 2012. Available at: http://leg islatie.just.ro/Public/DetaliiDocument/144470. Accessed 30 March 2021

33. Legislatie.just.ro (2007) HOTARARE nr. 674 din 28 iunie 2007. Available at: http://legislatie. just.ro/Public/DetaliiDocumentAfis/83733. Accessed 30 March 2021
34. Legislatie.just.ro (2005) HOTARARE nr. 321 din 14 aprilie 2005. Available at: http://legisl atie.just.ro/Public/DetaliiDocumentAfis/61215. Accessed 30 March 2021
35. Legislatie.just.ro (2002) Regulamentul General De Urbanism din 27 iunie 1996 (republicat). Available at: http://legislatie.just.ro/Public/DetaliiDocument/40188. Accessed 30 March 2021
36. McLeod I, Latimore M (2014) Challenges in producing an Australian noise exposure forecast. InterNoise 2014, Melbourne, Australia, 16–19 November
37. TO70 (2018) Review of Perth Airport New Runway Project—Preliminary Draft Major Development Plan. T070 Aviation Australia for City of Canning, North Melbourne, Australia
38. UK Civil Aviation Authority (CAA) (2014) CAP 1165 Managing Aviation Noise, CAA. https:// publicapps.caa.co.uk/modalapplication.aspx?appid=11&mode=detail&id=6251. Accessed 23 Mar 2021
39. Who.int (1997) WHOQOL Measuring Quality of Life. Available at: https://www.who.int/men tal_health/media/68.pdf. Accessed 30 March 2021

Beyond Flying Machines, Human Beings

Impact of Aircraft Noise on Health

Sarah Benz[ID]**, Julia Kuhlmann**[ID]**, Sonja Jeram**[ID]**, Susanne Bartels**[ID]**, Barbara Ohlenforst**[ID]**, and Dirk Schreckenberg**[ID]

Abstract Aircraft noise exposure is an environmental stressor and has been linked to various adverse health outcomes, such as annoyance, sleep disturbance, and cardiovascular diseases. Aircraft noise can trigger both psychological (annoyance and disturbance) and physiological stress responses (e.g. activation of the cardiovascular system and release of stress hormones). People are usually able to deal with this kind of stressor. However, a constant exposure to aircraft noise can cause a continuous state of stress. This in turn can constrain a person's ability to regenerate and restore its resources to cope with the noise situation. As a consequence, the risk for certain negative health outcomes can be increased. Within the ANIMA project, literature reviews on the effects of aircraft noise on health outcomes have been performed.

Explaining how far recent works from WHO and beyond showed that noise-induced annoyance and awakening are likely to mediate to more severe health impact

S. Benz (✉) · J. Kuhlmann · D. Schreckenberg
ZEUS GmbH, Centre for Applied Psychology, Environmental and Social Research, Sennbrink 46, 58093 Hagen, Germany
e-mail: benz@zeusgmbh.de

J. Kuhlmann
e-mail: kuhlmann@zeusgmbh.de

D. Schreckenberg
e-mail: schreckenberg@zeusgmbh.de

S. Jeram
National Institute of Public Health (NIJZ), Environmental Health, Trubarjeva 2, 1000 Ljubljana, Slovenia
e-mail: sonja.jeram@nijz.si

S. Bartels
Sleep and Human Factors Research, Institute of Aerospace Medicine, German Aerospace Center (DLR e.V.), Linder Höhe, 51147 Cologne, Germany
e-mail: Susanne.Bartels@dlr.de

B. Ohlenforst
Royal Netherlands Aerospace Centre NLR, Anthony Fokkerweg 2, 1059 CM Amsterdam, The Netherlands
e-mail: Barbara.Ohlenforst@nlr.nl

L. Leylekian et al. (eds.), *Aviation Noise Impact Management*,
https://doi.org/10.1007/978-3-030-91194-2_7

This chapter gives an overview of the relevant health outcomes affected by aircraft noise and summarises the results of different reviews and studies on these outcomes. Additionally, the underlying mechanisms of how noise impacts health are explained for daytime as well as night-time aircraft noise exposure (i.e. while sleeping). Further, the relevance of considering not only the general population, but vulnerable groups as well (such as children and elderly people) is described. Lastly, open questions for further studies are presented and discussed.

Keywords Aircraft noise exposure · Health outcomes · Noise annoyance · Sleep disturbance · Cardiovascular diseases · Mechanism · Stress

What Are the Health Impacts of Aircraft Noise Exposure

Aircraft noise exposure has been associated with various adverse health outcomes. In the ANIMA project the impact of aircraft noise on human health and well-being was reviewed for several health outcomes: cardiovascular diseases, sleep disturbance, annoyance, cognition, mental health, hearing impairment and other adverse effects, including adverse birth effects and metabolic diseases. Together, these are the critical and important health outcomes affected by environmental noise as mentioned by the World Health Organisation's (WHO) Environmental Noise Guidelines for the European Region [76]. Within the ANIMA project a literature review was carried out, including publications after the year 2014. We focused on very recent articles as earlier publications are already evaluated by the WHO (see https://www.mdpi.com/journal/ijerph/special_issues/WHO_reviews). The outcomes from the literature review are published in the report 'Recommendations on noise and health (Deliverable D2.3, [41]).

The WHO reviews as well as the ANIMA literature review demonstrate associations between long-term aircraft noise exposure and ischemic heart disease, annoyance, reading and oral comprehension in school children as well as sleep disturbance during the night. In the ANIMA review, associations were made between sleep disturbance, annoyance and certain long-term health outcomes, indicating that self-reported sleep disturbance and annoyance may be mediators of adverse health outcomes. In the following sections new findings on the effects of aircraft noise exposure on different health outcomes are summarised.

Cardiovascular Diseases

Several cardiovascular health effects were investigated, such as hypertension (high blood pressure), ischaemic heart disease (coronary artery disease) and stroke. New studies show that aircraft noise exposure may increase the risk of hypertension, especially if exposure is high during the night time. Evidence on heart diseases needs cautious interpretation and further research. It was not shown that an increased risk for stroke is associated with increased aircraft noise exposure. The lack of statistical significance could be related to the small number of persons which are exposed to the highest levels of aircraft noise. The studies investigated either the prevalence or the incidence of diseases associated with aircraft noise exposure. *Prevalence* describes the occurrence of a disease in a higher aircraft noise exposed population relative to the occurrence of the disease in a less exposed population. The aircraft noise induced *incidence* of a disease, however, refers to the occurrence of new cases of this disease in a high exposed population compared to new cases in an unexposed or less exposed population.

Circulatory System and Hypertensive Heart Diseases

The WHO review did not show a significant increase of the risk for hypertension associated with increased aircraft noise exposure. However, such risk was confirmed in case of road traffic noise [76]. The ANIMA literature review included ten publications on the circulatory system and hypertensive diseases.

The exposure to aircraft noise at night was frequently included in new studies of the circulatory system and hypertensive heart diseases. An association between the risk of hypertension and exposure to night-time aircraft noise reached 34% increase related to 10 dB increase in L_{night}. Additionally, two recent studies show a significant impact of aircraft noise on hypertensive heart disease. The importance of exposure during the night is evident [58, 59]. However, even when significantly increased risk for hypertension incidence was observed for individuals exposed to aircraft noise at levels of 50-54 dB LAeq24h the conclusion in a large case–control study was that there is no association between air traffic noise and hypertension [77].

Overall, new studies show and confirm the WHO statement on the association between aircraft noise and hypertension and add evidence on the importance of also considering the night-time noise exposure. Studies with additional methodological improvements would be needed to further reduce inconsistencies and improve the quality. Still, the findings of new cohort studies seem to point toward a harmful effect caused by aircraft noise exposure [29].

Ischaemic Heart Disease and Other Forms of Heart Diseases

The WHO review showed a significant but small increased risk for ischaemic heart disease incidence associated with increased aircraft noise exposure [76]. The ANIMA review included five publications on ischaemic heart diseases, myocardial infarction, cardiac arrhythmia and heart failure.

New studies use different approaches to examine the association between aircraft noise exposure and the occurrence of a disease. For example, different noise indicators were considered such as the intermittency ratio. The intermittency ratio, which is a noise parameter that describes how strongly a noise event emerges from the background, at night showed stronger relation to myocardial infarction hazard than continuous noise levels of the same average level. With respect to myocardial infarction, the most sensitive time of noise exposure was between 5.00 and 6.00 a.m. However, this was not confirmed by other researchers. Another approach is to use the mortality rate ratio (MRR). The MRR from cardiovascular disease resulted in an increase of 18% per 10 dB of the overall day exposure L_{den} to aircraft noise. For coronary heart disease and myocardial infarction, the MRR increased with 24% and 28% per 10 dB, respectively, and was higher for men compared to women [23]. For other forms of heart disease, an arrhythmia was observed during the night and heart failure or hypertensive heart disease was reported to be associated with aircraft noise exposure.

The association observed between aircraft noise exposure and risk of myocardial infarction or mortality from ischaemic heart disease or other forms of heart diseases needs cautious interpretation. The heart diseases are all in all multi-factorial determined and the impact of aircraft noise is relatively small. However, it becomes relevant given that in a population even health effects of small size sum up to a considerable number of people suffering from severe health problems.

Stroke

The WHO review did not show a significant increase of risk for stroke associated with increased aircraft noise exposure [76]. The ANIMA review included four publications on cerebrovascular disease including different types of stroke.

Overall, there is no conclusive evidence with respect to an association between aircraft noise exposure and stroke.

The findings of the recent Swiss National Cohort around Zurich Airport between 2000 and 2015 suggest that night-time aircraft noise can trigger acute cardiovascular mortality with a similar association found in previous studies for long-term aircraft noise exposure [58]. The RIVM review [70] and the review on aviation noise and public health [29] confirm the WHO conclusion that there is no evidence of risk for stroke associated with aircraft noise.

Key Message

Several studies on cardiovascular diseases show an association with aircraft noise exposure. However, they lack conclusive results. New studies add information on the importance of the night-time exposure to noise, and also the number and the level of individual noise events, therefore they should be considered in more detail.

Sleep Disturbance

The WHO review showed a significant increase in the probability of additional awakenings due to aircraft noise related to noise indicator L_{Smax} and an increase in percentage of persons reporting to be highly sleep disturbed (%HSD) in relation to noise indicator for night-time L_{night} [76]. The literature review conducted within ANIMA identified 24 publications comprising journal papers as well as conference proceedings.

The vast majority of the studies refer to cross-sectional studies in the field. Three studies have undertaken a pre-post-comparison for sleep disturbance/sleep quality before and after the change of a night flight regime [48, 61, 66]. In eight publications, disturbance was assessed by physiological measurements [4, 7, 19, 37, 47, 48, 49, 63]. Twenty-one studies used self-reports to assess different sleep outcomes for aircraft noise, such as insomnia, awakenings, and sleep quality. Only six of these studies specifically referred to aircraft noise as the source for disturbance [13, 17, 48, 51, 57, 60], whilst thirteen others did not [4, 7, 15, 34–37, 39, 43, 50, 56, 59, 63, 66]. Two studies applied both neutral sleep quality questions but also questions referring to aircraft noise as the source for sleep disturbance [52, 55].

Results of recent studies are generally in line with the findings of the WHO review. Physiologically measured disturbances of sleep quality, represented by an increase of the time to fall asleep and wake time, number of awakenings or increased motility were found for an increase in the exposure represented by higher average night levels or a higher number of (loud) aircraft noise events. Studies using physiological measurements confirmed the significant impact of the maximum sound pressure level on the probability for awakening reactions. In one study [63], however, results were not statistically significant, most probably due to the small sample of participants. The benefit of the implementation of a night curfew from 23:00 to 05:00 at a large German airport (Frankfurt Airport) was demonstrated with regard to the number of awakenings per night, total sleep time and also the time spent in deep sleep [48]. However, residents reported higher sleep disturbance and an increased number of awakenings in the early morning coinciding with the end of the night curfew at 05:00 [60]. Thus, the benefit of this night curfew is rather ambiguous.

A laboratory study compared the impact of the three major traffic noise sources—air, railway and road traffic—and revealed that the probability to wake up from equal maximum levels increased in the order aircraft < road < railway noise. This order is reversed to that usually found for self-reported long-term sleep-disturbance, and annoyance [19].

Since the assessment of both exposure variables and sleep outcomes differed considerably between the eight studies included in this review, a comparison between results is not possible.

Effects of aircraft noise exposure were also shown for self-reported sleep disturbances, decreased sleep quality or similar sleep outcomes. Eighteen of the twenty-one publications report an effect of aircraft noise on participants' self-reported sleep outcomes. Studies not showing an effect all referred to general sleep outcomes not mentioning aircraft noise as the source for sleep disturbances. The conclusion that effects of aircraft noise exposure on self-reported sleep disturbance are higher when aircraft noise was mentioned as a source for the sleep disturbances, was already drawn in the recent WHO evidence review on the impact of environmental noise on sleep [6]. The magnitude of effect also depended on the assessment methods for aircraft noise exposure. The magnitude of effect was enhanced in comparisons between exposure groups vs. control groups and low exposure vs. high exposure (e.g., [38, 43]). In contrast, when aircraft noise exposure was represented by average sound pressure levels or the number of aircraft noise events during the night, not all studies revealed a significant association to sleep disturbances or sleep quality (e.g., [37, 63]). It was concluded that average noise levels were not sufficient predictors for sleep disturbances and the number of events and maximum level should be taken into account as well. The Intermittency Ratio has been shown to be a relevant predictor of self-reported sleep disturbance and adding important information to average noise levels [13, 57].

Overall, self-reported sleep disturbance or decreased sleep quality do not necessarily reflect the physiologically measured sleep quality or sleep disturbances due to aircraft noise (e.g., [4, 48]). The studies included in this review using measures of self-reports differed considerably with regard to the assessed sleep outcomes (sleep disturbances (e.g., in Brink [13]), insomnia (e.g., in Kwak et al. [43]) etc.) and questions used for this assessment, e.g. specifically if sleep disturbances are attributed to aircraft noise in the wording of questions (e.g., in Röösli et al., [57]) or not (e.g.,in Janssen et al. [37]). Therefore, the possibility to compare the results of the various studies is limited.

Key Message
Physiological measurements reveal sleep disturbances due to aircraft noise exposure, mainly represented by awakenings. Self-reported measures of sleep outcomes are affected by aircraft noise exposure, too, but do not necessarily reflect physiologically measured sleep outcomes. The magnitude of the effect of aircraft noise exposure on sleep is influenced both by the assessment of exposure variables and sleep outcomes. Average sound pressure levels are insufficient predictors of both physiologically-measured and self-reported sleep outcomes. The number of noise events and maximum levels should be considered, too.

Cognitive Impairment

The WHO review on cognition showed that most of the studies focus on the impact of aircraft noise on children. Children exposed to aircraft noise above 55 dB L_{den} have a higher risk of experiencing cognitive constraints related to their reading skills and oral comprehension [57]. The ANIMA review included only one new publication on cognitive impairment in children [39]. In this new study, a 20 dBA increase in aircraft noise exposure was associated with a 2-month delay in reading abilities for the whole sample, and with a 3-month delay in the subsample of non-migrant children. For the evaluation of the noise effect, other factors impacting reading should also be considered, especially socioeconomic status and the number of books at home.

The ANIMA review supports the WHO conclusion on the negative association between aircraft noise exposure and reading comprehension in children. The review on aviation noise and public health [29] confirms the WHO conclusions.

Key message
Aircraft noise has an effect on cognitive functioning in children related to reading skills and oral comprehension. These effects are important to be considered in protection of children's health.

Mental Health and Well-Being

In the WHO review it was emphasised that consistent conclusions for the effects of aircraft noise exposure on mental health and well-being could not be drawn. This is due to the small number of studies, the differences in the experimental design of the studies and a variation of methods for noise metrics and outcome measurements. Further, no estimates of risks could be drawn from the results [77]. In newer studies included in the ANIMA review as well as resulting from new searches there is new but still inconsistent evidence for a relationship between aircraft noise and mental health.

In studies with short-term measures of well-being and quality of life, referring to a momentary time period, higher aircraft sound levels were associated with lower levels of happiness [28] and well-being [44]. Further, small but significant effects of aircraft noise on quality of life were found in children [39] and adults [61], that is, with increasing aircraft noise levels reported quality of life decreased.

In a study on the impact of aircraft noise exposure on well-being and health in children it was shown that noise exposure had no direct effects on child-reported physical well-being and parents' reports of children's health [64].

No association was found for aircraft noise exposure and the use of medication indicated for mental health issues [10], and psychological distress measures [9]. However, in another study significant differences regarding depression scores between high exposure groups and the control group were observed [35].

In two studies examining the long-term effect of aircraft noise on diagnoses of depression no direct association was found for depression diagnoses one year later in a German study [11] and ten years later in a Swiss study [25].

Finally, for diagnoses of manifest disorders a study analysing insurance data found a positive relationship between aircraft noise exposure and diagnosed unipolar depression when socioeconomic status was taken into account, i.e. with increasing aircraft noise levels an increase in risk for diagnoses of depression was investigated [62].

Results of the ANIMA review and new literature support the findings of the initial WHO review indicating inconsistent evidence for the influence of aircraft noise on mental health outcomes. 3 of 5 studies showed weak but significant associations of quality of life and well-being measures with noise exposures showing health-related quality of life to be impaired by aircraft noise, while only two other studies found significant evidence for the impact of aviation noise on psychological distress [9] and diagnosed depressions [62].

However, further analyses support the assumption that there might not be a direct effect of aviation noise on mental health measures but effects may be mediated by annoyance, i.e. an increase in noise levels leads to an increase in annoyance ratings which further contributes to other health effects [11, 25].

Key Message
New studies are available indicating a negative effect of aircraft noise exposure on well-being, quality of life and diagnosed depression, but overall findings on mental health are still inconsistent and scarce.

Hearing Impairment

The WHO review found no evidence of an association between aircraft noise exposure and hearing impairment and tinnitus [77]. The ANIMA review has not identified new studies that would investigate the association between aircraft noise exposure and hearing impairment outcomes or tinnitus.

Key Message
There is no evidence that aircraft noise would cause hearing impairment in the general public.

Adverse Birth Outcomes

The WHO review identified a knowledge gap and a need for long-term studies on adverse birth outcomes (pre-term delivery, low birth weight and congenital anomalies) and other adverse effects from exposure to environmental noise, to inform future recommendations properly [77]. The ANIMA review did not identify any new study investigating the association between aircraft noise and adverse birth outcomes.

Key Message
There is a need for further research on the adverse birth and reproductive outcomes due to the importance of long-term morbidity that they can cause [29].

Metabolic Diseases

The WHO identified a research gap on the impact of aircraft noise exposure on metabolic diseases, which is why they could not draw firm conclusions [77].

The ANIMA review could only identify very few studies examining the impact of aircraft noise exposure on different metabolic diseases. Two studies found a significant increase in waist circumference per 10 dB L_{den} increase in aircraft noise exposure.

Three studies are available analysing the relationship between aircraft noise exposure and obesity. Two of these studies found a significant association between aircraft noise exposure and an increase in waist circumference [21, 54]. Results by Pyko et al. [54] further showed a significant weight gain of 0.03 kg per 10 dB L_{den} increase in aircraft noise exposure. In another study, no such associations were observed between aircraft noise and adiposity markers as well as the development of obesity [27].

Three studies that looked at the impact of aircraft noise exposure on the incidence or prevalence of diabetes could be identified. Two studies did not show a significant relationship between aircraft noise and diabetes incidence and prevalence of diabetes [21, 72], whereas Eze et al. [24] describe a significant association between aircraft noise and incidence of diabetes, indicating that the risk of diabetes increases with increasing noise levels.

Key Message

Overall, there are very few studies available investigating the effect of aircraft noise exposure on metabolic diseases. Therefore, no firm conclusions can be drawn from the current evidence. More research is needed on this topic.

Noise Annoyance—A Mediator of Aircraft Noise Effects on Health?

Annoyance is one of the most studied and established effects of noise and is, therefore, often used as a noise impact measure for estimation and regulation purposes. As increasing aircraft noise exposure levels are linked to an increase in aircraft noise annoyance, it can further be hypothesised that increasing annoyance levels might contribute to other adverse health outcomes.

Annoyance as a stress response and health outcome itself is described as "a relation between an acoustic situation and a person who is forced by noise to do things he/she does not want to do, who cognitively and emotionally evaluates this situation and feels partly helpless" (Guski et al. [30], p. 525). Due to the multi-dimensional structure of annoyance with its cognitive, emotional and behavioural aspects, it might be related to, or even contribute to, various health outcomes or even to disorders.

Health outcomes can also be discussed as contributing to the manifestation of noise annoyance. The relation of noise annoyance and health outcomes thus leads to two questions: Do high ratings of annoyance play a role in the development and maintenance of diseases? Are people, who are suffering from any form of disease, more bothered, annoyed or disturbed by aircraft noise?

New studies support an indirect role of annoyance in the relationship of aircraft noise exposure and health outcomes. That is, noise exposure influences noise annoyance, which in turn affects the health outcome. Supporting evidence for this theory results from studies on the effect of aircraft noise exposure on cardiovascular diseases, sleep outcomes, and mental health measures.

There has been a vast amount of studies on the impact of aircraft noise on cardiovascular diseases, but only few of them also examined the relationship to annoyance. Eriksson et al. [20] found the relative risk for hypertension among participants reporting annoyance to be higher than in participants who are not annoyed. This is supported by findings from Babisch et al. [2] and Baudin et al. [8]. By contrast, in a small study from Italy, no association between blood pressure and annoyance was shown [15].

A few studies showed a link between mental health and well-being-related measures and noise annoyance. Spilski et al. [64] reported indirect effects of aircraft noise on physical well-being in children via noise annoyance, i.e. with an increase in noise annoyance children's self-reported physical well-being decreased. Similarly, a higher risk for psychological distress was observed for people being extremely annoyed by noise in comparison to a lower risk for people being less annoyed [9]. Baudin et al. [10] found an association between aircraft noise annoyance and the use of anxiolytics (medication for anxiety disorders), implying a mediating role of annoyance for the link of aircraft noise exposure to mental health outcomes.

As shown earlier, aircraft noise did not have a direct effect on mental-health related quality of life [61] and diagnoses of depression [11], but in both studies an indirect effect via annoyance was found. The results suggest that aircraft noise exposure decreases mental health-related quality of life [61] and predicts the development of depression one year later [11] via noise annoyance. Both studies further indicate that there is a reciprocal association, i.e. that diagnoses of depression and poorer mental-health related quality of life also contributed to higher ratings of annoyance a year later. In addition, whereas the absolute aircraft noise level was not directly associated with mental health-related quality of life, in one of the studies [61] it turned out that the change in noise exposure due to the opening of a new runway lead directly and indirectly via noise annoyance to a poorer mental health-related quality of life. This indicates the importance of communication and engagement in airport noise management, particularly in situations of change (see also Chaps. 8 and 9).

Moreover, effects of annoyance were also observed for sleep quality [3] and physical activity [26]. Better rated sleep quality was accompanied with a lower rating of long-term aircraft noise annoyance [3], while noise annoyance was negatively associated with physical activity, i.e. higher ratings of transportation noise annoyance predicted reduced physical activity [26].

Some studies even showed signs of reversed causality, i.e. the health outcome also predicted a future increase of noise annoyance [11, 61]. This indicates that vulnerability due to physiological and/or psychological health issues may limit resources to cope with noise which can contribute to higher annoyance.

Due to different methods to assess noise annoyance as well as different health outcomes and measures, it is difficult to draw consistent conclusions. However, evidence indicates that annoyance contributes to adverse mental health outcomes.

Key Message

Several studies showed a link between noise annoyance and various mental health outcomes. As stress responses (i.e. annoyance) are considered to be an important element in the development of some diseases it is recommended to further investigate the relationship between annoyance and health outcomes.

Overall, the reviews highlight the importance of addressing aircraft noise annoyance and sleep disturbance as the most critical outcomes. The assumption is that interventions aiming at the reduction of noise annoyance might in turn at least partly reduce negative health outcomes. Figure 1 gives a summary of the different health outcomes associated with aircraft noise exposure and depicts the underlying mechanisms and role of noise annoyance.

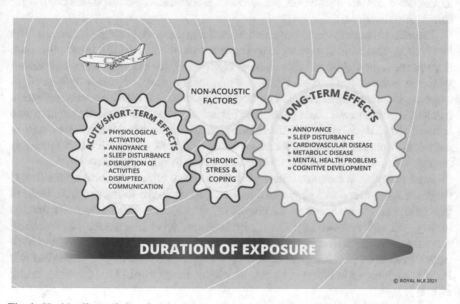

Fig. 1 Health effects of aircraft noise exposure and the role of annoyance

Why Does Aircraft Noise Exposure Have an Impact on Health?

General Health-Related Mechanisms of Aircraft Noise Exposure

We are constantly surrounded by sound. In a general assumption, sound can be evaluated as positive, neutral, or negative. When sound is characterised as unpleasant or unwanted, the notion is changed to noise. Environmental noise can cause disturbances, e.g. in daily activities such as watching television, concentrating or in conversations with people. Further, environmental noise can be considered as an environmental stress factor that challenges the human system [12]. Stress is a reaction of the body and mind in demanding situations. In particular, stress occurs in situations of uncertainty and unpredictability [40]. Stress can be considered as a reaction in and to specific situations and stimuli, but also as a process, as stress responses can trigger subsequent reactions.

The human stress system is vital and essential to tackle demanding situations. The human body and mind always try to maintain a state of balance [16]. In case this state is threatened, the fight-or-flight response is activated, for example when confronted with a dangerous wild animal. From an evolutionary perspective, this fight-or-flight response is crucial for survival. The body shuts down current irrelevant activation (such as digestion etc.) while enhancing all acutely vital processes to stay alert and focused. Nowadays, acute life-threatening situations or stimuli are rather rare in industrial countries. However, there are other situations or stimuli that trigger this response, e.g. when environmental exposures such as noise challenge us.

In an established reaction scheme for the adverse effects of noise on health [1] the chain is described as follows: In a hierarchical chain the sound exposure leads to psychological effects in terms of cognitive, emotional and behavioural reactions (annoyance and disturbance). On a physiological level it activates physiological processes, e.g. the activation of stress hormones, that are considered as stress indicators. Prolonged physiological activation through noise can trigger biological risk factors, e.g. change in blood pressure, which are directly linked to long-term health effects [1].

An individual evaluation of the situation or stimulus further plays a key role in the stress mechanism. Psychological models suggest that for a stress response it is also important whether a situation or stimulus is perceived as demanding. A second important aspect is whether an individual feels like he/she is able to cope with the situation with the available resources [45]. Thus, when an individual feels threatened by a noise situation, the first reaction is usually trying to cope with it, e.g. by closing the windows.

As suggested by Stallen [65], coping and perceived control are essential concepts influencing the degree of annoyance. Coping is an individual's capability of having resources to deal with the noise (or other situations) that are perceived as demanding and being able to apply strategies. This process of coping with noise events is dynamic and is associated with reappraisals of the noise situation and the success of coping. Strategies can refer to how the noise is valued, the mindset or direct adjustments of the behaviour to the noise conditions. This can be closing the windows when the noise situation is perceived as disturbing. Coping directly relates to perceived control, which can be described as an individual's belief of being capable to influence certain events in its life. To be able to cope means that one perceives having control over the situation (and vice versa), this in turn can decrease annoyance. Hence, when one is able to cope, one experiences less stress, i.e. less annoyance (See also Chap. 8).

Furthermore, the resources to cope with demanding situations from time to time require restoration. Being in nature, having access to green areas and doing physical activities outdoors allow recovery [18, 32, 31]. Studies show that transportation noise is not only a stressor in itself, but hinders outdoor physical activities and constrains the restorative environmental qualities, thus hampering recovery from daily stress [26, 73, 74]. This means, if an environmental demand lasts over a long period of time, resources for coping decline, and restoration is constrained. Long-term stress can be harmful and the degree of annoyance can vary.

It is important to note that a physiological stress response is not independent from a psychological stress response as they are naturally linked and contribute to each other, e.g. being annoyed by noise and the release of stress hormones are reciprocally related. Further, research suggests that annoyance plays a central role in the relationship between aircraft noise exposure and health effects. In recent years, researchers have started to study annoyance as a precursor/mediator for further physical and mental health issues. That means that part of the effect of noise exposure on a health outcome is explained by the effect of noise exposure on noise annoyance, which in turn contributes to health effects (examples are described in the section on annoyance as a mediator).

Whether an individual evaluates a certain situation or stimulus as stressful, and is annoyed by it, is also determined by other factors. Acoustical features and noise characteristics (e.g. harmfulness, intensity, duration, loudness etc.) are critical when it comes to annoyance as some features can be more annoying than others, but also non-acoustical factors are crucial, e.g. attitudes, concerns, expectations etc. Their impact is elaborated on further in Chap. 8.

To summarise, noise as a stressful stimulus can affect the regulation of the human organism [16]. Short-term reactions of the stress system to noise might not be problematic. Stress responses only become unhealthy when activation of the stress system is prolonged while the resources for coping are limited. If the process of coping with noise is not successful in the long run and recovery is not possible, it is likely that this environmental demand together with the lack of control and ability to cope can lead to stress-related long-term health effects.

Health-Related Mechanisms of Aircraft Noise Exposure During Sleep

Undisturbed sleep of sufficient length is a vital process for human beings, providing the necessary daytime alertness, performance ability and health (see [5]). According to the WHO [75], environmental noise-related self-reported sleep disturbance is widespread in the European Region leading to the highest number of noise-related loss of healthy life years compared to other noise health outcomes, followed by noise annoyance. With regard to the number of healthy life years lost in the European Union, noise-induced sleep disturbance is therefore regarded as "the most deleterious non-auditory effect of environmental noise exposure." (Basner et al. [5], p. 5). Deterioration of sleep quality and interruptions of sleep together with noise annoyance [8, 10] are regarded as belonging to the possible key variables in the causal pathway of noise-induced cardiovascular and metabolic diseases [77]. The consequences of interrupted sleep from transport noise can be classified as immediate reactions, short-term reactions, and long-term consequences [53, 75].

Nocturnal noise affects the human organism in a rather direct way. Even during sleep in an unconscious state, the hearing system has an alerting function to prevent harm from possible ambient threats and therefore continually inspects the environment acoustically. Sudden noise events that emerge from the background have hinted at threats in the early history of humankind. It was necessary that humans were able to awake and react quickly if necessary. Noisy events are subliminally perceived and evaluated even during sleep and can provoke physiological reactions that enhance the alertness of the individual. Therefore, meaningful noise events (e.g. one's own crying child) are more likely to cause reactions than less meaningful noise events appearing at the same sound level [6].

(a)   Immediate reactions to nocturnal noise

Acute noise exposure affects the function of multiple organs and systems, including an increase in blood pressure and heart rate. These reactions are most likely induced by the release of stress hormones, such as adrenaline and noradrenaline. These reactions helped humans in their early history to react adequately to the threat, i.e. to remove the threat by actively confronting it (= fight reaction) or to flee from it (= flight reaction). These stress reactions even appear when noise is not consciously perceived, e.g. during sleep. Stress reactions due to intruding noise can disturb the balance in the organism. These changes refer to blood pressure, blood flow, blood lipids, carbohydrates (glucose) regulation, electrolytes, and thrombosis/fibrinolysis [75].

As a consequence of the above described stress responses, immediate reactions in sleeping behaviour may occur. These comprise short arousal responses, changes from a deeper to a lighter sleep stage, awakenings, body movements, and in consequence, an increase in total wake time, a reduced time in deep sleep, and more general sleep loss [75].

(b)   Short-term reactions to nocturnal noise

As a consequence of the sleep loss and the reduced restoration during the night, sleepiness in the morning and the following day and a reduced well-being as well as a decrease in cognitive performance during the daytime can occur [75].

(c)   Long-term reactions

Chronic sleep loss and recurring interruptions of sleep are a major risk factor for cardiovascular and metabolic diseases. This link set up the assumptions that recurring noise-induced awakenings and the resulting sleep loss may account for the higher risks of negative health outcomes (see above) after a longer period of aircraft noise exposure. However, the relationship between the immediate and long-term effects of noise is not completely clear, yet, in particular as mediators such as noise annoyance seem to play a relevant role for long-term health effects as well [8, 9]. The assumption that noise-induced sleep disturbance is part of the causal pathway from nocturnal exposure to increased risks for cardiovascular and metabolic diseases is often replicated (e.g. WHO, [76]), evidence for the mediating effect is scarce [56] or contradictory [21]. Very recently it was concluded that nocturnal aircraft noise exposure increases the risk of developing hypertension via a direct effect on blood pressure as well as via a mediated effect as a consequence of chronic sleep disturbance [56].

The appearance of acute reactions to nocturnal aircraft noise do not differ from natural reactions, such as spontaneous awakenings. However, a considerable increase in the number of these immediate reactions is assumed to constitute a health issue as it reduces the restorative power of sleep [6, 75]. Healthy adult individuals briefly awaken approximately 20 times during an 8 h night and most of these awakenings are too short to be remembered the next day. Up to now, it is not clear how many additional noise-induced awakenings are needed to cause adverse effects on restoration and health. Large differences between individuals are assumed with regard to an acceptable number of additional noise-induced awakenings depending on the vulnerability of the individual towards noise effects and the presence of additional, non-acoustical risk factors for health outcomes [6]. Several vulnerability factors are discussed below.

Individual Risk Factors in the Causal Chain from Noise to Health Outcomes

The vast majority of studies included in the WHO reviews and the ANIMA review considered noise effects in the general population. Only few studies have focused on so-called vulnerable groups, who are considered as more susceptible to adverse effects of aircraft noise and at a higher than expected risk for developing particular diseases [68]. In the context of noise-induced health effects, vulnerability factors comprise physical and mental health parameters, phase in life, lifestyle factors and habits, educational and socioeconomic status as well as characteristics of the environment. Groups considered to be at higher risks are for instance children, elderly people, shift-workers, chronically ill including mentally ill people, noise sensitive people as well as people with a low socioeconomic status [68]. The mechanisms of the vulnerability factor are not fully understood yet and are not necessarily the same for different vulnerable groups. Children, for instance, are in a sensitive developmental life stage and do not yet possess adequate strategies to cope with the noise, although not being per se more vulnerable. They are regarded as being less annoyed than adults, having lower risks for sleep disturbances but are more susceptible for cognitive impairments and cardiovascular diseases [68]. However, recent research in primary school children showed an aircraft noise-induced reduction of deep sleep due to noise that is comparable to the deep sleep reductions due to obstructive sleep apnea syndrome [4], which is regarded as a risk factor for the development of mental, metabolic and cardiovascular dysfunctions. Elderly people are likewise not considered to be at higher risk for annoyance. But they are regarded as more susceptible to noise-induced cardiovascular dysfunctions while their susceptibility for sleep disturbance due to aircraft noise is not completely clear yet [68].

One crucial factor resulting in higher susceptibility, in particular for annoyance [46] but also for cardiovascular disease [8], psychological distress as well as psychiatric disorders [71], is noise sensitivity. Noise sensitivity has a genetic component but can also be the result of mental or physical illness [68]. Moreover, noise sensitivity has been found to be highest in middle age, attributed to a higher workload and caring for children or other family members [46]. In summary, it is concluded that future research should focus on subgroups to understand the effects and mechanisms of aircraft noise for health effects in populations at higher risks and to provide group-specific exposure response relationships. Moreover, attention should be given to potential accumulation of environmental as well as social risk factors across the lifespan but also during specific life stages resulting in higher susceptibility to noise, poorer capacities to cope with the noise and consequently to higher risks for adverse health effects. Research in this field could shed more light in starting points for intervention measures.

Current Gaps in the Relationship Between Aircraft Noise Exposure and Health Outcomes

Despite the vast amount of studies on the impacts of aircraft noise on health, a couple of open questions remain. Study results have shown the association between aircraft noise exposure and aircraft noise annoyance as well as aircraft noise induced sleep disturbances (https://www.mdpi.com/journal/ijerph/special_issues/WHO_reviews). Sleep disturbance and aircraft noise annoyance highly correlate with each other [67] and are thought to be relevant in the causal pathway from aircraft noise exposure to cardiovascular diseases [22]. However, the causal link between noise annoyance and sleep disturbance is still unclear, i.e. whether aircraft noise annoyance facilitates sleep disturbances or whether sleep disturbances foster aircraft noise annoyance. A reciprocal relationship between these two factors is also conceivable. That is, sleep disturbance caused by aircraft noise can cause tiredness and reduce one's resources during daytime. This in turn could contribute to aircraft noise annoyance. On the other hand, if one feels annoyed by aircraft noise, this could affect one's sleep and one is more easily disturbed by aircraft noise at bedtime. Results from one study indicate the former: a higher self-reported sleep quality was linked to lower long-term aircraft noise annoyance [3]. However, empirical evidence on this relation is still scarce.

An analysis done by Schreckenberg et al. [61] using longitudinal data from the NORAH study focused on the relationship between aircraft noise annoyance and mental health-related quality of life (HQoL) and results indicate a reciprocal relationship between these two factors. A detailed overview of studies related to annoyance and mental health is provided by van Kamp and Davies' review [69]. Overall, there are mixed results regarding a direct relationship between aircraft noise exposure and mental health. However, noise annoyance seems to be a relevant mediator. Further, Kamp and Davies [69] reviewed studies including noise sensitivity as one additional important modifying factor.

Many studies examine the relationship between aircraft noise exposure and various health outcomes, such as sleep disturbances or cardiovascular diseases. However, there are limitations in some of those studies that need further consideration and should be addressed in future studies accordingly [42]. Limitations can encompass the study design itself, including the question of causality, noise exposure assessment and outcome operationalisation, aspects concerning the response rates and the disregard of potential confounding factors (i.e. other factors that might additionally influence the health outcome).

To tackle the question of causality, i.e. whether factor A causes B or vice versa, prospective, long-term studies with (at least) two measurements should be conducted to establish a potential causal relationship. Another limitation for some studies is a selective non-response, i.e. people with certain characteristics do not participate or participate to a much lesser extent. For example, in online surveys, elderly people are often underrepresented. Aspects such as these need to be considered.

The L_{den} is probably the most frequently used noise metric assessing noise exposure. However, there is a debate about whether this is always the best option to choose. Other metrics may be better suited to reflect the actual sound environment or better fit examining the research question at hand (e.g. [33]).

When looking at the relationship between noise exposure and different diseases, the operationalisation of the outcome is very important. One can rely on self-report health measures or use health data provided by insurances or other institutions that contain medical diagnoses. However, both data sources can lead to an under- and overestimation of a disease. Further, health data may lack relevant information on specific characteristics or behaviours of a person that might additionally influence the health outcome (confounding factors).

Future studies should aim to address the above-mentioned limitations to improve data quality and generate valid and robust evidence.

Conclusions

Aircraft noise exposure poses a worldwide health issue. Literature reviews conducted by WHO and within ANIMA have identified a relationship between aircraft noise exposure and adverse health outcomes such as annoyance, sleep disturbance, and cardiovascular diseases. Further, more recent studies found evidence for an impact of aircraft noise exposure on mental health measures. It has been found that noise annoyance and sleep disturbance also play a role as mediators of adverse health effects. A continuous experience of aircraft noise annoyance has been linked to adverse health effects through stress mechanisms. Further sleep disturbance has been found to promote adverse health effects, e.g. for cardiovascular diseases. Consequently, reducing noise annoyance and sleep disturbance can help to decrease adverse health effects and to improve people's well-being and their quality of life. To reduce aircraft noise annoyance there are two important aspects. First, decreasing aircraft noise exposure should be the main focus to reduce annoyance and related health effects. Second, the possibility to recover from the noise exposure should be accomodated, e.g., by providing access to recreational and green areas and areas of reduced noise exposure [34].

The number of existing studies on the various health outcomes differs enormously. However, only a few studies are available on, e.g., metabolic diseases and as the existing results are not consistent they do not allow for drawing firm conclusions. There is an urgency for future research to further investigate the impact of aircraft noise exposure on health for different populations such as vulnerable groups like children and elderly people. As annoyance and sleep disturbance are also mediators to health outcomes, it is essential to better understand and fully identify the underlying mechanisms to efficiently minimise further adverse health effects.

Assessing noise exposure, its impact and related research results grants the evaluation of noise interventions that can be improved and harmonised after evaluation. In the WHO Environmental Noise Guidelines for the European Region [76] this is already acknowledged as for the first time the noise guidelines include recommendations on noise interventions. The systematic WHO review on the health impact of noise interventions [14] provides a study protocol to be followed in future research that allows for evaluating the impact of noise interventions. This allows for the development of proper treatment and health care, prevention of health effects and mitigation strategies. A consistent monitoring system could provide a comprehensive approach to establish how interventions affect annoyance and interrelated outcomes. With this knowledge mitigation strategies for noise annoyance can improve people's well-being as well as quality of life and mitigate potentially adverse long-term health effects. A better understanding of these underlying mechanisms and the impact of aircraft noise exposure on health can serve as an essential guidance for developing the specific design and targeted implementation of successful noise interventions and mitigation strategies.

References

1. Babisch W (2002) The noise/stress concept, risk assessment and research needs. Noise Health 4:1
2. Babisch W, Pershagen G, Selander J, Houthuijs D, Breugelmans O, Cadum E, Vigna-Taglianti F, Katsouyanni K, Haralabidis AS, Dimakopoulou K, Sourtzi P, Floud S, Hansell AL (2013) Noise annoyance—a modifier of the association between noise level and cardiovascular health? Sci Total Environ 452–453:50–57. https://doi.org/10.1016/j.scitotenv.2013.02.034
3. Bartels S (2014) Aircraft noise-induced annoyance in the vicinity of Cologne/Bonn Airport— The examination of short-term and long-term annoyance as well as their major determinants, Technische Universität Darmstadt.
4. Bartels S, Quehl J, Aeschbach D (2019) Effects of nocturnal aircraft noise on objective and subjective sleep quality in primary school children. In: International Congress on Acoustics (ICA), 09.-13.09.2019, Aachen, Deutschland
5. Basner M, Babisch W, Davis A, Brink M, Clark C, Janssen S et al (2014) Auditory and non-auditory effects of noise on health. Lancet 383:1325–1332
6. Basner M, McGuire S (2018) WHO environmental noise guidelines for the European region: a systematic review on environmental noise and effects on sleep. Int J Environ Res Public Health 15:519
7. Basner M, Witte M, McGuire S (2019) Aircraft noise effects on sleep: results of a pilot study near Philadelphia International Airport. Int J Environ Res Public Health 16(17):3178
8. Baudin C, Lefèvre M, Babisch W, Cadum E, Champelovier P, Dimakopoulou K, Houthuijs D, Lambert J, Laumon B, Pershagen G, Stansfeld S, Velonaki V, Hansell AL, Evrard A-S (2020) The role of aircraft noise annoyance and noise sensitivity in the association between aircraft noise levels and hypertension risk: results of a pooled analysis from seven European countries. Environ Res 19:110179. https://doi.org/10.1016/j.envres.2020.110179
9. Baudin C, Lefèvre M, Champelovier P, Lambert J, Laumon B, Evrard AS (2018) Aircraft noise and psychological ill-health: the results of a cross-sectional study in France. Int J Environ Res Public Health 15(8):1642. https://doi.org/10.3390/ijerph15081642
10. Baudin C, Lefèvre M, Babisch W, Cadum E, Champelovier P, Dimakopoulou K, Houthuijs D, Lambert J, Laumon B, Pershagen G, Stansfeld S, Velonaki V, Hansell AL, Evrard A-S (2021a) The role of aircraft noise annoyance and noise sensitivity in the association between aircraft noise levels and medication use: results of a pooled-analysis from seven European countries. BMC Public Health, 21(300). https://doi.org/10.1186/s12889-021-10280-3
11. Benz S, Schreckenberg D (2019) Examination of the causal relationship between aircraft noise exposure, noise annoyance and diagnoses of depression using structural equation modelling. In: *Proceedings of the 23rd International Congress on Acoustics*. Aachen, Germany
12. Bodenmann G, Gmelch S (2009) Stressbewältigung. In: Margraf J, Schneider S (Hrsg.), Lehrbuch der Verhaltenstherapie (S. 617–629). Springer Berlin Heidelberg, Berlin, Heidelberg
13. Brink M, Schäffer B, Vienneau D, Pieren R, Foraster M, Eze IC, Rudzik F, Thiesse L, Cajochen C, Probst-Hensch N, Röösli M, Wunderli JM (2019) Self-reported sleep disturbance from road, rail and aircraft noise: exposure-response relationships and effect modifiers in the SiRENE study. Int J Environ Res Public Health 16(21):4186. https://doi.org/10.3390/ijerph16214186
14. Brown AL, van Kamp I (2017) WHO environmental noise guidelines for the European Region: A systematic review of transport noise interventions and their impacts on health. Int J Environ Res Public Health 14:873. https://doi.org/10.3390/ijerph14080873
15. Carugno M, Imbrogno P, Zucchi A, Ciampichini R, Tereanu C, Sampietro G, Consonni D (2018) Effects of aircraft noise on annoyance, sleep disorders, and blood pressure among adult

residents near the Orio al Serio International Airport (BGY). Italy La Medicina Del Lavoro 109(4):253–263

16. Chrousos GP (2009) Stress and disorders of the stress system. Nat Rev Endocrinol 5:374–381. https://doi.org/10.1038/nrendo.2009.106

17. Douglas O, Murphy E (2016) Source-based subjective responses to sleep disturbance from transportation noise. Environ Int 92–93:450–456

18. Dzhambov AM, Markevych I, Hartig T, Tilov B, Arabadzhiev Z, Stoyanov D, Gatseva P, Dimitrova DD (2018) Multiple pathways link urban green- and bluespace to mental health in young adults. Environ Res 166:223–233. https://doi.org/10.1016/j.envres.2018.06.004

19. Elmenhorst E-M, Griefahn B, Rolny V, Basner M (2019) Comparing the effects of road, railway, and aircraft noise on sleep: exposure-response relationships from pooled data of three laboratory. Studies 16(6):1073

20. Eriksson C, Bluhm G, Hilding A, Ostenson CG, Pershagen G (2010) Aircraft noise and incidence of hypertension-gender specific effects. Environ Res 110(8):764–772

21. Eriksson C, Hilding A, Pyko A, Bluhm G, Pershagen G, Östenson CG (2014) Long-term aircraft noise exposure and body mass index, waist circumference, and type 2 diabetes: a prospective study. Environ Health Perspect 122:687–694

22. Eriksson C, Pershagen G, Nilsson M (2018) Biological mechanisms related to cardiovascular and metabolic effects by environmental noise. Copenhagen: WHO Regional Office for Europe. http://www.euro.who.int/en/health-topics/environment-and-health/noise/publications/2018/biological-mechanisms-related-to-cardiovascular-and-metabolic-effects-by-environmentalnoise. Accessed 8 Aug 2018

23. Evrard AS, Bouaoun L, Champelovier P, Lambert J, Laumon B (2015) Does exposure to aircraft noise increase the mortality from cardiovascular disease in the population living in the vicinity of airports? Results of an ecological study in France. Noise Health 17:328–336

24. Eze IC, Foraster M, Schaffner E, Vienneau D, Heritier H, Rudzik FP, Hensch N (2017) Long-term exposure to transportation noise and air pollution in relation to incident diabetes in the SAPALDIA study. Int J Epidemiol 46(4):1115–1125

25. Eze IC, Foraster M, Schaffner E, Vienneau D, Pieren R, Imboden M, Wunderli JM, Cajochen C, Brink M, Röösli M, Probst-Hensch N (2020) Incidence of depression in relation to transportation noise exposure and noise annoyance in the SAPALDIA study. Environ Int 144:106014. https://doi.org/10.1016/j.envint.2020.106014

26. Foraster M, Eze IC, Vienneau D, Brink M, Cajochen C, Caviezel S, Héritier H, Schaffner E, Schindler C, Wanner M, Wunderli J-M, Röösli M, Probst-Hensch N (2016) Long-term transportation noise annoyance is associated with subsequent lower levels of physical activity. Environ Int 91:341–349. https://doi.org/10.1016/j.envint.2016.03.011

27. Foraster M, Eze IC, Vienneau D, Schaffner E, jeong A, Heritier H, Rudzik F, Thiesse L, Pieren R, Brink M, Cajochen C, Wunderli J, Röösli M, Probst-Hensch N (2018) Long-term exposure to transportation noise and its association with adiposity markers and development of obesity. Environ Int 121:879–889

28. Fujiwara D, Lawton RN, MacKerron G (2017) Experience sampling in and around airports, momentary subjective wellbeing, airports, and aviation noise in England. Transp Res Part D: Transp Environ 56:43–54

29. Grollman C, Martin I, Mhonda J (2020) Aviation noise and public health. Rapid evidence assessment. NatCen Social Research that works for society. Prepared for: The Independent Commission on Civil Aviation Noise (ICCAN)

30. Guski R, Felscher-Suhr U, Schuemer R (1999) The concept of noise annoyance: How international experts see it. J Sound Vib 223:513–527

31. Hartig T, Evans GW, Jamner LD, Davis DS, Gärling T (2003) Tracking restoration in natural and urban field settings. J Environ Psychol 23:109–123

32. Hartig T, Johansson G, Kylin C (2003) Residence in the social ecology of stress and restoration. J Soc Issues 59(3):611–636

33. Haubrich J, Benz S, Isermann U, Schäffer B, Schmid R, Schreckenberg D, Wunderli J-M, Guski R (2020) Leq+X—Lärmexposition, Ereignishäufigkeiten und Belästigung: Re-Analyse

von Daten zur Belästigung und Schlafstörung durch Fluglärm an deutschen und Schweizer Flughäfen (*Leq+X—Noise exposure, number of events and annoyance: Re-analysis of aircraft noise annoyance and sleep disturbance data at German and Swiss Airports.* Hauptbericht Bochum: Ruhr-Universität Bochum. https://doi.org/10.46586/rub.164.139

34. Haubrich J, Burtea NE, Flindell I, Hooper P, Hudson R, Rajé F, Schreckenberg D, et al (2019) ANIMA D2.4—Recommendations on annoyance mitigation and implications for communication and engagement. Zenodo. https://doi.org/10.5281/zenodo.3988131

35. Hiroe M, Makino K, Ogata S, Suzuki S (2017) A questionnaire survey on health effects of aircraft noise for residents living in the vicinity of Narita International Airport: the results of physical and mental health effects. In: Proceedings: ICBEN 2017. Zurich

36. Holt JB, Zhang X, Sizov N, Croft JB (2015) Airport noise and self-reported sleep insufficiency, United States, 2008 and 2009. Prev Chronic Dis 12:E49

37. Janssen SA, Centen MR, Vos H, van Kamp I (2014) The effect of the number of aircraft noise events on sleep quality. J Appl Acoust 84:9–16

38. Kim SJ, Chai SK, Lee KW, Park JB, Min KB, Kil HG (2014) Exposure–response relationship between aircraft noise and sleep quality: a community-based cross-sectional study Osong public health and research perspectives 5(2):108–114

39. Klatte M, Spilski J, Mayerl J, Möhler U, Lachmann T, Bergström K (2017) Effects of aircraft noise on reading and quality of life in primary school children in Germany: results from the NORAH study. Environ Dev 49(4):390–424

40. Koolhaas JM, Bartolomucci A, Buwalda B, de Boer SF, Flügge G, Korte SM et al (2011) Stress revisited: a critical evaluation of the stress concept. Neurosci Biobehav Rev 35:1291–1301

41. Kranjec N, Benz S, Burtea NE, Hooper P, Hudson R, Jeram S, Schreckenberg D et al (2019) ANIMA D2.3—Recommendations on noise and health. Zenodo. https://doi.org/10.5281/zenodo.2562748

42. Kranjec N, Kuhlmann J, Benz S, Schreckenberg D, Rajé F, Hooper P, Jeram S (2021) Aircraft noise health impacts and limitations in the current research. In: Proceedings of the 13th ICBEN congress on noise as a public health problem. Stockholm, Sweden, 14–17 June 2021

43. Kwak KM, Ju YS, Kwon YJ, Chung YK, Kim BK, Kim H et al (2016) The effect of aircraft noise on sleep disturbance among the residents near a civilian airport: a cross-sectional study. Annals Occup Environ Med 28:38

44. Lawton R, Fujiwara D (2016) Living with aircraft noise: airport proximity, aviation noise and subjective wellbeing in England. Transp Res Part D: Transp Environ 42:104–118

45. Lazarus RS, Folkman S (1984) Stress, appraisal, and coping. Springer Publishing Company, New York

46. Miedema HM, Vos H (2003) Noise sensitivity and reactions to noise and other environmental conditions. J Acoust Soc Am 113(3):1492–1504. https://doi.org/10.1121/1.1547437

47. Müller U, Elmenhorst E-M, Mendolia F, Quehl J, Basner M, McGuire S, Aeschbach D (2017) A comparison of the effects of night time air traffic noise on sleep at Cologne/Bonn and Frankfurt Airport after the night flight ban. In: Proceedings of the 12th international congress on noise as a public health problem (ICBEN). 2017, ID 3726

48. Müller U, Elmenhorst E-M, Mendolia F, Quehl J, Basner M, McGuire S, Aeschbach D (2016) The NORAH-Sleep Study: effects of the night flight ban at Frankfurt airport. In: Proceedings of Inter-noise 2016. Hamburg, Germany, pp 7782–7786. ISBN 978-3-939296-11-9

49. Nassur A-M, Léger D, Lefèvre M, Elbaz M, Mietlicki F, Nguyen P, Ribeiro C, Sineau M, Laumon B, Evrard A-S (2019) The impact of aircraft noise exposure on objective parameters of sleep quality: results of the DEBATS study in France. Sleep Med 54:70–77

50. Nassur AM, Lefevre M, Laumon B, Leger D, Evrard AS (2017) Aircraft noise exposure and subjective sleep quality: the results of the DEBATS study in France. Behavioral sleep medicine 1–12. 46. Kageyama T, Yano T, Kuwano S, Sueoka S, Tachibana H

51. Nguyen TL, Nguyen TL, Yano T, Nishimura T, Sato T, Morinaga M, Yamada I (2017) The opening of a new terminal building and its influences on community response around Hanoi Noi Bai International airport: comparison between arrival and departure sides

52. Perron S, Plante C, Ragettli MS, Kaiser DJ, Goudreau S, Smargiassi A (2016) Sleep disturbance from road traffic, railways, airplanes and from total environmental noise levels in Montreal. Int J Environ Res Public Health 13(8):809. https://doi.org/10.3390/ijerph13080809
53. Porter ND, Kershaw AD, Ollerhead JB (2000) Adverse effects of night-time aircraft noise (Rep. No. 9964). UK Civil Aviation Authority, London, UK
54. Pyko A, Eriksson C, Lind T, Mitkovskaya N, Wallas A, Ögren M, Östenson C, Pershagen G (2017) Long-term exposure to transportation noise in relation to development of obesity—a cohort study. Environ Health Perspect 125(11):117005
55. Rocha S, Smith MG, Witte M, Basner M (2019) Survey results of a pilot sleep study near Atlanta international airport. Int J Environ Res Public Health 16:4321
56. Rojek et al (2021) The relation of nocturnal exposure to aircraft noise and aircraft noise-induced insomnia with blood pressure
57. Röösli M, Vienneau D, Foraster M, Eze I, Héritier H, Schaffner E (2017) Short and long term effects of transportation noise exposure (SiRENE): an interdisciplinary approach. In: Proceedings: ICBEN 2017. Zurich
58. Saucy A, Schäffer B, Tangermann L, Vienneau D, Wunderli JM, Röösli M (2021) Does night-time aircraft noise trigger mortality? A case-crossover study on 24 886 cardiovascular deaths. Eur Heart J 42:835–843
59. Schmidt FP, Herzog J, Schnorbus B, Ostad MA, Lasetzki L, Hahad O, Schäfers G, Gori T, Sørensen M, Daiber A, Münzel T (2020) The impact of aircraft noise on vascular and cardiac function in relation to noise event number: a randomised trial. Cardiovasc Res 1–9. https://doi.org/10.1093/cvr/cvaa204
60. Schreckenberg D, Belke C, Faulbaum F, Guski R, Möhler U, Spilski J (2016) Effects of aircraft noise on annoyance and sleep disturbances before and after expansion of Frankfurt Airport—results of the NORAH study, WP1 'Annoyance and quality of life'. In: Proceedings: InterNoise 2016. Hamburg
61. Schreckenberg D, Benz S, Belke C, Möhler U, Guski R (2017) The relationship between aircraft sound levels, noise annoyance and mental well-being: An analysis of moderated mediation. In: Proceedings of the 12th ICBEN congress on noise as a public health problem. Zurich, Switzerland 18–22 June 2017. http://www.icben.org/2017/ICBEN%202017%20Papers/Subjec tArea03_Schreckenberg_0326_3635.pdf
62. Seidler A, Hegewald J, Seidler AL, Schubert M, Zeeb H (2019) Is the whole more than the sum of its parts? Health effects of different types of traffic noise combined. Int J Environ Res Public Health 16:1665. https://doi.org/10.3390/ijerph16091665
63. Smith MG, Rocha S, Witte M, Basner M (2020) On the feasibility of measuring physiologic and self-reported sleep disturbance by aircraft noise on a national scale: a pilot study around Atlanta airport. Sci Total Environ 718:137368
64. Spilski J, Rumberg M Berchterhold M, Bergström K, Möhler U, Kurth D, Lachmann T, Klatte M (2019) Effects of aircraft noise and living environment on children's wellbeing and health. In: Proceedings of the 23rd international congress on acoustics. Aachen, Germany
65. Stallen PJM (1999) A theoretical framework for environmental noise annoyance. Noise Health 3:69–79
66. Trieu BL, Nguyen TL, Bui TL (2019) Assessment of health effects of aircraft noise on residents living around Noi Bai International Airport. In: Madrid inter-noise conference 2019: noise control for a better environment. Madrid, Spain
67. van den Berg F, Verhagen C, Uitenbroek D (2014) The relation between scores on noise annoyance and noise disturbed sleep in a public health survey. Int J Environ Res Public Health 11(2):2314–2327
68. van Kamp I, Davies H (2013) Noise and health in vulnerable groups: a review. Noise Health 15(64):153–159. https://doi.org/10.4103/1463-1741.112361
69. van Kamp I, Davies H (2008) Environmental noise and mental health: five year review and future directions. In: Proceedings of the 9th ICBEN congress on noise as a public health problem. Mashantucket, Connecticut, U.S.A.

70. Van Kamp I, van Kempen EEMM, Simon SN, Baliatsas C (2019) Review of the evidence relating to environmental noise exposure and annoyance, sleep disturbance, cardio-vascular and metabolic health outcomes in the context of ICGB (N), RIVM Report 2019–0088 institute for public health and the environment (RIVM). Bilthoven, The Netherlands. https://doi.org/10.21945/RIVM-2019-0088

71. van Kamp I, Job SRF, Hatfield J, Haines M, Stellato RK, Stansfeld SA (2004) The role of noise sensitivity in the noise-response relation: a comparison of three international airport studies 2004. J Acoust Soc Am 116:3471–3479

72. Van Poll R, Ameling C, Breugelmans O, Houthuijs D, van Kempen E, Marra M, Swart W (2014) Gezondheidsonderzoek Vliegbasis Geilenkirchen (Desk Research) I. Hoofdrapportage: Samenvatting, Conclusies en Aanbevelingen Gezondheidsonderzoek Vliegbasis Geilenkirchen; National Institute for Public Health and the Environment: Bilthoven, The Netherlands (In Dutch)

73. Von Lindern E, Hartig T, Lercher P (2014) Assessing the relationship between perceived disturbances from traffic, restorative qualities of the living environment, and health. In: Proceedings of the 43rd international congress on noise control. Melbourne, Australia, 16–19 Nov 2014

74. Von Lindern E, Hartig T, Lercher P (2016) Traffic-related exposures, constrained restoration, and health in the residential context. Health Place 39:92–100. https://doi.org/10.1016/j.healthplace.2015.12.003

75. World Health Organisation (2011) Burden of disease from environmental noise. Quantification of healthy life years lost in Europe. WHO Regional Office for Europe, Copenhagen

76. World Health Organisation Regional Office for Europe (2018) WHO environmental noise guidelines for the European region. Copenhagen, Denmark: World Health Organisation, WHO Regional Office for Europe

77. Zeeb H, Hegewald J, Schubert M, Wagner M, Dröge P, Swart E, Seidler A (2017) Traffic noise and hypertension—results from a large case-control study. Environ Res 157:110–117

Coping with Aviation Noise: Non-Acoustic Factors Influencing Annoyance and Sleep Disturbance from Noise

Susanne Bartels⑩, Isabelle Richard⑩, Barbara Ohlenforst⑩,
Sonja Jeram⑩, Julia Kuhlmann⑩, Sarah Benz⑩, Dominik Hauptvogel,
and Dirk Schreckenberg⑩

Abstract Annoyance and sleep disturbances due to aircraft noise represent a major burden of disease. They are considered as health effects as well as part of the causal pathway from exposure to long-term effects such as cardiovascular and metabolic diseases as well as mental disorders (e.g. depression). Both annoyance and sleep disturbance are not only determined by the noise exposure, but also to a considerable extent by non-acoustic factors. This chapter summarises the most relevant

Detailing and exemplifying notions of Quality of Life, of possible compensation, of the impact of ability to cope with as well as the key factor of trust in authorities

S. Bartels (✉) · D. Hauptvogel
Sleep and Human Factors Research, Institute of Aerospace Medicine, German Aerospace Center (DLR e.V.), Linder Höhe, 51147 Cologne, Germany
e-mail: susanne.bartels@dlr.de

D. Hauptvogel
e-mail: dominik.hauptvogel@dlr.de

I. Richard
Environnons, 302 route de Mende, 34090 Montpellier, France
e-mail: isabelle.richard@environnons.com

B. Ohlenforst
Royal Netherlands Aerospace Centre NLR, Anthony Fokkerweg 2, 1059 CM Amsterdam, The Netherlands
e-mail: barbara.ohlenforst@nlr.nl

S. Jeram
National Institute of Public Health, Environmental Health, Trubarjeva 2, 1000 Ljubljana, Slovenia
e-mail: sonja.jeram@nijz.si

J. Kuhlmann · S. Benz · D. Schreckenberg
ZEUS GmbH, Centre for Applied Psychology, Environmental and Social Research, Sennbrink 46, 58093 Hagen, Germany
e-mail: kuhlmann@zeusgmbh.de

S. Benz
e-mail: benz@zeusgmbh.de

D. Schreckenberg
e-mail: schreckenberg@zeusgmbh.de

non-acoustic factors and briefly explains their mechanisms on annoyance and sleep as well as the potential to address these factors via intervention methods aiming at the reduction of adverse noise outcomes and an increase in the quality of life of airport residents. Here, the focus is on airport management measures that are considered to help improve the residents' coping capacity. Findings from the ANIMA case studies with regard to main aspects of quality of life in airport residents around European airports are briefly reported and recommendations for a community-oriented airport management are derived.

Keywords Noise annoyance · Sleep disturbance · Quality of life · Coping · Non-acoustic factors · Interventions

Annoyance as the Most Common Psychological Effect of Noise and Its Non-Acoustic Influence Factors

Annoyance due to aircraft noise includes behavioral, emotional and cognitive elements. These are (a) the feeling of disturbance due to noise combined with behavioral responses in order to minimise the disturbance, e.g. closing the window to reduce the noise from outdoors, (b) an emotional/attitudinal response like anger about the noise and a negative attitude towards the noise source, and (c) a cognitive response like the distressful insight that one cannot do much against this unwanted situation.

The multifaceted definition of annoyance already hints at several factors influencing an individual's reaction to noise. Following the research outcomes on noise effects, only about one third of the variations in long-term annoyance judgments can be explained by variables representing the average noise exposure such as L_{den}, L_{dn}, or L_{Aeq} [1]. Another third of the variation in noise annoyance ratings can be explained by non-acoustic factors, whilst the last third of variance has remained unexplained so far. Non-acoustic factors can be roughly described as those factors "which are not directly connected to the nature of the sound" [2]. Most researchers have defined non-acoustic factors as being those variables that modify, moderate or co-determine responses to noise, but not being part of the causal chain from sound via disturbances to annoyance and further health-related outcomes [e.g., 1]. In order to categorise the manifold non-acoustic factors, it seems plausible to discriminate between factors referring to attitudes and traits of the individuals exposed to noise (*personal and social factors*) and factors referring to the context of the noise situation (*contextual and situational factors*) [3]. Moreover, factors exist that refer to the *social aspects of the noise management* at the noise source. In the following, these factors are listed up and briefly explained with the focus on insights on individual strategies to cope with noise that turn out to be more or less successful. Results from ANIMA [4] underline the importance of considering coping strategies and possibilities more deeply to get a better understanding of the association between aircraft noise exposure and annoyance.

Personal and Social Factors

Attitudes, Concerns, and Expectations

Attitudes, concerns, and expectations belong to the most important non-acoustic factors influencing annoyance [e.g., 5, 6]. Positive evaluations of the noise source, such as the belief that the noise source is important for the local economy, reduce noise annoyance [7] whilst negative attitudes and concerns about the negative health outcomes and in particular the fear of aircraft crashes increase annoyance [e.g., 7, 8]. Aviation-related fears and negative attitudes can contribute even more to aircraft noise annoyance during the past 12 months than the average indicators of noise level such as L_{den}, L_{dn}, or L_{Aeq} [e.g., 7, 8]. Annoyance is also enhanced in individuals who believe that *the noise situation will worsen in the future* [e.g., 8, 9] and in individuals who generally *prioritise environmental and silence aspects* to economic issues when it comes to airport-related decisions [3, 9].

Noise Sensitivity and Personality Traits

Noise sensitivity is considered as a stable personality trait with regard to an individual's general susceptibility to noise [10] that may be associated with a more general disposition for experiencing negative emotions, such as anger, tension or anxiety [11, 12]. Noise sensitivity is considered as one of the most influential variables of noise annoyance besides the above-mentioned attitudes [e.g., 5].

Coping Strategies

The ability to cope with a noise situation depends on possibilities for control and can differ from one person to another. There are three main ways for a person to deal with noise exposure:

1.　Adopt a short-term coping strategy referring to the "here and now", such as strategies that focus on problem-solving or emotions at a given moment. These strategies target the adaptation to the noise exposure via actions to reduce the discomfort induced by the noise, e.g. by cognitive rationalisation, escaping from noise or at least attempting to decrease it, and by covering the noise source [13].
2.　Adopt a long-term coping strategy that refers to taking actions against the annoyance and its resolution, for example, by participating in citizens' groups and developing collective strategies to change regulations, by moving to another life place [13] or by simply complaining to the airport.
3.　Do not adopt any coping strategies, which can be explained either by the fact that the exposed individuals do not perceive any discomfort, that they are able to delegate the responsibility, or that they feel not to have control over the noise situation and/or that they experience so-called learned helplessness [14]. The

latter is an adverse mental outcome developed by people who repeatedly feel a real or perceived absence of control over the outcome of a stressful situation. As a consequence, they have learned that they are seemingly unable to control or change anything, and hence are helpless in this situation. Residents who resigned to the stressful noise situation are at higher risk to develop health problems.

In view of this, the ability to cope and coping strategies of an individual are obviously important non-acoustic factors that determine the way of living with noise exposure. And this goes far beyond the mere feeling of annoyance. Coping strategies are therefore considered as a factor involved in the causal pathway from noise exposure to health impacts [15]. However, several questions are not fully resolved: Does the inability to cope have an impact on annoyance as suggested by [16]? Does learned helplessness tend to reinforce negative emotions that would foster annoyance? Future research should bring light in this circle of exposure, annoyance, learned helplessness and health risks. What are the options to break it, e.g. by providing adequate possibilities to improve coping capacities and by having control over a noise situation? There is already evidence that personal control and learned helplessness are relevant intervening factors in the causal chain from noise exposure to annoyance and further health outcomes [e.g. 8, 17]. However, systematic knowledge about successful interventions improving the process of coping with noise is lacking.

Trust in Authorities and Perceived Fairness

The aforementioned ability to cope and the psychological aspects of perceived control are not only dependent on individual abilities and convictions. The behavior of the airport managers also plays a role including to what extent it is perceived as trustworthy and fair in regard to the affected residents [1, 18]. In this context, many aspects impact on the perception of trust and fairness, but among them is the perception that the airport authorities do their best to avoid unnecessary noise and a perception of an airport communicating honestly and taking the concerns of noise-affected residents seriously. Fairness aspects are strongly related to trust in authority and seem, thus, to be able to reduce annoyance by implementing a consistently fair communication. Fairness aspects have been known to play an important role in various fields (e.g. in the organisational and judicial context) since the 1970s and recently gained attention also in the context of aircraft noise exposure [19, 20]. Research on fairness has identified a variety of different factors that are important for establishing an environment that allows building trust through fairness. These factors are considered in detail in Chap. 11 and recommendations are made how these can be taken up by airport management.

Residential Satisfaction

Residential satisfaction is a frequently studied factor in the socio-psychological field. There is evidence showing that feeling well in the neighbourhood and being content with its (acoustic) appearance and infrastructure impacts annoyance [3, 6, 9]. However, since residential satisfaction also has a strong subjective component and noise effect studies including residential satisfaction are mostly cross-sectional, it is also conceivable that residential satisfaction is rather a consequence of noise affected by annoyance [21].

Demographics

Although often examined in annoyance research, significant effects of demographic factors such as age, gender, occupational status, educational level, homeownership, dependency on the noise source, and use of the noise source, on annoyance were only seldomly found and if so, they were small [5, 6]. Age has a rather curvilinear effect; i.e. relatively young and relatively old people are less annoyed [5] whilst gender seems to have no influence on annoyance at all. Slightly higher annoyance is reported for people with higher educational level and occupational status, for homeowners, and for people who neither are dependent on the noise source nor use it [5].

Contextual and Situational Factors

Degree of Urbanisation and Background Noise Exposure

A typical factor that is lying outside an individual and that persists across different noise situations is the surrounding of an individual. There is at least some evidence that the type of the neighborhood has an influence on annoyance. Noise annoyance seems to be highest in rural areas, followed by suburban, urban, commercial, and industrial areas in decreasing order [3, 22]. Indeed, in an urban area, residents' expectations are congruent with the noise in contrast to rural environments representing a much more peaceful place for people where noise is not expected. The specific situation in rural areas needs to be taken into account for airports surrounded by rather rural areas.

Access to Greenery and Recreational Areas, Appearance of Neighborhood

Coping with environmental stress as produced by noise requires an individual's resources and coping capabilities to be restored. Access to nature or green areas is regarded as allowing such recovery [23]. For example, the availability of vegetation and green spaces as well as the perceived neighborhood greenery can reduce annoyance due to road traffic noise [24, 25] and railway noise [25]. However, green areas per se do not necessarily reduce aircraft noise annoyance. The availability of residential green areas was also shown to increase aircraft noise annoyance for those that were still exposed to aircraft noise [25]. The reason for this might be that aircraft noise is more alien and intrusive in residential areas than road traffic noise [25]. Moreover, the more the neighborhood is perceived as green, the more residents expect a quieter residential environment and might regard aircraft noise as even more intrusive. However, green spaces as compensation strategies can encourage people living under the flight paths to adopt more healthy coping strategies and, thus, improve their quality of life [15]. Moreover, literature shows that the presence of vegetation presented simultaneously with moving water can reduce annoyance as it improves the soundscape [26, 27].

Access to a Quiet Side of the Dwelling

In the context of annoyance due to railway and road traffic, it was shown that the possibility to escape from noise has an annoyance-reducing effect. For example, people residing along a very busy road perceive lower annoyance when their bedroom or living room is directed to a quiet façade [28, 29] that also represents a facet of perceived control over the noise situation. Besides the mere exposure reducing effect, the visual quality of the space (a courtyard or greenery) had an impact on annoyance [30]. Even though aircraft noise cannot be compared to road traffic noise and a quiet façade is per se not feasible, the beneficial effect of the opportunity to escape from noise, e.g. via a quiet room in the building or access to nearby quiet recreational areas, is assumed to be transferable as it offers a measure to cope with the noise.

Differences in Annoyance in Changing Versus Stable Exposure Situations

A fundamental change in the noise exposure, i.e. an abrupt reduction or increase that is not only due to temporal changes, causes different levels of annoyance than would be expected at airports with a stable exposure. For so-called high-rate change airports that were announced to have or that actually experienced a step change in exposure, for instance because of the opening of a new runway, and, thus, an increase of flight numbers, community annoyance is usually higher than at airports without an experienced or announced change (so-called low-rate change airports) at the same exposure level [31]. Evidence for a change effect exists both for a step-increase and step-decrease in the noise exposure [e.g., 32]. The mechanisms of this effect are not fully understood, yet. Several explanations including a change in residents' attitudes or their retaining of previous and no longer appropriate coping strategies have been discussed [33]. Also, the time it takes the change effect to extinguish is only roughly estimated ranging from several months to several years [33].

A step change in noise exposure can also occur due to (unforeseen) events other than operational expansions or re-organisations. In this context, the abrupt reduction of flight numbers during the COVID-19-pandemic seems worth mentioning. The effect of this unforeseen period of decreased aircraft noise exposure on future annoyance after resuming regular operation schemes is currently not clear and warrants consideration in future research on aircraft noise impacts. Relevant questions are, for instance, whether and to what extent airport residents have got "used" to the reduced noise exposure and, thus, have changed their expectation towards acceptable noise levels in their neighbourhood area, to what extent they have changed their habits, for instance with regard to outdoor activities and window-opening behaviour as well as, whether residents' general noise sensitivity and noise tolerance have changed during the lockdown. With regard to the perception of the noise (relief) during the pandemic and future expectations, at least some of these questions are currently targeted in [34].

Temporal Factors of a Noise Situation

As a result of a busy and noisy working environment, a demand for quiet and restful periods, in particular during the evening and night, has been established in modern industrialised civilisations [35, 36]. The time of day when the noise occurs is, therefore, a relevant situational influence factor [3, 37]. Certain times of a day coincide with specific activities like communication including conversation, socialisation, listening to the radio, and watching TV, prevailing in the afternoon and evening. During the night, the need for recreation and sleep prevails [38]. The early morning hours are likewise regarded as very susceptible to noise due to an individual's psycho-physiological adaptation process to the rhythm of the day [36]. Higher annoyance was reported for the weekend [37] most likely due to the fact that the weekend coincides with noise-susceptible activities, above all recreation. In general, the activities contribute to the explanation of why at the same noise exposure level people differ in their short-term annoyance ratings (e.g. per event, hour, or day [3]). In noise effect studies assessing long-term noise responses (e.g. over a period of the past 12 months), activities are specifically addressed in questions on how often or how much these activities were disturbed by noise. In these studies, activity disturbances are regarded as primary reactions to noise preceding annoyance and not as a non-acoustic factor modifying the annoyance without itself depending on noise exposure [e.g. 39].

This last point leads to the method of annoyance assessment. The standardised, general one-item question and scale recommended by the International Commission on the Biological Effects of Noise [ICBEN; 40] is the most common method to assess transportation noise annoyance. Notwithstanding the huge benefits of this internationally standardised method, some deficiencies exist [41, 42] since it is not capable of considering the time of the day the annoyance is experienced (morning, night, etc.). The ICBEN-question refers to the past twelve months, but there is also a bias in memory capacity. Individuals generally tend to remember mostly the very recent or very early experiences [recency and primacy effect, 43]. With regards to the self-assessment of annoyance, research has shown that individuals tend to refer to recent situations [44] or the worst situations [45]. Moreover, the semantics a respondent puts under the term of annoyance (loudness, fear, anger, depression) may have an effect as well. As mentioned in the beginning of this chapter, the concept of annoyance covers behavioral, emotional and cognitive elements. Acknowledging this, the ICBEN question makes an attempt to capture the multidimensionality of the concept via only one single-item question that combines the aspects "annoyance", "disturbance", and "bother". Whether these concepts can really be subsumed under the general construct of annoyance, is disputable. It could be valuable in further research to complement the ICBEN-question by additional standardised questions to better match the concept of annoyance [46].

Social Aspects of the Noise Management

As described, trust in authorities and perceived fairness play a major role in coping with aircraft noise exposure. Their role can be generally extended to the experience of a neighbourly relationship with the airport and its activities. Social aspects of noise management and noise exposure are non-acoustic factors that may additionally affect noise annoyance. Examples of such social, non-acoustic factors are:

- The effect of noise insulation on noise-induced annoyance is evident [6]. Although exposure from outdoor noise diminished, evidence for an annoyance-reducing effect of noise insulation is mixed. The mere fact that noise insulation has been installed at home did not affect annoyance [9], but being satisfied with the sound attenuation of insulation windows at home had an effect [47]. Results from ANIMA show that insulation scheme procedure is not considered by residents as a relevant solution depending on the season and the regional climate. It is rather perceived as a necessary measure but not sufficient to reduce their discomfort regarding aircraft noise [4].
- A shift or redistribution of noise exposure across populations. Such operational measures change the exposure of the affected people, but they are also related with non-acoustic factors described above, such as the perception that noise authorities care for the residents' needs and health. The distribution of costs and benefits that fall across the population might be uneven and perceived as unfair [4]. To solve this problem, rigorous and accepted methods for balancing uneven costs and benefits of aviation are required [48].

Given the nature of those social non-acoustic factors, fairness beliefs and trust in authorities are potentially addressable by a community-oriented airport management, meaning that interventions that are undertaken are accompanied by consultation, education and communication as well as community engagement.

Sleep Disturbance as the Most Common Neuro-Behavioral and Physiological Effect of Noise and Its Non-Acoustic Influence Factors

Disturbances of sleep are a major adverse reaction to nocturnal noise and can be manifested by changes in the sleep depth and sleep continuity and, as a consequence of this less restorative sleep, by fatigue and a reduced cognitive performance during the following day. As already described in Chap. 9, sudden noisy events have hinted at threats in the early history of humankind. It was necessary that humans were able to react quickly, including when asleep. Noisy events are, therefore, subliminally perceived and evaluated even during sleep and can provoke physiological reactions such as an awakening, bringing an individual back to a conscious state. The extent of the sleep disturbance per night depends on the number of noise events and each one's acoustic properties, such as the level of the noise event, or the speed that the level rises, which is an indicator of how fast the event is approaching. Potentially threatening noise events consequently cause a reaction with higher probability [49]. Whether a noise event causes an awakening, however, does not only depend on its acoustic properties but also on situational and personal factors.

Situational Factors

In the framework of the ANIMA-project, a standardised sleep model has been developed that describes the probability for an aircraft noise-induced awakening from sleep that does not only include acoustic features of the aircraft noise event (e.g. maximum level, duration, speed of the level rise) but also situational factors [50]. Whether an individual wakes up is influenced by the duration it has already been asleep. With increasing sleep time, the internal biological drive for sleep decreases and sleep becomes more susceptible for disturbances, as it is particularly the case in the early morning hours. When sleep pressure is low, the human organism is not only prone for an awakening but also has greater problems to fall asleep again [49]. The time spent asleep is also connected to the prevailing sleep stages, which are again influencing the probability for an aircraft noise to evoke an awakening reaction. It is less likely to awake from deep sleep stages than from light sleep stages. Moreover, whether an aircraft noise event causes an awakening or not is influenced by the background sound pressure level at the time when the noise event occurs. The probability to awake is enhanced when the aircraft noise event stands out from the background noise.

Personal Factors

Personal factors such as age can have an influence on sleep. Sleep rhythm and the duration spent in the different sleep stages change during a lifespan. With increasing age, the amount of deep sleep decreases [51]. Children are less likely to awake than adults at the same sound pressure level [52] whilst the elderly are assumed to be at higher risks for aircraft noise induced-sleep disturbances. Age is therefore included in the ANIMA standardised sleep model [50]. However, the effect of aircraft noise on elderly people has sparsely been investigated. An effect of age on self-rated sleep disturbances due to aircraft noise has recently been shown [53]. Gender is not a relevant influence factor of the probability to awake due to aircraft noise according to the ANIMA standardised sleep model [50] and other studies [54]. Additionally, self-rated sleep disturbance due to aircraft noise seems not to be remarkably affected by gender [53].

Noise sensitivity is assumed to be an important non-acoustic factor for both self-rated sleep disturbance and acute physical reactions to transportation noise [55]. However, the evidence of differences between noise sensitive versus non-sensitive individuals is scarce and systematic analyses of the effect of noise sensitivity on physiologically measured sleep are lacking [54].

The extent of the impact of personal factors may also depend on the measurement of sleep disturbance. Asking for self-rated sleep disturbances due to, e.g. aircraft noise, leaves room for personal attitudes, emotions and knowledge similarly as it is the case for annoyance judgments. The relation between self-rated sleep disturbance assessed without any reference to aircraft noise and the exposure level is weaker than when aircraft noise is explicitly mentioned as a potential source of the sleep disturbance [49]. But also for physiologically measured sleep disturbances and awakenings, the effect of aircraft noise exposure differs remarkably between individuals. These differences are not fully explainable by gender, age, and noise sensitivity [54]. Attitudes towards aviation may be a source for these differences [56].

Relation Between Sleep Disturbance and Annoyance

As described above, annoyance can evolve from the repeated disturbance of intended activities including sleeping. Annoyance is therefore sometimes regarded as one secondary effect of sleep disturbance as well as the resulting perception of fatigue, decreased cognitive functioning and changed mood [35]. Evidence exists that (short-term) annoyance assessed in the morning and with regard to the past night is related to the past night's aircraft noise exposure [57]. Likewise, an association between nocturnal aircraft noise exposure and cognitive functioning [58] and with self-rated tiredness in the morning [59] was found. Nevertheless, whether annoyance judgments are covering experiences of sleep disturbance or whether noise annoyed residents suffer from sleep disturbance more often, is not clear yet [60]. Presumably, the relation between sleep disturbance and annoyance is reciprocal. Poor sleep from nocturnal noise can affect the next day's mood, performance and well-being and as a consequence promote annoyance. However, feeling annoyed due to aircraft noise during the day may affect sleep as well since annoying/upsetting situations that were experienced during the day can be processed during sleep.

Role of Non-Acoustic Factors for Aircraft Noise Interventions

A crucial conclusion that can be drawn from the research on the effect of non-acoustic factors so far, is that those factors have the largest impact on the noise response that tackle the individual's capacity to cope with the noise. As described above, control over the noise situation or aviation-related decisions can strengthen one's coping capacity. The perception of having control covers two components: On the one hand, direct and immediate measures can be undertaken, e.g., controlling a noise situation via closing windows, or, having access to respite locations and periods. On the other hand, people can have indirect control via being represented in noise-related decision-making processes and, thereby, having voice either personally or through trusted authorities that act and decide on the residents' behalf. In that sense, it is important to distinguish long-term from short-term coping strategies in the same way that we differentiate between long-term and short-term annoyance. Short-term annoyance is associated with short-term coping strategies and can strongly influence the discomfort felt in a specific situation. In contrast, long-term annoyance is associated with long-term coping strategies and can have on the long run a strong impact on residents' health. By improving the ability of residents to cope with their noise exposure in the long-term and not only at a given time, interventions could reduce residents' annoyance. Residents' coping capacity is assumed to be enhanced by giving them decision-making power, by acknowledging them as knowledgeable of their experiences in the field of noise exposure, by acknowledging and considering the environmental and the health costs of each planned project, etc. Thus, enhancing the coping capacity may tackle certain outcomes that are related to annoyance, such as improving residents' health, creating trust and transparency between airport operators, authorities and residents, and saving more time and energy in raising aviation-related projects.

Further to the importance of perceived control for coping with noise, the possibility of recreation is crucial as it allows for restoring coping capacities that are diminished when dealing with ongoing stress. Besides continuously reducing aircraft noise exposure as well as implementing interventions targeting non-acoustic factors, noise management should focus on increasing people's perceived control and coping capacities and allow for space and time for recreation.

Therefore, some authors have categorised non-acoustic factors with regard to the strength of importance for aircraft noise responses, in particular annoyance, and the modifiability by means of interventions [2, 61, 62]. A review and summary were produced within ANIMA [63]. For this chapter, we revised and enhanced this system of categorising non-acoustic factors by adding the assessment of the effect of non-acoustic factors on sleep-disturbance and exemplifying how the factors can be addressed in aircraft noise management (see Table 1).

Even when aircraft noise interventions such as those mentioned in Table 1 aim at reducing adverse noise effects addressing both acoustic as well as non-acoustic factors of, for example, annoyance or sleep disturbance, airport operators and aviation-related authorities have to be aware that the aircraft noise management activities have an impact on communities' quality of life in general. This is true for all airport activities. Therefore, in the next section, we will describe main areas of people's quality of life and link them to airport/aviation activities that potentially address these quality of life facets in an airport region.

Role of Airport Interventions on Quality of Life in Airport Regions

There are a variety of definitions and frameworks relating to quality of life. The WHO defines quality of life (QoL) as "an individual's perception of their position in life in the context of the culture and value systems in which they live and in relation to their goals, expectations, standards and concerns" [65, p 1], which is a broad conceptualisation that emphasises a person's motives and perceptions. Most of previous research has been conducted on the interaction between noise exposure and annoyance or health-related effects. The innovative approach within the ANIMA project addressed the relationship between aircraft noise exposure, noise annoyance, and QoL. It is essential to understand the impact of aircraft noise management and noise-related interventions that can influence different dimensions of QoL next to health-related outcomes. Currently, little is known about the quality of life dimensions and their interactions that are important for residents around airports. For the purpose of addressing issues that communities in the vicinity of airports face, one existing framework, the EUROSTAT approach [66], was considered best fit and adapted in the ANIMA project [67]. The EUROSTAT framework considers nine QoL dimensions with specific indicators for measuring QoL: material living conditions, productive or main activity (job), health, education, leisure, economic security and physical safety, governance and basic rights, natural and living environment, and overall experience of life. To some extent the different dimensions overlap and interdependencies are likely to occur.

Within the ANIMA project, a study was conducted in four different European airport regions in an attempt to capture crucial aspects of residents' quality of life [4]. A re-analysis of recently collected data from a survey conducted around Schiphol revealed that participants living near Schiphol Airport are generally satisfied with

Table1 Role of non-acoustic factors in aircraft noise management

Factor	Effect on …		Address-ability	Starting points for modification/improvement
	Annoy-ance	Sleep		
Personal and social factors				
Attitudes, concerns, expectations	+	+	+ −	Communication and engagement campaigns accompanying interventions and projects (e.g. operational changes) and raising awareness in both airport management and the affected communities; dialogue forum and information service; acknowledgement of adverse health and environmental outcomes by airport management; transparent complaint management; compensation programs with the engagement of communities, e.g., via compensation that is selectable by the residents
Trust in authorities, perceived fairness	+	+ ?	+	
Coping strategies	+	?	+ −	Communication and engagement campaigns; dialogue forum and information service; transparent complaint management; education on coping possibilities; respite times with dynamic noise maps with information on planned air traffic around airports making noise exposure predictable and enabling residents to plan noise-susceptible (outdoor) activites
Residential satisfaction	+ −	+ ?	+	Community engagement in design of neighborhood appearance; land-use planning; provision of access to greenery and quiet recreational areas
Demographics	+ −	+ −	+ −	Age: Land-use planning regarding vulnerable groups (children, elderly, i.e. placement of kindergartens, schools, nursing homes) House ownership: House purchase scheme, noise exposure schemes

(continued)

Table1 (continued)

Factor	Effect on …		Address-ability	Starting points for modification/improvement
	Annoy-ance	Sleep		
Contextual and situational factors				
Degree of urbanisation & background noise	+	+	−	Improved noise-sensitive land use planning legislation; access to recreational, calm and green areas; operational interventions
Access to greenery & recreational areas, appearance of neighborhood	+ −	?	+	
Access to a quiet side of the dwelling	+	+	+ −	Insulation programs; indoor design of dwellings and arrangement of rooms (e.g. bedrooms); sound adaptive buildings and urban design
Time/day when noise occurs	+	+	+	Operational interventions (respite time, night curfews); dynamic noise maps with information on planned air traffic around airports making noise exposure predictable and enabling residents to plan noise-susceptible (outdoor) activites; raising awareness for the requirement of noise respite times
Social aspects of noise management				
Insulation	+	+	+	Consultation and community engagement; insulation programs engaging communities, e.g., via selectable insulation measures; dynamic noise maps with information for planned air traffic around airports
Operational intervention	+	+	+	

Adapted from [2, 61, 62] and modified according to the review presented in this chapter and results obtained from ANIMA studies [4, 64]

Note + evidence for an effect or addressability is given, − evidence for an effect or addressability is not given, + − evidence for an effect or addressability is ambivalent, ? evidence for an effect or addressability has not yet been examined, + ? an effect is assumed but has not yet been examined.

their residential area. However, there is a difference in people's residential satisfaction depending on the noise contour they live in. Although all groups were on average either "satisfied" or "very satisfied" with their residential area, the results suggest that the exposure to higher levels of aircraft noise may negatively affect people's residential satisfaction. Similar results were found conducting focus groups around Marseille Airport. Overall, participants were satisfied with their living environment, but aircraft noise exposure and other negative effects of air traffic were mentioned as aspects of their residential area negatively affecting their QoL. People living around Frankfurt Airport were asked about their understanding of quality of life. They mentioned aspects such as family, health, nature, the social and living environment, and security. Most participants stated that their living environment has a crucial impact on their quality of life, which can be both positive (nature) and negative (noise). When asked about their living environment, residents living near Heathrow Airport mentioned, for example, being concerned about air quality and noise pollution. With respect to quality of life, they rated the existing infrastructure and nature as positive aspects and mentioned negative aspects such as road traffic and an increase in aircraft noise.

At the end of 2020, a survey of various factors related to QoL and the impact of aviation during Covid-19 was carried out to gain a better understanding of the effect of less air traffic on residents' perception of aviation. To this end, a group of people living close to Schiphol was compared with a group living around the city of Utrecht. The results suggest that the group living around Utrecht (placed nearer to the local airport) takes a more negative view of aviation than the group living near Schiphol (living further away from the local airport). This negative affinity with aviation is also supported by other recent Dutch studies [e.g., 68].

In the section above, it was explained that non-acoustic factors of annoyance referring to attitudes, concerns, personal traits, residential satisfaction and demographics of the person exposed to aircraft noise, affect the perception of annoyance. Results from case studies conducted within ANIMA suggest an interaction of several of those non-acoustic factors and a direct impact on people's QoL. If people were exposed to higher noise levels they reported lower residential satisfaction and other negative effects of air traffic seemed to negatively affect their QoL. When noise exposure increases, the balance between advantages and disadvantages of the noise exposure may shift. People living closest to airports do not necessarily have more benefits such as mobility, or economic safety.

Natural and Living Environment

High exposure to aircraft noise may have a negative impact on residents' natural and living environment as it disturbs people's daily activities. People might be more concerned about the noise exposure, increasing air traffic, lowering housing prices, lower air quality, negative developmental effects in children and the feeling of shame for the noise. When people feel ashamed in front of their friends or families for the noise exposure at their homes, a direct negative impact on their social life is likely to result. Non-acoustic factors such as increased concerns may alter people's well-being, impact their health and their social life and in that way affect their QoL.

Productive or Main Activity

It seems that an economic relationship with the airport creates fewer negative associations with aviation. Most people working for the aviation industry trust their company and they identify themselves with aviation and their work. Their residential satisfaction may be high and together with less concerns their perception of noise annoyance may be limited. In this way, non-acoustic factors related to noise annoyance are intertwined with the residents' QoL.

Education and Leisure

There are dimensions where airports can have a direct positive impact on their neighborhood. Airports can have a beneficial take on improving quality of life in adding value to certain aspects such as structural improvements (e.g. infrastructure), supporting leisure activities in sponsoring training gear for local sports clubs, providing access to education, knowledge, public activities, leisure and cultural places, general aviation or drone areas. Research indicated that children living in the vicinity of airports show impaired reading abilities, explained by aircraft noise exposure [69]. Creating access to education, knowledge, libraries and learning facilities may be a way to respond to impaired reading abilities or even counteract adverse effects. Regarding leisure time, as mentioned in the section on the non-acoustic factor of "Access to greenery and recreational areas" the possibility to recreate is able to minimise annoyance. In addition, it is also an important aspect of QoL, both for children, and adults in general as it allows physical activities, learning by playing, it improves health and allows for recovery from daily stress.

Using a concept like QoL to assess the impact of an airport on its surroundings allows for observing both positive and negative consequences that are linked to the airport. Next to decreasing people's stress response to noise, i.e. annoyance, by reducing the noise exposure, it can be useful to improve their QoL as this can help increase people's coping capacities. While many interventions aimed at reducing or minimising noise responses in the past, only few interventions additionally focused on improving quality of life specifically. Although not specifically addressed, existing interventions can have unintended (positive) side effects on residents' QoL. Several studies have linked aircraft noise annoyance to health-related QoL [e.g. 16].

In addition, QoL factors might be positively influenced in an indirect way. By improving residents' QoL, airports can affect non-acoustic factors of annoyance such as concerns, negative attitudes and residential satisfaction. Building on the relationship and creating trust requires frequent and transparent communication. Well-developed community engagement strategies should strongly focus on establishing a good relationship and not only be applied when an operational change or procedure is planned. Another important aspect is commitment to engage residents. When carrying out community engagement actions, people's opinions and the local situation need to be considered. Feedback should be provided and information on the progress of investigated topics should be reported back to the residents. When airport management engages and establishes good relationships with their neighborhood the feeling of being in control and empowerment might be granted to the residents. Trust in aviation and the related authorities can improve people's QoL. When considering QoL and aiming at improvements in QoL dimensions/aspects through interventions, it is necessary to focus on local characteristics and tailor interventions to the conditions and local issues of the targeted area.

Conclusions: How to Deal with Non-Acoustic Factors in Aircraft Noise Management?

This chapter has highlighted the importance of considering non-acoustic factors in aviation-related decision-making and management. The Balanced Approach to aircraft noise management of the International Civil Aviation Organisation (ICAO) has already adopted the idea of considering non-acoustic factors. Consideration of people's views (via improved communication and information access as well as consultation) has been emphasised to be involved in each of the four existing elements (a) reduction of noise at source, (b) land-use planning and management, (c) noise abatement operational procedures, and (d) operating restrictions [70]. The goal is to address specific noise problems at individual airports in order to identify the noise-related measures that achieve cost-effective maximum environmental benefits using objective and measurable criteria. Within the ANIMA project, several case studies at small and large airports have been carried out in order to investigate the potential of best practice interventions in aircraft noise management that address non-acoustic factors.

The case studies revealed the importance of involving all stakeholders, such as representatives of airport operators, local communities, civil aviation authorities and policy makers. Transparent policy of noise management and community engagement showed to be of crucial importance in reaching promising interventions aiming at reducing annoyance and enhancing quality of life of citizens.

The results of the case studies confirmed what is theoretically derived: community engagement has to be understood as a possibility for residents to not only have a voice, but above all, to take part in the decision-making process. Their voice has to count rather than be only consultative. This is confirmed by experienced aircraft noise campaigners such as [71] when he stated in his reflection of 20 years of campaigning: "The key thing for airports to address is the local area-specific issues which concern the campaigners. And communities need to be involved in shaping these solutions, particularly at times of change" [71, p 3].

For instance, some results of ANIMA [64] showed that people describe an ideal relationship with the airport as sharing information, fairness, mediation, trust as well as recognition. On a more practical level, options like open and transparent dialog and communication, benefits, being involved, performance and compensation can help to create this ideal relationship. Specifically, these ideas translate into improvements directly affecting noise reduction, noise compensation, and communication. All in all, communication and engagement are crucial elements of aircraft noise management for addressing non-acoustic factors of aircraft noise annoyance and sleep disturbances. They are also good starting points to enhance residents' quality of life as they have the capacity to lower annoyance.

This chapter has demonstrated the immense effect that non-acoustic factors have on sleep disturbance, annoyance and more broadly on quality of life. Overall, it appears necessary to revise the manner in which annoyance is assessed by adding more standardised questions, and to better investigate the role of coping strategies and the ability to cope with noise and on the capacity to reduce health impacts. The proposed starting points of interventions that address the non-acoustic factors (Table 1) are derived from evidence mainly on findings of non-acoustic contributors to annoyance. In part, it is assumed that most of these non-acoustic factors also affect sleep disturbances. Whether the suggested interventions listed in Table 1 actually have a beneficial effect on citizens' quality of life including a decrease in sleep disturbance and annoyance has to be systematically examined in evaluation studies. Such studies should be a fixed part of the implementation of aircraft noise interventions.

Chapter 11 elaborates more on this and presents criteria and standards to guide the establishment of successful and beneficial communication and engagement. At the end of this chapter, we point out that the evaluation and its results themselves address non-acoustic factors as, once communicated, the evaluation contributes to the building of trust in authorities, perceived fairness and self-regulation (being part of a—hopefully successful—story).

References

1. Guski R (1999) Personal and social variables as co-determinants of noise annoyance. Noise Health 3:45–56
2. Asensio C, Gasco L, de Arcas G (2017) A review of non-acoustic measures to handle community response to noise around airports. Curr Pollution Rep 3:230–244. https://doi.org/10.1007/s40 726-017-0060-x
3. Bartels S (2014) Aircraft noise-induced annoyance in the vicinity of Cologne/Bonn Airport. The examination of short-term and long-term annoyance as well as their major determinants. Technische Universität Darmstadt (Dissertation). http://tuprints.ulb.tu-darmstadt.de/4192/
4. Kuhlmann J, Rajé F, Richard I, Ohlenforst B (2020) Deliverable 3.6 evaluations of previous interventions in improving quality of life. Zenodo. https://doi.org/10.5281/zenodo.4288282

5. Miedema HME, Vos H (1999) Demographic and attitudinal factors that modify annoyance from transportation noise. J Acoust Soc Am 105(6):3336–3344
6. Fields JM (1993) Effect of personal and situational variables on noise annoyance in residential areas. J Acoust Soc Am 93(5):2753–2763
7. Schreckenberg D, Meis M, Kahl C, Peschel C, Eikmann T (2010) Aircraft noise and quality of life around Frankfurt Airport. Int J Environ Res Public Health 7:3382–3405. https://doi.org/10.3390/ijerph7093382
8. Kroesen M, Molin EJE, van Wee B (2008) Testing a theory of aircraft noise annoyance: a structural equation analysis. J Acoust Soc Am 123(6):4250–4260
9. Wirth K (2004) Lärmstudie 2000-Die Belästigungssituation im Umfeld des Flughafens Zürich. Dissertation [Swiss Noise Study 2000: The status quo of annoyance in the vicinity of the airport Zurich (doctoral dissertation)], Universität Zürich, Schweiz
10. Zimmer K, Ellermeier W (1999) Psychometric properties of four measures of noise sensitivity: a comparison. J Environ Psychol 19:295–302
11. Stansfeld SA (1992) Noise, noise sensitivity and psychiatric disorder: epidemiological and psychophysiological studies. Psychol Med Monogr Suppl 22:1–44
12. Persson R, Björk J, Ardö J, Albin M, Jakobsson K (2007) Trait anxiety and modeled exposure as determinants of self-reported annoyance to sound, air pollution and other environmental factors in the home. Int Arch Occup Environ Health 81(2):179–191
13. Levy-Leboyer C, Moser G (1987) Individual differences in noise annoyance: four explanations. In: Koelega HS (ed) Environmental annoyance: characterisation, measurement and control. Elsevier, Amsterdam, The Netherlands, pp 293–299
14. Seligman MEP (1975) Helplessness: on depression, development, and death. W. H. Freeman, San Francisco, USA
15. Föllmer J, Moore G, Kistemann T (2020) Urban parks as nature-based solutions for improved well-being under the flight paths: a soundscape analysis in the vicinity of Heathrow Airport, EGU General Assembly 2020, Online, 4–8 May 2020, EGU2020–17661. https://doi.org/10.5194/egusphere-egu2020-17661
16. Schreckenberg D, Benz S, Belke C, Möhler U, Guski R (2017) The relationship between aircraft sound levels, noise annoyance and mental well-being: an analysis of moderated mediation. In: Proceedings of 12th ICBEN congress on noise as a public health problem, Zurich, Switzerland
17. Hatfield J, Job RFS, Hede AJ, Carter NL, Peploe P, Taylor R, Morrell S (2002) Human response to environmental noise: the role of perceived control. Int J Behav Med 9(4):341–359
18. Stallen PJM (1999) A theoretical framework for environmental noise annoyance. Noise Health 1(3):69–79
19. Maris E (2008) The social side of noise annoyance (De sociale kant van geluidhinder). Doctoral thesis, Department of Social en Organisational Psychology/Cognitive Psychology, Faculty of Social and Behavioural Sciences, Leiden University, The Netherlands
20. Liebe U, Preisendörfer P, Bruderer Enzler H (2020) The social acceptance of airport expansion scenarios: a factorial survey experiment. Transp Res D Transp Environ 84:102363. https://doi.org/10.1016/j.trd.2020.102363
21. Kroesen M, Molin EJE, Miedema HME, Vos H, Janssen SA, van Wee B (2010) Estimation of the effects of aircraft noise on residential satisfaction. Transp Res Part D: Transp Environ 15(3):144–153. https://doi.org/10.1016/j.trd.2009.12.005
22. Lercher P, de Greve B, Botteldooren D, Rüdisser J (2008) A comparison of regional noise-annoyance-curves in alpine areas with the European standard curves. Paper presented at the 9th International Congress on Noise as a Public Health Problem, July 21–25, 2008, Foxwoods, CT, USA
23. Hartig T, Mitchell R, de Vries S, Frumkin H (2014) Nature and health. Annu Rev Public Health 35:207–228
24. Van Renterghem T (2019) Towards explaining the positive effect of vegetation on the perception of environmental noise. Urban Forestry Urban Greening 40:133–144
25. Schäffer B, Brink M, Schlatter F, Vinneau D, Wunderli, JM (2020) Residential green is associated with reduced annoyance to road traffic and railway noise but increased annoyance to aircraft noise exposure. Environ Int 143:105885

26. Lugten M, Karacaoglu M, White K, Kang J, Steemers K (2018) Improving the soundscape quality of urban areas exposed to aircraft noise by adding moving water and vegetation. J Acoust Soc Am 144(5):2906–2917. https://doi.org/10.1121/1.5079310

27. Yu L, Kang J (2010) Factors influencing the sound preference in urban open spaces. Appl Acoust 71(7):622–633

28. Bodin T, Bjork J, Ardö J, Albin M (2015) Annoyance, sleep and concentration problems due to combined traffic noise and the benefit of quiet side. Int J Environ Res Public Health 12:1612–1628

29. De Kluizenaar Y, Janssen SA, Vos H, Salomons EM, Zhou H, van den Berg F (2013) Road traffic noise and annoyance: a quantification of the effect of quiet side exposure at dwellings. Int J Environ Res Public Health 10:2258–2270

30. Van Kamp I, Brown AL, Schreckenberg D (2019) Soundscape approaches in urban planning: implications for an intervention framework. In: Proceedings of the 23rd international congress on acoustics, 9–13 Sept 2019. Aachen, Germany, pp 405–410

31. Guski R, Schreckenberg D, Schuemer R (2017) WHO environmental noise guidelines for the European Region: a systematic review on environmental noise and annoyance. Int J Environ Res Public Health 14:1539. https://doi.org/10.3390/ijerph14121539

32. Brink M, Wirth KE, Schierz C, Thomann G, Bauer G (2008) Annoyance responses to stable and changing aircraft noise exposure. J Acoust Soc Am 124(5):2930–2941. https://doi.org/10.1121/1.2977680

33. Brown AL, van Kamp I (2009) Response to a change in transport noise exposure: competing explanations of change effects. J Acoust Soc Am 125(2):905–914. https://doi.org/10.1121/1.3058636

34. Marshall B, Xypolia K, Walford A (2020) Aviation noise during lockdown. Ipsos MORI Report for the ICCAN, report no. 20–036970–01. https://iccan.gov.uk/wp-content/uploads/2020_10_08_Aviation_noise_during_lockdown_Ipsos_survey_report_for_ICCAN-min.pdf

35. Porter ND, Kershaw AD, Ollerhead JB (2000) Adverse effects of night-time aircraft noise (Rep. No. 9964). UK Civil Aviation Authority, London, UK

36. Hoeger R (2004) Aircraft noise and times of day: possibilities of redistributing and influencing noise exposure. Noise Health 6(22):55–58

37. Schreckenberg D, Meis M (2006) Gutachten Belästigung durch Fluglärm im Umfeld des Frankfurter Flughafens–Endbericht [Annoyance due to aircraft noise in the vicinity of Frankfurt Airport—Final report]. ZEUS GmbH, Hörzentrum Oldenburg, Bochum, Oldenburg, Germany

38. Fields JM (1985) The timing of noise-sensitive activities in residential areas, NASA Contractor Report 177937 (Rep. No. NASA Contractor Report 177937). The Bionetics Corporation, Hampton, Virginia, USA

39. Flindell I, Le Masurier P, Le Masurier H (2021) Resolving uncertainties in understanding community attitudes to aircraft noise. Appl Acoust 178:108032

40. Fields JM, de Jong RG, Gjestland T, Flindell IH, Job RFS, Kurra S et al (2001) Standardized general-purpose noise reaction question for community noise surveys: research and a recommendation. J Sound Vib 242(4):641–679

41. Sparrow V, Gjestland T, Guski R, Richard I, Basner M, Hansell A, de Kluizenaar Y, Clark C, Janssen S, Mestre V, Loubeau A, Bristow A, Thanos S, Vigeant M, Cointin R (2019) Aviation noise impacts. White Paper. State of the science 2019: aviation noise impacts. Chapter 2 in ICAO Environmental report. https://www.icao.int/environmental-protection/Documents/Scientific Understanding/EnvReport2019-WhitePaper-Noise.pdf

42. Richard I (2017) Aircraft annoyance: a psychosocial approach. Presentation in the ISG Aviation Noise Impacts workshop. OACI, November 1–3, 2017, Montréal, Canada

43. Ebbinghaus H (1913) On memory: a contribution to experimental psychology. Teacher College, New York

44. Brink M, Schreckenberg D, Vienneau D, Cajochen C, Wunderli JM, Probst-Hensch N, Röösli M (2016) Effects of Scale, question location, order of response alternatives, and season on self-reported noise annoyance using ICBEN Scales: a field experiment. Int J Environ Res Public Health 13(11):1163. https://doi.org/10.3390/ijerph13111163

45. Hallmann S, Guski R, Schuemer R (2001) Cognitive Processes in global noise annoyance judgments. Paper presented at the 30th international congress and exposition on noise control engineering, August 28–30, 2001. The Hague, The Netherlands

46. Schreckenberg D, Belke C, Spilski J (2018) The development of a multiple-item annoyance scale (MIAS) for transportation noise annoyance. Int J Environ Res Public Health 15(5):971. https://doi.org/10.3390/ijerph15050971

47. Kastka J (1999) Untersuchung der Fluglärmbelastungs-und Belästigungssituation der Allgemeinbevölkerung der Umgebung des Flughafen Frankfurt [Examination of aircraft noise exposure and aircraft noise-induced annoyance in the vicinity of Frankfurt Airport], Report. Heinrich-Heine-Universität Düsseldorf, Germany

48. Upham P, Maughan J, Raper D, Thomas C (eds) (2003) Towards sustainable aviation. Earthscan Publication, New York, USA

49. Basner M, McGuire S (2018) WHO environmental noise guidelines for the European region: a systematic review on environmental noise and effects on sleep. Int J Environ Res Public Health 15:519

50. Müller U (2019) ANIMA D3.2 - Development of indicators for night noise protection zones. Zenodo. https://zenodo.org/record/5517783

51. Ohayon MM, Carskadon MA, Guilleminault C, Vitiello MV (2004) Meta-analysis of quantitative sleep parameters from childhood to old age in healthy individuals: developing normative sleep values across the human lifespan. Sleep 27(7):1255–1273. https://doi.org/10.1093/sleep/27.7.1255%JSleep

52. Bartels S, Quehl J, Berger M, Aeschbach D (2021) Exposure response-relationships between nocturnal aircraft noise and sleep disturbances in primary school children. In: Proceedings of 13th ICBEN congress on noise as a public health problem. Stockholm, Sweden, virtual congress

53. Brink M, Schäffer B, Vienneau D, Pieren R, Foraster M, Eze IC, Rudzik F, Thiesse L, Cajochen C, Probst-Hensch N, Röösli M, Wunderli JM (2019) Self-reported sleep disturbance from road, rail and aircraft noise: exposure-response relationships and effect modifiers in the SiRENE study. Int J Environ Res Public Health 16(21):4186. https://doi.org/10.3390/ijerph16214186

54. McGuire S, Müller U, Elmenhorst EM, Basner M (2016) Inter-individual Differences in the Effects of Aircraft Noise on Sleep Fragmentation. Sleep 39(5):1107–1110. https://doi.org/10.5665/sleep.5764

55. Marks A, Griefahn B (2007) Associations between noise sensitivity and sleep, subjectively evaluated sleep quality, annoyance, and performance after exposure to nocturnal traffic noise. Noise Health 9(34):1–7. https://doi.org/10.4103/1463-1741.34698

56. Elmenhorst E-M, Müller U, Mendolia F, Quehl J, Aeschbach D (2016) Residents' attitude towards air traffic and objective sleep quality are related. In: Proceedings of the 45th international congress and exposition on noise control engineering (InterNoise 2016), pp 7744–7746

57. Quehl J, Müller U, Mendolia F (2017) Short-term annoyance from nocturnal aircraft noise exposure: results of the NORAH and STRAIN sleep studies. Int Arch Occup Environ Health 90(8):765–778. https://link.springer.com/article/10.1007%2Fs00420-017-1238-7

58. Elmenhorst EM, Elmenhorst D, Wenzel J, Quehl J, Mueller U, Maass H, Vejvoda M, Basner M (2010) Effects of nocturnal aircraft noise on cognitive performance in the following morning: dose-response relationships in laboratory and field. Int Arch Occup Environ Health 83(7):743–751. https://doi.org/10.1007/s00420-010-0515-5

59. Nassur AM, Lefèvre M, Laumon B, Léger D, Evrard AS (2019) Aircraft noise exposure and subjective sleep quality: the results of the DEBATS study in France. Behav Sleep Med 17(4):502–513. https://doi.org/10.1080/15402002.2017.1409224

60. Schreckenberg D (2018) Knowledge gaps concerning health impacts of environmental noise. In: Proceedings of Euronoise 2018. Heraklion, Crete–Greece, pp 985–991

61. Sánchez D, Naumann J, Porter N, Knowles A (2015) Current issues in aviation noise management: a non-acoustic factors perspective. In: Proceedings of the 22nd international congress of sound and vibration. Florence, Italy

62. Vader R (2007) Noise annoyance mitigation at airports by non-acoustic measures. Inventory and initial analysis. LVNL, Air Traffic Control the Netherlands, Amsterdam, The Netherlands
63. Haubrich J, Burtea NE, Flindell I, Hooper P, Hudson R, Rajé F, Schreckenberg D (2019) ANIMA D2.4—Recommendations on annoyance mitigation and implications for communication and engagement. Zenodo. https://doi.org/10.5281/zenodo.3988131
64. Hauptvogel D, Richard I, Haubrich J, Kuhlmann J, Heyes G, Benz S, Hooper P, Bartels S, Schreckenberg D (2021). ANIMA D3.9 - Engagement guideline. Zenodo. https://zenodo.org/record/5517797
65. World Health Organiation (1997) Division of mental health and prevention of substance abuse. WHOQOL: measuring quality of life. World Health Organiation. https://apps.who.int/iris/handle/10665/63482
66. EUROSTAT (2017) Final report of the expert group on quality of life indicators. http://ec.europa.eu/eurostat/web/products-statistical-reports/-/KS-FT-17-004
67. Roosien R, Schreckenberg D, Benz S, Kuhlmann J, Hooper P (2018) ANIMA-Public version of D3.1 Study to identify the gaps—Quality of Life indicators. Zenodo. https://doi.org/10.5281/zenodo.1549205
68. Lelij B, van der Bos L, Roelofs S (2020) Luchtvaart in Nederland, Draagvlakonderzoek onder het Nederlands publiek, meting 2020 (derde meting, tijdens COVID-19-pandemie) https://www.rijksoverheid.nl/documenten/rapporten/2020/11/20/bijlage-6-rapport-motivaction-luchtvaart-in-nederland-draagvlakonderzoek-meting-2020
69. Klatte M, Spilski J, Mayerl J, Möhler U, Lachmann T, Bergström K (2017) Effects of aircraft noise on reading and quality of life in primary school children in Germany: results from the NORAH study. Environ Dev 49(4):390–424
70. International Civil Aviation Organisation (ICAO) (2008) Guidance on the balanced approach to aircraft noise management (2nd edn). Doc 9829 AN/451
71. Stewart J (2021) Think Local. Act Local. How airports can deliver for local communities. www.ukna.org.uk/uploads/4/1/4/5/41458009/delivering_for_local_airport_communities.pdf

Engaging Communities in the Hard Quest for Consensus

G. Heyes⬤, D. Hauptvogel⬤, S. Benz⬤, D. Schreckenberg⬤, P. Hooper⬤, and R. Aalmoes⬤

Abstract Mistrust, negative attitudes and the expectation of not having any voice against airport authorities can considerably impact on the perception of aircraft noise exposure, lead to increased annoyance and can even influence sleep quality of the noise affected residents. As a result, quality of life can reasonably be assumed to be reduced. This chapter focuses on measures to engage airport communities in aviation-related decision making by improving the information and communication of airports in order to enhance residents' 'competence' and also trust in the airport noise authorities. The role of non-acoustical factors, including aviation-related media coverage in this process, is discussed and results from a media coverage analysis conducted in the ANIMA project are presented. Based on research on perceived fairness in communication, recommendations are given as to how to communicate and engage residents with the aim of building a neighbourly relationship between airport authorities and residents on an even footing and, thus, enable an improved exchange leading to deeper understanding and comprehension by both parties. Results from the ANIMA review on airport management strategies (including communication and engagement aspects) of several European airports are presented and conclusions are drawn about what characterises good (or bad) communication and community

---

Explaining successful methods employed to engaged communities and the prerequisite for such success

G. Heyes (✉) · P. Hooper
Ecology and Environment Research Centre, Department of Natural Sciences, Manchester Metropolitan University, Chester Street, Manchester M1 5GD, UK
e-mail: g.heyes@mmu.ac.uk

D. Hauptvogel
Sleep and Human Factors Research, Institute of Aerospace Medicine, German Aerospace Center (DLR e.V.), Linder Höhe, 51147 Cologne, Germany

S. Benz · D. Schreckenberg
ZEUS GmbH, Centre for Applied Psychology, Environmental and Social Research, Sennbrink 46, 58093 Hagen, Germany

R. Aalmoes
Royal Netherlands Aerospace Centre NLR, Anthony Fokkerweg 2, 1059 CM Amsterdam, Netherlands

© The Author(s) 2022
L. Leylekian et al. (eds.), *Aviation Noise Impact Management*,
https://doi.org/10.1007/978-3-030-91194-2_9

engagement strategies for the purpose of a neighbourly relationship between the airport and its residents.

Keywords Communication · Engagement · Non-acoustic factors · Fairness · Evaluation · Noise management

Introduction

Aircraft noise has been shown to cause adverse effects on human health (see Chapter 7). It is assumed that this is partly mediated by the effect of annoyance. Previous chapters have taught us the important role of non-acoustic factors for the levels of annoyance.

As described in Chapter 8 there are different non-acoustical factors, i.e., situational and personal/social factors that contribute to how noise is perceived and processed, and that these non-acoustic factors can affect the impact of environmental hazards. Non-acoustical factors can be summarised as factors that are not directly connected to the sound [1], but modify or co-determine the response to it. In this way managing non-acoustical factors can be seen as a crucial and essential opportunity to minimise annoyance reactions and reduce the adverse effects of noise. This is also due to the fact that noise reduction alone has not resulted in corresponding reductions in annoyance. Hence, non-acoustic factors are seen as having a critical impact on noise effects, and are equally important to consider when tackling annoyance and other noise responses. In Chapter 8, several categories of non-acoustical factors were identified, some of which can be more influenced than others, with some especially important. One's general sensitivity to noise, personal (mis-)trust in responsible authorities, attitudes towards the airport and aviation in general as well as expectations and fears, e.g., for health risks and aircraft crashes, have been identified as the most important non-acoustic factors. Besides these factors, socially shared information is also relevant for how we perceive or what we experience [2]. Particularly vital and influential factors are those related to communication and social exchange. Communicating with other people about issues informs our knowledge and shapes our expectations. For example, a neighbour complaining about something can directly affect how we perceive the topic of complaint.

By studying the influence of discourse on people's experience of aircraft noise, research provides evidence that the discourse in an airport region not only originates the degree of annoyance but also how policy discourse resonates in private discourses [3, 4]. One finding was that people are influenced by policy in that they refer to policies when talking about noise experience. Private discourse often directly reflects the story lines of annoyance policies. This was shown by comparing two airport regions, Amsterdam Schiphol and Zurich Kloten. Further, it seems that noise experience was influenced by the discourse in that annoyance ratings were supported by policy discourse arguments. Taken together this can mean that when people are engaged in the process of policy definition it can contribute positively to their experience of noise.

The Role of Media Coverage

Another factor able to contribute to the impacts of noise is public discussion and how the media reports noise. One could summarise it by saying that dominant policy discourses can shape our experience [4], which suggests that media coverage is at least able to shape our experience and perception of noise. Interest of the media focuses on coverage on deviating opinions and events, such as demonstrations, not on activities that influence existing practices.

Findings from research on other sources of environmental exposures suggest that the way the media frames information in its reports shapes the expectations of people around the exposure source. Moreover, expectations around possible adverse health outcomes of exposure sources can contribute to the occurrence of negative health outcomes [5], which has shown that the framing of information influences how participants perceive the noise. Studies on wind turbine noise for example suggest that when participants saw negative framed material on the effects of infrasound it affected the number and intensity of health complaints people reported when exposed to [6] or annoyed by [7] infrasound. The same was true for a group that were shown positive information about infrasound, they reported less health complaints and even some positive effects while exposed to infrasound. It is assumed that media coverage about potential adverse health effects builds up expectations on the consequences of the exposure and this, for various reasons, increases or even causes potential health outcomes.

As highlighted in Chapter 8, the perception of avoidability, unpredictability, uncontrollability, and procedural unfairness increases stress responses and annoyance, and reduces the perception of being able to cope. In particular, when the exposure situation is likely to change, e.g. due to re-allocation of flight paths, increase in air traffic, and/or an expansion of the airport, questions arise from new or increasingly exposed residents, such as: Are these changes necessary or could they be avoided? Is the new situation predictable? Are the changes in exposure fair and are the way these changes are established and the decisions made fair? Information is essential in residents having answers to these questions, and thus having some form of perceived control of the situation. This is particularly the case in ambiguous situations where residents already dislike situations, and are expected to dislike future situations. Hence, this is where socially shared information such as from the media comes into play. Another point in line with this is that it facilitates the adoption of an attitude, intention, or behaviour, if this follows a social norm, that is, if relevant persons (family, neighbours, friends) expect such an attitude, intention, or behavior from someone [8]. If family members or neighbours are annoyed by aircraft noise, it is easier to be annoyed too, or to regard this as confirmation of one's own annoyance (in terms of 'I am not alone with my annoyance'). In addition, other people talking about noise issues can raise awareness or draw attention to a noise issue. The latter has become particularly relevant in the era of the internet and social-media in which the networks or individuals, be them formal or informal, have extended significantly. Social-media in particular enables one's views to be potentially influenced by a large number of people who may not be experts on a subject but who nonetheless can play an informing role in the development of an individual's perceptions on a given subject (the role of social media is discussed in more detail in Chapter 10).

The role of media, in particular local media, is not so much that it produces noise responses such as annoyance, but that it can reflect socially shared knowledge, opinions, and perceptions of noise which particularly become relevant in situations of change. In this sense, it is hypothesised that the way aircraft noise is covered in media articles influences the way that noise exposure is perceived and processed, e.g. resulting in expectations and contributing to noise annoyance and further health issues.

Analysis of media reports showed that the motivation for participation in the NORAH study around Frankfurt Airport was influenced by media coverage about Frankfurt Airport: a higher number of reports about the study were related to a higher number of completed interviews in the study (Guski, Peschel, Wothge, 2014). This indicates that the media articles seem to have contributed to the residents' awareness of the importance of the study and that it would be useful perhaps even for one's own residential quality of life to participate—in terms of 'if the study is on health effects of aircraft noise and repeatedly described in the media, it has a point there and I should be part of it'.

Results of the ANIMA Media Coverage Analysis

The media analysis conducted within ANIMA deals with how media reports about aircraft noise and how related topics may influence annoyance ratings assessed in the NORAH study on health effects of aircraft noise.

This was done by linking media reports around Frankfurt Airport during the NORAH study to the annoyance ratings over the same time periods and examining if annoyance ratings are influenced by media reports. Looking at the content of news reports headlines, categories of topics were derived from the reports, such as "night flight", "noise exposure","protest", among others, and analyses were conducted to find out how reports with certain content may affect the annoyance ratings. For each participant, reports from 180 days prior to the study interviews were taken into account as it was assumed that some time was needed to process media reports and assimilate information.

First results indicate that media coverage about certain noise-related topics have an impact on annoyance ratings, in that stronger annoyance has been reported when media articles more frequently reported about these noise-related topics (for further details see Hauptvogel et al., 2021).

First and foremost, the focus of most media is reporting about existing issues and to shed light on problems that are relevant and/or pending. It is about displaying and focusing attention on e.g. local noise issues around an airport. The frequency of reporting about certain topics reflects the relevance of these topics.

So when it is assumed that annoyance ratings can partly be explained by media reports, it is not based on the assumption that reporting explains the annoyance but that media coverage can have an impact on how exposures are perceived and to be aware of how framing of certain health issues are delivered. Media reporting can have an effect itself in providing attention to the specific problem. When airports or other authorities work on improvements to the problem, media coverage adapts accordingly. Media coverage can therefore cause but also extend existing discourse about the topic, which in turn shapes opinions and can influence the perception of the noise itself.

Therefore the focus of any intervention should not be to change information but to change the problem, which in turn changes the information about the change, topic or issue. This is a dynamic process. Thus, when communication and engagement measures are trying to determine the discourse in the region around the topic itself, this cannot be simply confined to changing the communication around it but to include interventions that focus on reduction of noise. In other words, meaningful communication and engagement is that relating to the reduction of the causes of negative impacts.

Change of communication and engagement strategy of an airport has to be accompanied by technical or operational changes and vice versa. Communication without implementing changes may even encourage higher mistrust in responsible authorities. The whole dynamic has its origin in the problem, the noise source, itself and how it is managed.

Transferring this to the Balanced Approach could mean that communication and engagement has to be built across all four pillars.

Communication and Engagement and Noise Management

Given the nature of the described non-acoustical factors, including media coverage, it is hardly surprising that researchers and the aviation industry have identified communication and engagement as key elements in the management of noise impacts.

The aviation industry has gone to considerable effort to reduce noise and noise impact over the past 50 years, mostly via significant reductions in noise from individual aircraft, driven by increasingly stringent certification regulations regarding aircraft design. These reductions have not, however, resulted in corresponding reductions in annoyance. Instead, public opinion is an increasing constraint to airport activity, despite fewer people being exposed to higher levels of noise than in previous years [9]. The ICAO Balanced Approach has looked to help address this by not just reducing noise at source (although this remains important), but also through other measures that are designed to better manage noise for the benefits of residents. As well as encouraging reductions in noise at source, the Balanced Approach also outlines actions that can be taken with regard to: land-use planning and management policies that seek to reduce noise exposure on the ground, either by keeping noise sensitive developments (i.e. conurbations) away from high-noise areas, or by managing sound on the ground, through insulation programmes, Operational Procedures, such as moving flight tracks so as to not over fly communities, and; operating restrictions, for instance night flight limits or absolute caps on aircraft movements.

And, finally, in 2007 the Balanced Approach Guidance was expanded to include 'People issues'. This added fifth pillar focuses on communication strategies, advocating the use of enhanced information that is easily accessible by the public and emphasises the role of consultation. Although not formally adopted through the Balanced Approach as a core pillar, the concept of communication and engagement as a noise management tool is now seen to be increasingly important.

Communication and engagement does not purely exist as an additional pillar through which noise can be managed—it can also help aid the successful implementation of other balanced approach measures. Successful noise management actions must be technically feasible or viable in order to be implemented, and together with a range of technical data, the industry has typically focused the development of Balanced Approach interventions on such data in order to develop interventions that are deemed to have the greatest potential impact and benefit for noise affected communities. Indeed, national noise policy is often focused on such considerations, leading airports to develop, for instance, new operational procedures based on aggregated noise metrics and success criteria such as the number of people exposed to certain levels of noise. This is a sensible approach, which can provide airports with confidence that the noise management actions they develop will be more likely to result in positive outcomes. As previously mentioned however, improvements in noise as measured through such approaches is not a guarantee that residents will perceive them as successful, or that there will be a positive impact on annoyance and complaints. The reason for this is that truly successful noise management interventions require a further consideration to technical feasibility and viability—desirability. Put simply, if a noise management intervention looks good on paper, but is not deemed to be effective or desirable in the eyes of those it is designed to serve (i.e. residents), then it is less likely to be perceived by those same residents as being an effective or appropriate response to the noise they experience. Through communication and engagement, airports are able to explain noise and noise management processes to residents, but also gain their feedback and insight into what success looks like in residents' own eyes. This information can be incorporated into decision making and help to produce noise outcomes that are more likely to be viewed as appropriate.

Why is Fairness so Important in this Context?

The operation of an airport inevitably leads to noise. Unfortunately and despite the application of the four "traditional" pillars of the ICAO Balanced Approach, it cannot be ruled out that the noise affects some people more than others. The nature of aircraft noise means that it has to be distributed in a certain way over parts of the population.

Logically, this distribution is inherently unfair—as some people get more noise than others. Aircraft noise is man-made and the exposure to it is often seen as a social conflict arising from the fact that residents view noise as the airport exposing them [10]. In order to come to a certain distribution of the noise, decisions have to be made. Procedures have to be applied to reach these decisions and the results of the decision-making process need to be communicated to affected people. For this reason, it makes sense to look at the exposure to aircraft noise from the perspective of fairness research, in particular research on procedural and interpersonal fairness, which offer some important starting points on how to deal with this inherently unequal distribution.

Fairness as the Overall Goal

An observation made since the 1970s is that people are more likely to accept and adopt unfavourable outcomes of decisions when the decisions are based on correct information, when the decision-making process is free from bias and applied consistently over time and, above all, when the affected people have been involved in the decision-making process [11]. This so-called "fair process effect" is based on the observation that giving people "voice" makes them more likely to accept decisions [12–14]. As described in detail in Chapter 8, noise annoyance is a stress response that depends on various factors such as how much coping opportunities and resources people perceive.

In evolutionary terms, procedural fairness is an extremely important indicator for a person to be an accepted and valued member of a group. It therefore fulfils the need for belonging and self-esteem [12, 14].

This means that airport management should apply procedures that are as fair as possible and recognised as such by the public. The assumption that giving voice leads to increased perceived fairness and reduced annoyance due to noise exposure has already been shown in studies [15] when people who could express their preference for a certain sound were significantly less annoyed than people who could not. However, annoyance was particularly high among people whose preference was actively ignored. A more recent study [16] also showed that many opportunities to participate led to a higher acceptance of a fictitious airport expansion. It also showed that the focus on the jobs created by the airport expansion had no effect on acceptance.

In sum, it can be said that procedural aspects of aircraft noise distribution have an enormous influence on how people perceive aviation, the airport and the noise and to what extent they are annoyed by the noise. Interventions that take these insights into account can therefore be very effective.

Despite the positive effect of having voice or control in the decision-making process, fairness research has also shown the critical impact of providing information and justification of a decision for the perception of the outcome of this decision. From the perspective fairness regarding informational aspects and regarding the interaction between two parties (so-called informational and interpersonal fairness), people may perceive unfairness, even though they consider the procedure and its result as fair, just because of an improper treatment or a lack of justification by the decision-maker [17]. But also in case of a negative outcome, the decision process may be recognised as fairer when an adequate justification or causal account is given by the authority who made the decision [18, 19]. These findings point to the need for a good communication strategy of the airport management and we will come back to the lessons learned from this branch of fairness research when we define criteria for good communication and information.

To give an overview, research has identified a set of criteria and standards relating to the fairness aspects mentioned above, which, taken together, can create a perception of fair process and fair interaction with the parties concerned. Research distinguishes between several facets of fairness. A distinction can be made between procedural, interpersonal and informational fairness. All of these main fairness standards comprise a number of criteria:

	Fairness standards	
Procedural	Process control	procedures provide opportunities for voice
	Decision control	procedures provide influence over outcome
	Bias suppression	procedures are neutral and unbiased
	Representativeness	procedures take into account concerns of subgroups
	Consistency	procedures are consistent across persons and time
	Accuracy	procedures are based on accurate information
	Correctability	procedures offer opportunities for appeals of outcomes
Informational	Truthfulness	explanations about procedures are honest
	Justification	explanations about procedures are thorough
Interpersonal	Propriety	enactment of procedures refrains from improper remarks
	Respect	enactment of procedures refrain from improper remarks

(Rules taken from [11, 13, 20, 21], after [22], Colquitt)

With these research-derived criteria, concrete recommendations can be derived on what constitutes good communication and engagement and how to build a neighbourly relationship with residents of local airport communities.

What is Communication and Engagement

At its core, communication refers to the dissemination of information from one person or organisation, to another person or organisation. For instance, governments may communicate information about certain changes to legislation, or about new laws or policies to the public—government health and safety warnings around the time of the Covid-19 pandemic being a good example. For aviation, airports may communicate for a range of reasons, for example sharing noise data or operational changes to their communities, or performing marketing activities regarding things like the promotion of noise management measures, reductions in noise levels as described through metrics such as L_{den}, or quality of life benefits afforded to residents as a result of airport activity as well as contributions to the national or regional economy. What really defines communication however is the one-way flow of information that it typically implies. That is, one actor passing on information to another. Typically, communication tools therefore include things like newspaper articles, radio advertisements, websites, mail and other printed media such as noise action plans, noise contour maps or other corporate reporting—with more recent innovations including the use of social media to, for example, communicate things like airport operating conditions. The intent of such activities is for a specific message, or messages, to be heard by a target audience, at a specific point in time, and with a targeted outcome. As such communication activities tend to lose meaning over time, and whilst their one-way flow of information and generic targeting can be helpful in explaining things to residents, they can also lead to disengagement from receptive audiences or confusion if messages are unclear, misunderstood or not trusted. This is particularly difficult for airports, who are tasked with explaining highly complicated, multi-faceted and technical data in simple and easily digestible formats. This is a significant challenge as simple communication measures can lack relevant information, whilst communication materials that show a range of information can be critiqued for being too complicated to understand. This is compounded by the fact that communicating noise through different metrics has a range of different advantages and disadvantages. Noise contours for example, do a good job at illustrating aggregated noise levels around an airport, however they fundamentally describe an audible factor, through a visual medium, and describe noise in a way that is not experienced by residents, who live through individual noise events. The result is that contour maps are often poorly understood by residents [23], despite legislation such as the Environmental Noise [24]/49/EC requiring airports to produce such contour maps and to disseminate them to the public. In worst case scenarios poor communication can lead to mistrust between airports and community groups who may begin to question the information that they are being told, thus raising the question of the value of the communication itself.

Engagement, on the other hand, refers not just to the provision of information to stakeholders, but to establishing a dialogue. Here the objective is to embark on a conversation with stakeholders to explain things to them, but importantly, to also listen. The concept is rooted in the fact that residents are the experts on their own lived experiences and can offer important insight that may otherwise remain unknown, and that could play an important role in decision making around the development of any noise management interventions that are likely to be perceived as acceptable. Hence, the aim is not only to pass information onto stakeholders, but to also listen to stories about their lives, their fears, the things they do in life, and to build empathy for them and their perspectives on given issues. Put simply, engagement implies not just talking, but also listening, and understanding and the need to tailor messages and information to different people, in so doing having the potential to become more meaningful interactions over time. The importance of engagement can be seen through concepts such as design thinking, which are applied in organisational settings to develop solutions to a range of operational challenges. The process is based on the idea that considering the needs of a given beneficiary of a service is essential in order to maximise the likelihood of the success of that service. The process is rooted in deep engagement with stakeholders, including the use of multi-stakeholder design teams, collecting qualitative data to complement quantitative information, and under-standing and addressing core challenges directly. Similar approaches are already set out in aviation noise through proposed processes in the United Kingdom's Civil Aviation Authority CAP 1616 [25] document and the United States Federal Aviation Authority Program 150 [26]. Both take iterative step processes to develop noise management interventions that include a focus on understanding resident needs and embed them as core principles in the development of noise management actions. Methods for engagement go beyond the mere dissemination of information as with pure communication, and involve more participatory methods such as consultation, focus groups, workshops or full collaborative and participative working groups. Hence communication and engagement can be seen as sitting on a spectrum, from the simple provision of information, through to more participatory levels that afford degrees of citizen empowerment through partnerships, delegation of control. This has been helpfully illustrated by [1], who, as illustrated in Table 1, created a Wheel

Table 1 Asensio el al. [1] types of public participation	Category	Sub-Category
	Information	Minimal communication
		Limited information
		Good quality information
	Consultation	Limited consultation
		Customer care
		Genuine consultation
	Participation	Effective advisory body
		Partnership
		Limited centralised decision making
	Empowerment	Delegated control
		Independent control
		Entrusted control

of Participation for airport noise management, adapted from the work of Arnstein's Ladder of Public Participation [27] to illustrate the types of public participation that exist.

Communication tools may still be used as part of engagement, but rather than as the primary output, they merely lay the framework on which a wider discussion can take place. At the same time, it should be stressed that engagement with stakeholders does not imply that good levels of communication have taken place. It is entirely possible for example, that an airport may be seeking to engage with residents, but communicating noise information to them poorly, or even in a manner that residents deem to be dishonest (such claims may be untrue, but if they are true in the eyes of residents they remain a relevant management concern). Likewise, processes of engagement do not necessarily mean success. Engagement has to be meaningful and with an honest intent to listen to and learn from stakeholders. Failure to do this can result in mistrust, which once lost can be almost impossible to win back.

Building on findings from case study research conducted in ANIMA, some of the characteristics and key principles of, and differences between, communication and engagement are outlined below:

- Communication typically sets out to describe what is happening, or what has happened, or to perform basic consultation regarding a set of predetermined interventions. Engagement on the other hand, explains why things are happening, and seeks to obtain the input of stakeholders regarding decisions that have not yet been taken, the aim being to produce fair outcomes.
- Communication describes one way dialogues between airports, speaking to residents. This means that communication methods more often than not include contour maps, noise reporting, noise action plans, or marketing information. Engagement on the other hand describes two-way flows of information, and therefore utilises methods such as consultation events, workshops, focus groups and Dialogue Forums. These require more effort and resources to operate but better reflect a more engaged and informative process that is more likely to lead to outcomes that are perceived to be successful in the eyes of stakeholders.
- Communication typically uses quantitative data to describe and communicate noise. This is useful in that it is an attempt to describe noise in the most accurate way possible. However it is also beset with difficulties of describing a complex and highly technical concept (noise) through simple metrics. Engagement may also use the same information, but its two-way flow of information also concerns qualitative data, i.e. how residents feel about noise and how noise is likely to affect them.
- In pure communication, the actor that is leading the communication typically takes on the role of expert. This can lead to hierarchical stakeholder relationships that can make establishing trust difficult, and can cause the lead communicator to discount other sources of information. Engagement on the other hand is typically based on levelled hierarchies in which all stakeholders are seen to have potentially valuable information to offer decision making processes. Empathy plays a key role and consensus is deemed more likely to be reached through understanding.

The above may suggest that engagement is a more comprehensive approach than communication, it should not be seen as necessarily being best practice in every scenario, as the level of engagement activity undertaken by an airport should be determined by the desired output of the interaction. Hence, both communication and engagement approaches should be used with an awareness of the attributes and benefits of each, and importantly, the circumstances surrounding the area in which they are to be implemented, for instance what is the ultimate desired outcome of the interaction, the understanding of which may itself require some form of engagement. That said, best practice dictates that engagement should at the least be considered whenever an airport is looking to communicate something to its residents, or to make operational changes or other modifications to airport activity. The importance of this can be appreciated through the fact that noise management, at its fundamental core, exists for the benefit of airport residents, be it due to direct pressure to manage noise as demanded by communities, or in response to legislation designed to protect noise affected communities from the potentially significant noise impact caused by noise exposure. It is therefore important to not just develop noise management actions or general airport operations that are technically feasible or viable, but to also consider what actions are desirable in the eyes of those residents.

A Tale of Communication and Engagement Gone Wrong

Vienna Airport is the largest airport in Austria and of major economic importance to the region. The airport built its second runway in 1972, however they projected that airport capacity would be reached by 2012 and that an additional runway would therefore be needed to continue airport growth.

Hence, the airport began plans for a third runway in 1998 to the south of existing airport infrastructure. However, the airport made such an announcement without effective consultation or dialogue with its communities.

The result was significant opposition to the runway by local community groups who felt aggrieved about the lack of consultation, and the health impacts that they would be subjected to from increased traffic, particularly for communities who would be newly overflown by aircraft arriving and departing from the new runway. By not being engaged with, trust was damaged and opposition campaigns proved so successful that approval for the third runway was not granted—indeed, some 20 years later, the runway has still not been built.

This is an example of an airport not engaging with its residents effectively, and demonstrated the potential impact to airport operations from doing so. However, Vienna Airport learned from this mistake, and as we demonstrate later in the chapter, they are now regarded as one of the best examples of an airport communicating and engaging with its residents.

How to Do 'Good' Communication and Engagement

Noise managers increasingly understand the human response to noise and the role of non-acoustic factors in driving annoyance. Addressing such factors is however complicated, and coupled with external pressure for absolute reductions in noise, has seen the majority of noise management actions focus on addressing acoustic factors. Although such an approach is understandable, doing so has not always led to successful outcomes—hence why noise (as measured through metrics such as noise level equivalents) has remained stable or fallen at many airports, against a background of increased levels of reported annoyance.

Despite the continuing trend that communication and engagement are recognised by airports as important, there is a lack of clear recommendations on what constitutes successful communication, how to implement it and how to evaluate it. So what needs to be emphasised here is that any kind of communication and engagement should be underpinned by certain quality criteria and theoretical principles. For this purpose, we suggest focusing on principles derived from research on fairness in social exchanges. This is the only way to achieve a long-term and sustainable trust and acceptance of the airport. Great progress in the ANIMA Project was achieved since not only theoretical recommendations were derived but their application in practice was assessed as well. So how have airports been performing in terms of communication and engagement? This has been a key question throughout the ANIMA Project. Airports have been communicating about noise for many decades, with approaches moving over time from a purely dissemination of information approach, towards processes more aligned to consultation and engagement that can aid airport decision making.

In a review of airport case studies across the European Union, ANIMA research came to the following conclusions about communication and engagement:

- There has been an evolution from communication towards more participative forms of discourse, notably an increase in consultation and the development of noise dialogue or community programs.
- However, communication and engagement tends to happen in a relative ad-hoc manner with data provision often following guidance to produce quantitative noise data only, and with such data often being disseminated in ways that publics find hard to comprehend.
- Communication and engagement tends to remain largely about information provision rather than leveraging the potential benefits of engagement in light of the role of non-acoustic factors.
- Communication and engagement often happens without an intended outcome that seeks to address given challenges or needs.
- There is rarely any evaluation as to the impact of any communication and engagement.
- Communication and engagement is generally seen as ancillary noise management activities, rather than as playing a key informing role in the success of other interventions, or as a management tool in their own right.

As with all aspects of noise management, it is important that airports do not follow prescribed advice based on 'best practices' from elsewhere, but rather base their actions on their own definitions of 'good practice' as appropriate for their own circumstances. That said, there are some core guidelines that can help to ensure that good communication and engagement is taking place between airports and their community stakeholders. In the Table below we set out a range of recommendations that airports should consider when looking to conduct 'IDEAL' communication and engagement with residents. One should also note however that as a two-way process, communication and engagement is not necessarily in the hands of airports in its totality. Communities too have a responsibility to engage with airports about noise, to learn about noise management and to understand noise data made available to them. That said, it has to be stressed that as the source of the noise, and with the agency to make change, it is airports who must play the lead role in facilitating engagement and in providing information that is both relevant to residents and that is produced in a way that is comprehensible to non-experts.

The 'IDEAL' characteristics of communication and engagement	
I	Inclusive and diverse: No communities or hard to reach groups should be left behind. This can include those who do not have a history of complaints, and those in deprived areas or those consisting of different nationalities
	Information provision: Residents should be provided with data relevant to them. This means taking the time to understand what those data are, how they can be illustrated or described, and what appropriate communication channels might be
	Impartial: Advanced communication and engagement is not an easy task as it can involve having difficult conversations with conflicting voices. Independent facilitation can help overcome these challenges whilst also providing access to experts in the facilitation of things like focus groups and workshops. Data provided by impartial experts can also help to build trust
	Interrogate: It is important to ask questions about any pre-held perceptions about noise problems and their likely solutions as what may appear to be a challenge to be solved (i.e. reducing complaints), may actually be triggered by something at a deeper level. Questioning such perceptions and gaining insight from residents can be a useful way to understand how core challenges can be addressed, to identify targeted outcomes, and to establish potential criteria on which such outcomes can be evaluated
D	Decisions: All stakeholders may have expert knowledge that has the potential to inform decision making, or to influence the potential success of a given intervention. It can be helpful therefore to perform stakeholder analysis or stakeholder mapping when performing any activities that are likely to influence noise to identify two factors: who has interest in the issue, and who can have influence over the issue. With this information it is possible to determine who should be engaged about noise—although it should be considered that sometimes there can be unintended consequences that could affect groups that were not expected. It can therefore be helpful to include all groups in engagements in order to develop well rounded understanding and to aid decision making
	Direct: Airports should be honest with the citizens. This means that airports should start communicating honestly, directly and transparently from the beginning of a decision process

(continued)

(continued)

	The 'IDEAL' characteristics of communication and engagement
E	Early: Communities should be communicated with early and often throughout any changes that may affect them. This is important to make them aware of what is happening, but also to understand their needs, preferences, fears and so on, and to communicate any potential changes to the noise they may be exposed to (be it on a trial or temporary basis)
	Easy: It is important that data is communicated and explained as clearly as possible and that information is easy to understand without any previous knowledge or expertise. Presenting complex information that people find difficult to grasp can lead to airports being accused of hiding data by purposely putting up barriers. Communication and engagement should be tailored to the characteristics of each airport and community groups and what the interaction sets out to achieve. This includes using appropriate language and data, both in terms of relevance to the subject of the communication or engagement, but also to the expertise and comprehension of the recipient
	Explain: Airports should not just be explaining what has happened and what the results of any changes have been. They should also articulate, clearly, why decisions have been made, whether other options were considered, why other options may not have been selected. Noise action plans can be a great way to demonstrate that noise has been addressed at a strategic level
	Empathy: Effective communication and engagement means going beyond numbers and thinking in qualitative terms by developing stories of the lived experiences of residents and developing and acknowledging empathy for those stories. Airports can also tell their own stories to help articulate the significant difficulties that they have in managing noise, thus helping to foster empathy for their own situation
A	Accessible: Information should be easy to find and not hidden in technical reports, or multiple clicks into a website. For communication to be received effectively its intended audience should be able to access that information as easily as possible. Hard to find information gives the impression of mis-intent, which can be harmful to trust in airport-stakeholder relationships
	Authentic: Communication that does not set out to convey a certain message or have some intended outcome should generally be avoided as it can be considered as communication for communication's sake. Rather any communication should have some targeted outcome or rationale for taking place. Meanwhile engagement should be based on concepts of empowerment, trust and learning—engagement without these factors is less likely to lead to socially-optimal outcomes
	Accurate: It is easy to begin any decision-making process with perceptions of the challenge and any likely solutions. It is no different for noise. What can be perceived by an airport to be an issue that needs to be solved by obvious operational solutions may not actually be the core issue that needs to be addressed. For instance, setting out merely to reduce complaints is not likely to be as effective as setting out to solve the 'triggers' to those complaints. Management interventions that seek to address challenges without going to these deeper levels can result in money and time being wasted, or worse—damaging a situation yet further. It can be important to spend time listening and speaking to stakeholders to try to better understand a given noise problem
	Amenable: If decisions are made that are wrong from the citizens' point of view or there is new knowledge, then there are possibilities to amend these decisions

(continued)

(continued)

The 'IDEAL' characteristics of communication and engagement	
L	Legitimacy: We all have our own internal maps about what the world looks like, and to each of us those maps are reality. It is important to respect those views. Treating stakeholders and their views with respect and dignity is important in building trust and building effective relationships with residents and campaign groups

Vienna Airport: Now an Example of Good Communication and Engagement

Following from the opposition campaigns that resulted from the third runway announcement, in 2001 the airport embarked on a formal mediation process with all their stakeholders in an attempt to heal the wounds from the conflict surrounding the third runway and to build a better relationship across all stakeholders. The mediation group counted as many as 50 parties including air traffic control, airlines, mayors from communities, and communities themselves.

The mediation process concluded in 2005 with two key outcomes. First, a mediation contract was established which agreed to put in place a number of noise regulations and limits to protect local communities, whilst acknowledging the importance of the airport to the local economy. It also saw a creation of an environmental fund for breaches of noise limits to be channelled back into community projects. These achievements were made possible because of an acknowledgement on the side of industry that they had a responsibility to protect communities from noise, whilst community groups also acknowledged the importance of airport growth to the local economy. This created a shared vision on which all parties could build.

Second, the group founded a Dialogue Forum with the purpose of handling issues and conflicts related to flight operations and to develop solutions to any conflicts of interest that may arise. The Forum comprises members of all stakeholders, including from all communities around the airport. All communities are included on any issues that are discussed, even if they are not directly impacted, with the aim of promoting fairness, whilst helping to ensure that unpredicted impacts could be accounted for. The group meets regularly and are independently chaired away from the airport to help to ensure accountability and levelled hierarchies of control. Meanwhile a member of AustroControl (the Austrian air navigation services provider) also sits on the forum and provides data for residents as requested, also explaining results to them, and thus helping to build trust and confidence in the data provided. To date, the airport has not implemented any major changes without prior approval from the Dialogue Forum, and the process has proved robust enough to mean that there have been no rejections of any management proposals made to date.

Evaluation

Evaluation is a vital instrument to assess, validate and rate the success of an communication, engagement or other noise management measure during the whole process of forming, implementing, and postprocessing an intervention, helping to keep track of each step of the process of implementation and to reflect on the process and derive implications for beneficial adjustments. Further, evaluation helps to assess intended and unintended outcomes and the impact the intervention has on the target group and its effectiveness in terms of cost–benefit analysis.

Fundamental aspects of an evaluation include defining the aim of an intervention (i.e. what do we want to achieve with the intervention), definition of a target group (i.e. who is to be addressed and/or involved), definition of success criteria (e.g. when is an intervention considered as successful—with decrease of complaints, with a measurable increase in Quality of Life or a measurable increase in perceived fairness?), defining the way of proceeding to achieve the goals as well as how to come to an agreement on the procedure of engagement. These aspects of evaluation should be defined in advance.

During the implementation process it is favourable to monitor the implementation according to definition criteria, e.g. is this the right target group? Are people responding as anticipated to the intervention? What preliminary outcomes are observed, both intended and unintended? Is there a need for fine-tuning?

The impact assessment is conducted after the intervention has been implemented. Corresponding to the success criteria it is to be assessed what has been achieved regarding the defined outcome. Was the engagement process carried out as outlined at the beginning of the process? The fairness questionnaire developed within ANIMA project can be a useful instrument to assess/evaluate the process of the implementation of an intervention and the intervention itself.

Results from the evaluation process can be used to tailor future interventions to the characteristics of an airport region and/or to adapt and thus improve already implemented interventions.

Evaluating Fairness in the Context of Aircraft Noise Management—Introduction of an Psychometric Instrument

Since fairness is considered to be a highly important part of effective communication and engagement a psychometric questionnaire has been developed in the framework of ANIMA.

Based on the latest research in the field of justice psychology and in accordance with findings that have emerged in the exchange with affected citizens, a psychometric questionnaire was developed by conducting a study in the proximity of various German airports. This questionnaire is able to empirically capture the quality and success of airport management strategies via focusing on the perceived procedural, informational and interpersonal aspects of the residents' perception of the airport management's actions. In summary, the questionnaire is able to capture the perception of a fair and neighbourly relationship.

Since different aspects of fairness are captured in a differentiated manner, concrete statements can be made on whether interventions that are intended to address certain aspects of neighbourliness (e.g. more involvement of citizens, comprehensible information provision) are also perceived as such.

With the survey of aspects of neighbourliness, not only can the effectiveness of interventions be assessed, but it can also be determined at which points an intervention is necessary at all. Thus, the questionnaire offers an empirical instrument that can be used in a versatile and economical way due to its proven psychometric quality.

ANIMA in Action

The ANIMA Project has approached noise on a theoretical and practical level, by conducting research to help understand and inform on future communication and engagement practice, but also by working directly with airports to disseminate what we have learned. Below, we present one such case study, carried out in collaboration between ANIMA researchers and Rotterdam The Hague Airport.

Rotterdam the Hague

Rotterdam The Hague Airport is a regional airport near the city of Rotterdam, with a maximum capacity of just over 50 thousand flights a year [28, 29]. It has one paved runway and it features mostly holiday traffic, general aviation flights and helicopter movements. Due to the location of the airport in the vicinity of the city of Rotterdam and surrounding villages, noise annoyance is an issue. A regional consultation committee, called "Commissie Regionaal Overleg" (CRO), deals with matters related to noise annoyance by aviation. The CRO consists of representatives of the airport, local government, and community representatives.

A project group consisting of representatives from the airport, the community, the local ANSP, experts and the NLR was formed to investigate the benefits of optimising the runway 06 take-off procedure. The aim is to reduce the overall aircraft noise annoyance. According to calculations of noise contours including L_{den} and L_{Amax} noise levels, a reduction in noise exposure for some areas was expected by adjusting the initial turn of the departure procedure. Overall the noise exposure would be shifted. This means that some regions initially exposed to higher noise levels would benefit with the alternative departure procedure. However, other regions which are initially exposed to lower noise levels would receive higher noise levels.

There is no secondary goal for this operational change, such as capacity increase or reduction of fuel. Therefore, the opinion of the community is leading in the final decision on accepting the alternative departure procedure or not. Due to the complex situation, NLR was asked to investigate the possible impact of this operational change for the whole community.

Within this study several challenges were discovered. It was difficult to explain the calculated noise levels to the community. Therefore, a simulation was set up to compare the aircraft noise between the original and the alternative procedure for five different locations around the airport.

For each location, the current and the alternative flyover sounds were played, and subsequently, they were intermittent played (with 4 s interval) for direct comparison. The locations were discussed within the initial project group. Five representative locations along the flight path that were the most and the least impacted were chosen for the simulations. The simulation was first tested by people from the CRO, and after that, evaluated by 15 people recruited by the community representatives. The key provision in the set-up is that the locations that were used were not known beforehand, and it was also not known which procedure was the original or the alternative one.

Results from the simulation will be gathered and presented to the community, together with the disclosure of the locations belonging to the evaluated fly-overs. Results should help to evaluate which changes in noise level are audible, and put them in relation to the noise report on this measure. It may also provide directions for future optimisations to address whether they may benefit the community or not. Key learning from this study are related to the community engagement strategy, the structure of the process, the way the information was presented, the feedback from the community and the evaluation on how perceptual data can be used next to traditionally applied calculations to form a fair decision-making process.

Closing Remark

This chapter has outlined the importance of communication and engagement and set out some core principles that can aid airports in improving speaking to, and listening to, their communities.

It is important that airports engage with citizens more effectively, throughout the entire process of developing and delivering an intervention in order to increase the potential for noise management interventions being successful. It is vital that stakeholders are communicated to and engaged with throughout the process of intervention development if we aspire to develop outcomes that are more likely to be deemed acceptable by all stakeholders.

This can be daunting for airport management who may come from more technical backgrounds and may lack the experience or expertise in qualitative forms of data capture, dissemination and decision making. However, such difficulties will not be resolved by inaction, but by being embraced and embedded in approaches to noise impact management.

References

1. Asensio C, Gasco L, de Arcas G (2017) A review of non-acoustic measures to handle community response to noise around Airports. Curr Pollut Rep 3(3):230–244. https://doi.org/10.1007/s40 726-017-0060-x
2. Crichton F, Chapman S, Cundy T, Petrie KJ (2014) The link between health complaints and wind turbines: support for the nocebo expectations hypothesis. Frontiers in Public Health 220(2):1–8. https://doi.org/10.3389/fpubh.2014.00220
3. Bröer C (2006) Policy annoyance how policy discourses shape the experience of aircraft sound (Dissertation). Publisher Aksant, Amsterdam
4. Bröer C (2008) Private trouble, policy issue: people's noise annoyance and policy discourse. Critical Policy Analysis 2(2):93–117
5. Benedetti F, Lanotte M, Lopiano L, Colloca L (2007) When words are painful: unraveling the mechanisms of the nocebo effect. Neuroscience 147:260–271. https://doi.org/10.1016/j.neuros cience.2007.02.020
6. Crichton F, Dodd G, Schmid G, Gamble G, Cundy T, Petrie KJ (2014) The power of positive and negative expectations to influence reported symptoms and mood during exposure to wind farm sound. Health Psychol 33(12):1588–1592. https://doi.org/10.1037/hea0000037
7. Crichton F, Dodd G, Schmid G, Petrie KJ (2015) Framing sound: using expectations to reduce environmental noise annoyance. Environ Res 142:609–614. https://doi.org/10.1016/j.envres. 2015.08.016
8. Cialdini RB, Kallgren CA, Reno RR (1991) A focus theory of normative conduct: a theoretical refinement and reevaluation of the role of norms in human behaviour. In: Zanna MP (ed) Advances in experimental social psychology, vol 24. Academic Press, San Diego, CA, pp 201–234
9. Guski R, Schreckenberg D, Schuemer R (2017) WHO environmental noise guidelines for the European Region: a systematic review on environmental noise and annoyance. Int J Environ Res Public Health 14(12):1539. https://doi.org/10.3390/ijerph14121539
10. Stallen PJM (1999) A theoretical framework for environmental noise annoyance. Noise Health 1(3):69–79
11. Leventhal GS (1980) What should be done with equity theory? In: Social exchange. Springer, pp 27–55
12. Lind EA, Tyler T (1988) The social psychology of procedural justice. Springer Science & Business Media
13. Thibaut J, Walker L (1975) Procedural justice: a psychological analysis. L. Erlbaum Associates
14. Tyler T, Lind EA (1992) A relational model of authority in groups. In: Advances in experimental social psychology, Vol 25. Elsevier, pp 115–191
15. Maris E, Stallen PJM, Steensma H, Vermunt R (2006) (Un) sound management. Three laboratory experiments on the effects of social non-acoustical determinants of noise annoyance. Paper presented at the INTER-NOISE and NOISE-CON Congress and Conference Proceedings
16. Liebe U, Preisendörfer P, Enzler HB (2020) The social acceptance of airport expansion scenarios: a factorial survey experiment. Transp Res D Transp Environ Planning A 84:102363
17. Bies RJ, Moag JS (1986) Interactional communication criteria of fairness. Res Org Behav 9:289–319
18. Bies RJ, Shapiro DL (1987) Interactional fairness judgments: the influence of causal accounts. Soc Justice Res 1(2):199–218
19. Bies RJ, Shapiro DL (1988) Voice and justification: their influence on procedural fairness judgments. Acad Manag J 31(3):676–685
20. Adams JS (1965) Inequity in social exchange. In: Advances in experimental social psychology Vol 2. Elsevier, pp 267–299
21. Leventhal GS (1976) The distribution of rewards and resources in groups and organisations. Adv Exp Soc Psychol 9:91–131
22. Bartels S (2014) Aircraft noise-induced annoyance in the vicinity of Cologne/Bonn Airport-The examination of short-term and long-term annoyance as well as their major determinants

23. Hooper P, Flindell I (2013) Exchanging aircraft noise information with local communities around airports: the devil is in the detail. In: 42nd international congress and exposition on noise control engineering 2013, INTER-NOISE 2013: Noise Control for Quality of Life

24. Directive 2002/49/EC of the European Parliament and of the Council of 25 June 2002 relating to the assessment and management of environmental noise. Official Journal of the European Communities 18.7.2002; L189/12

25. CAA (2021) Airspace Change. Guidance on the regulatory process for changing the notified airspace design and planned and permanent redistribution of air traffic, and on providing airspace information. CAP 1616. Civil Aviation Authority. March 2021. 4th edn. Available on-line: Guidance on the regulatory process for changing the notified airspace design and planned and permanent redistribution of air traffic, and on providing airspace information. CAP 1616. Available at: https://publicapps.caa.co.uk/docs/33/CAA_Airspace%20Change%20Doc_Mar2021.pdf (Accessed: 30 March 2021).

26. FAA (2015) Fact Sheet—The FAA Airport Noise Program. Federal Aviation Authority. Available at: https//https://www.faa.gov/news/fact_sheets/news_story.cfm?newsId=18114. Accessed 30 March 2021

27. Arnstein SR (1969) A Ladder of Citizen participation. J Am Inst Plann 35(4):216–224. https://doi.org/10.1080/01944366908977225

28. EHR –ROTTERDAM/Rotterdam (2021) AIP from AIS the Netherlands, effective 25 February 2021. On-line. https://www.lvnl.nl/eaip/2021-01-14-AIRAC/html/eAIP/EH-AD-2.EHRD-en-GB.html. Accessed 30 March 2021

29. Rotterdam The Hague Airport (2021) Annual Statistics 2010–2019. https://www.rotterdamthehagueairport.nl/content/uploads/2020/01/Totaal-per-jaar-januari-2020.pdf. Accessed 17 Mar 2021

Towards Innovative Ways to Assess Annoyance

Catherine Lavandier⬤, Roalt Aalmoes⬤, Romain Dedieu⬤,
Ferenc Marki⬤, Stephan Großarth, Dirk Schreckenberg⬤,
Asma Gharbi⬤, and Dimitris Kotzinos⬤

Abstract Technological changes have driven the developments in the field of noise annoyance research. It helped to increase knowledge on the topic substantially. It also provides opportunities to conduct novel research. The introduction of the internet, the mobile phone, and miniaturisation and improved sensor technology are at the core of the three research examples presented in this chapter. The first example is the use of a Virtual Reality simulation to evaluate aircraft flyovers in different environments, and it examines how visual perception influences noise annoyance. The second example describes the use of a mobile application applying an Experience Sampling Method to assess noise annoyance for a group of people living near an airport. The third and final example is a study over social media discussions in relation to noise annoyance and quality of life around airports. These three examples demonstrate how novel

C. Lavandier (✉) · R. Dedieu · A. Gharbi · D. Kotzinos
CY Cergy Paris University, Cergy-Pontoise, France
e-mail: catherine.lavandier@cyu.fr

R. Dedieu
e-mail: romain.dedieu1@cyu.fr

A. Gharbi
e-mail: asma.gharbi@ensea.fr

D. Kotzinos
e-mail: dimitrios.kotzinos@cyu.fr

R. Aalmoes
Royal Netherlands Aerospace Centre NLR, Amsterdam, The Netherlands
e-mail: roalt.aalmoes@nlr.nl

F. Marki
Budapest University of Technology and Economics, Budapest, Hungary
e-mail: marki@hit.bme.hu

S. Großarth · D. Schreckenberg
ZEUS GmbH, Hagen, Germany
e-mail: grossarth@zeusgmbh.de

D. Schreckenberg
e-mail: schreckenberg@zeusgmbh.de

technologies help to collect and analyse data from people who live around airports, and so improve our understanding of the effect of noise on humans.

Keywords Virtual reality · Social media · Experience sampling method · Mobile application · Soundscape · Sound perception · Audio-visual interaction · Sound environment · Quality of sound · Quality of life

Introduction

Noise annoyance is a known health-related societal problem for a long time, and cannot be solved in the near future. But the conditions and environment in which it occurs do change. Significant changes have taken place in the twentieth century that affect how noise annoyance occurs and how it is perceived, and this trend continued in the twenty-first century. If we look at the forces that come into play, we can discriminate four interrelated factors (Fig. 11.1).

First, the noise sources change: if we look at the transportation domain, new vehicles appear (such as drones), or existing vehicles go through a disruptive cycle, such as the movement from petrol-based towards electric-powered automobiles. Research on noise mitigation measures has also reduced the impact by noise at the source. For instance, by increasing the by-pass ratio of jet engines, a significant noise reduction is achieved, making aircraft much more quiet than earlier generations (see Chap. 5 and Fig. 11.2).

Second, human perception and attitude, and consequently how people react to noise, have changed. Noise nuisance that was previously accepted and considered part of the environment is, with a more vocal community, noticed much more and may lead to complaints that are more significant. In one way, the improved democratic instruments or government protection measures enhance the ability to complain about noise issues, but new and more efficient ways of communication cannot be ignored

Fig. 11.1 Noise sources that cause annoyance have changed during the course of time. Human perception and the attitude towards noise annoyance have changed as well. Technological developments and expanded knowledge have both driven these changes on both noise sources and human perception and attitude

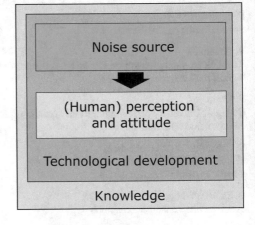

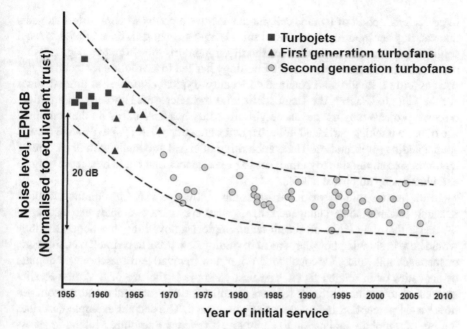

Fig. 11.2 Trends of noise reduction through the twentieth century. *Source* Le livre blanc de l'acoustique en France en 2010 (edited by SFA, the French Acoustical Society)

as well. With the internet, people are more organised to complain via email, websites, and social media, and authorities are also more organised to communicate with them and enhance people's engagement.

Indeed, aircraft noise annoyance per given intensity of average sound levels have increased over the last decades although single airplanes got quieter at the same time [14]. As described in Chapt. 9, there is evidence that annoyance mediates the impact of noise exposure on further long-term health risks, indicating that the increase in annoyance over time would in the long run also affect long-term health effects of aircraft noise.

The third reason is the gain in knowledge on noise annoyance on humans. The vast amount of research undertaken improved both psycho-acoustic knowledge and the impact on health. It created standardised exposure–response curves and standardised noise annoyance research questionnaires, such as the ICBEN scales [10]. And, the improvements in noise effect research led to human health reports on noise impact such as that by the WHO [46].

The fourth reason for the change of environment is the change of technology: technological advances reduced noise at the source, made people more aware of noise nuisance, and helped increase knowledge on the topic of noise annoyance. But it will also provide researchers with new means to conduct noise research. The introduction of the internet with email, websites, and, subsequently, social media creates a larger audience for conducting large-scale evaluations. The introduction and large-scale adoption of (smart) cell phones creates a potential pool of test subjects that can register their location, record sounds, and answer questions related to local soundscapes and noise annoyance. Finally, miniaturisation, improved processing power, and development of sensor technology has led to revolutionary technologies such as Virtual Reality and Augmented Reality. Typical examples of these devices are the Google Glasses, the Oculus Rift, and the Microsoft Hololens. In addition, research projects may use advanced virtual-reality headsets based on the combined use of head-tracking and Head Relay Transfer Functions (HRTFs) to provide natural spatial audio representation. The combination of visual and audio stimuli can create an immersive simulation to mimic real-life experiences that could otherwise only be examined using empirical studies.

In the next sections, we focus on some innovative ways to conduct acoustic research towards noise annoyance. In no way are these examples exhaustive, in the sense that they cover all recent technological innovations; but hopefully, they would help in showing how they can help finding new ways to characterise, mitigate, or manage annoyance; Or conducting other new innovative research in the domain of acoustics or in related fields. The first example is the use of a Virtual Reality simulation to evaluate aircraft flyovers in different environments, and it examines how visual perception influences noise annoyance. The second example describes the use of a mobile application to apply an Experience Sampling Method to assess noise annoyance for a group of people living near an airport. The third and final example is a study over social media discussions in relation to noise annoyance and quality of life around airports. The last two examples could also be combined with the dynamic population maps that are described in the following chapter, to correlate people's location and their annoyance, a novel approach not seen previously in aircraft annoyance research.

Immersive Simulation to Mimic Real-Life Experiences

Communication and engagement by airport authorities, local government, or local planners with communities is important and is discussed in other sections of this book. With respect to communication on noise impact, a more difficult task is at hand to explain predicted noise levels and what they mean for the affected communities. There are different ways to present changes to the noise on paper, and those used often are the 24h annual noise level, single-event peak-levels, or number of (highly) annoyed people near the airport. But to make these numbers better comprehensible for the layman, a demonstration that simulates aircraft flyovers at the predicted sound level would clarify what these numbers mean. Novel approaches that use aural and visual stimuli can be used for this purpose. There are some virtual reality applications (auralisation and visualisation) which can be used by residents to give them better understanding of the impact of future airport scenarios in land-use planning, as the virtual reality creates a higher immersion for the user. Virtual Reality headsets feature a greater field of view than projector or TV screen. Additionally, using head-tracking sensors, they allow the system to change the visuals in a way that allows the user to look in all directions. The same is true for the (spatial) sounds that are produced and provide audio-directivity. But this immersion has to be ecologically valid if authorities want to be trusted by residents during communication campaigns.

Validation of a Virtual Reality Application for Aircraft Noise

In this section, a validation study [8] is presented to evaluate the Virtual Community Noise Simulator (VCNS) for the perception of aircraft noise (Fig. 11.3). The VCNS has been developed by the Netherlands Aerospace Centre (NLR) based upon earlier work done at NASA Langley [38]. The current set-up makes use of the shelf hardware, such as an Oculus Rift CV1 VR headset, supported by a powerful laptop computer, and separate headphones for the audio.

Participants of the study take part in a perceptual experiment. Two landscapes are presented in which the participants evaluate the flyover sounds and visuals of three distinct aircraft. These three aircraft are the Airbus A320neo, the Airbus A380, and a revolutionary design called the "BOLT" (Blended wing body with Optimised Low-noise Technologies, Fig. 11.4, see also Chap. 6). The influence of the visuals is also measured by presenting the sound of one aircraft with the visuals of the other aircraft as well. In one additional condition, the visual is not visible and this is represented by an overcast sky. To prevent influence from different background noises, a single background recording was used in both landscapes.

In order to test if the size of the aircraft (or the absence of the aircraft vision because of clouds) has an influence on the perception of the audio-visual situation, all synthesised sounds were crossed with all visual situations. So, twenty-four environmental situations were created (three types of aircraft sound x four types of visual

Fig. 11.3 Artist impression of the virtual community noise simulator by NLR

Fig. 11.4 The "BOLT" architecture

aircraft source x two types of landscape). The four types of visual aircraft source correspond to the three aircraft plus one situation with clouds in the sky. The two landscapes correspond to one green park, and one urban situation.

Sixty participants were immersed in these twenty-four situations. After each flyover, they had to use the joystick on the touch controller to give answers to a questionnaire which appeared in the virtual world (Fig. 11.5). It consisted of four ratings:

(1)   Overall, does this situation seem more or less

Unpleasant/Unbearable ….. Pleasant/ Bearable?

(2)   Does the association of sound with visual seem more or less

Unrealistic/Non credible/Incoherent ….. Realistic/Credible/Coherent?

(3)   Is the sound of this aircraft more or less

Unpleasant/Unbearable ….. Pleasant/Bearable?

Fig. 11.5 Question about the audio-visual situation in the virtual world

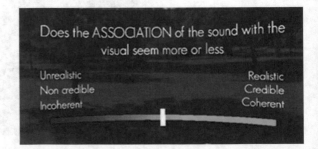

(4)   Does the noise level of this aircraft seem more or less

Strong/Loud ….. Weak/Quiet?

After assessing the twenty-four situations, participants had to fill a final questionnaire concerning the simulator's overall efficiency in creating a feeling of reality, and concerning personal information such as their noise sensitivity, or their quality of life. As the experiment was conducted by the Cergy Paris University (CYU, France), it was previously submitted to this ethical committee which approved it.

For the overall pleasantness measure, three groups of participants rated the audio-visual situations differently. One group rated all the situations in the "negative" part of the scale, which means that they found the environment less pleasant than the other groups. Generally, participants of this group are more noise sensitive than the other participants and a little bit older. Another group of less noise sensitive people rated the overall pleasantness in the middle of the scale. The last group rated the overall pleasantness in the "positive" part of the scale. In this group, participants are a little bit younger, whatever their noise sensitivity.

If we have a look on the influence of the landscape, it seems that the majority of participants were not influenced by the landscape, but some of them preferred the flyover in the park, because the situation is greener [45] and some of them disliked the flyover in the park because the flyover disturb the quietness of this environmental situation [4]. The size of the aircraft has no influence on the sound perception nor on the overall pleasantness.

For the sound pleasantness, the aircraft sound of the A380 is the most unpleasant because it is the loudest one ($L_{Amax,30s} = 76.1$ dB(A)). This sound also has the lowest pitch. Then the A320neo and the new aircraft are perceived as more pleasant as they are both less noisy ($L_{Amax,30s} = 72.1$ dB(A) and $L_{Amax,30s} = 71.3$ dB(A) respectively). People react globally in agreement with results of the literature about sound perception [12, 26, 30]. The study about realism can explain why the results are so close to scientific literature about aircraft noise.

In total, 78% of the participants found the virtual audio-visual environment very or extremely realistic (Fig. 11.6). All sound syntheses are of similar real-

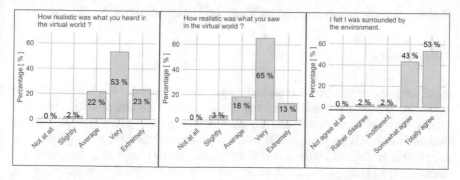

Fig. 11.6 Distribution of the answers given by participants at the end of the experiment about realism and immersion

istic/credible/coherent quality. Nevertheless, participants noticed that the simulation of the A320neo is the most credible from a visual point of view. The odd architecture of the "BOLT" aircraft renders its visual simulation less credible, the A380 is also less credible than the A320neo because people are less familiar with this large aircraft. It has also been found that the realism of the visuals are not only deteriorated by the unknown nature of the aircraft but also by clouds. Moreover, presence of clouds leads to an overall more unpleasant situation, and the sound of the new aircraft is judged louder under clouds compared to the same sound coming from any aircraft that is seen in the sky.

To conclude, the quality of the application has been validated by our perceptual experiment. 96% of the participants felt surrounded by the environment and the results, which are in line with literature about aircraft noise, show that this virtual reality tool can be used for communication with residents around airports in a fair approach. If we want to improve the quality of the application, the efforts should focus on the visualisation of the artificial clouds.

Effectiveness of This Virtual Reality Application for Better Communication

In order to test the effectiveness of this simulation tool, an in-situ experiment will be organised in a city around an airport where an operational change of the aircraft route is planned. If people feel that they understand this change better with the virtual reality tool than with classical maps (noise maps and aircraft trajectories), in theory they should feel in more control and thus should be more confident with the airport authorities. The hypothesis behind this in-situ experiment is that the noise annoyance could be reduced with the use of such a tool, reducing fear about what will happen in the future.

Mobile Application to Assess People's "Annoyance"

Introduction

For about two decades, airport residents' long-term annoyance has been studied by asking the ICBEN question, "Thinking about the last 12 months or so, how much did aircraft noise as a whole bother, disturb or annoy you?" [10]. Typically, and in line with the ICBEN recommendations, annoyance ratings have been given on 5-point verbal scales and 11-point numerical scales. By dichotomising the answers in values of high (1) and not high (0) annoyance, the percentage of respondents highly annoyed related to computed average noise levels provide exposure–response curves

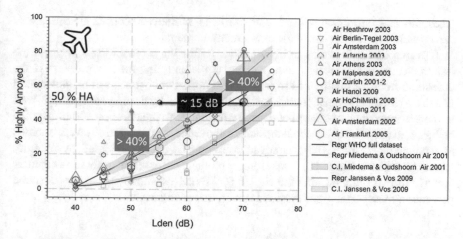

Fig. 11.7 Variance in exposure–response curves for the percentage of persons highly annoyed by aircraft noise [15]. *Source* Guski et al. [15], adapted and modified

that inform noise policy. Unfortunately, various study exposure–response curves deviate significantly (Fig. 11.7).

During these field studies, researchers also asked participants other questions to find reasons for their specific level of annoyance. However, here again, significant differences occurred. While there have been found many important driving factors to get annoyed, after decades of aircraft noise annoyance research, we are still not in the position to be able to set mathematical models, which would estimate with acceptable accuracy the annoyance due to its contextual embedment. It turned out that non-acoustic factors play an important role in the judgements [13], but the weight of one or the other factor is hugely different from place to place, and situation to situation, because airport residents and their living conditions are different. Those who live near a huge hub, like Heathrow or Frankfurt, behave differently from those who live near medium or small sized airports. Whether the airport has night traffic or not is also important. Over the years it has also been found that people in a changing situation (e.g. a new runway or a new flight route) behave completely differently as opposed to those living around a steady state airport. All this leads to the conclusion that while there have been quite a lot of field studies conducted already, because of the diversity of situations, we still don't have enough information to predict annoyance.

However, these classical field studies are quite costly and furthermore it has increasingly become difficult to get participants to take part. So, we need new ways to make test procedure and annoyance estimation easier, thus less costly and to get "access" to a lot of people, to find the necessary number of volunteers to take part. And therefore mobile phones come into play [24, 34, 37]: through them, people are very easily be accessed with regard to their short-term annoyance, the test procedure can be much more flexible (e.g. asking participants not just once but at several random times is easily done by an appropriate mobile application) and data exchange between researcher and participant is also fast and reliable. Moreover, acoustic measurements can be stored on the device, at the assessment moments.

It seems that it is worth trying to modify also the methodology, because the importance of non-acoustic factors means, at the end, that the key problem is the deterioration of the quality of life of people. Further, annoyance can be understood as a stress response to noise (see Chaps. 9 and 10) driven by the noise exposure as well as the perceived lack of control or capacity to cope with the noise. In addition, when being asked retrospectively about the noise annoyance over a period of several months as recommended by ICBEN, annoyance judgments could be biased by memory capacity (see also Chap. 10). Therefore, a momentary assessment, which captures feelings in real time, became the choice for ANIMA's pilot study to assess sound perception in airport regions closer in time to the sound events.

The ANIMA Mobile Application (AnimApp)

Experience Sampling Method

The method proposed—in order to assess acute perception of sound events—is called Experience Sampling Method (ESM; sometimes also called Ecological Momentary Assessment). In contrast to the assessment of retrospective long-term judgments as it is done in most of the 'classical' socio-acoustic surveys, this method allows to assess the experiences in-situ repeatedly on different (consecutive) days at different times of day. Hence, the ESM approach can be characterised as "capturing life as it's lived" [3]. By installing a survey-software on participants' devices, researchers are able to prompt for several assessments, whenever it appears to be necessary. Data is then submitted to a server and is available as soon as the upload is finished. Although ESMs have been found to be useful in many scientific disciplines they have just shortly found their way into modern noise impact assessments [7, 11, 28]. Here, we have found a promising way to get a realistic insight into people's everyday noise experience, which we regard as essential when examining sound perception and its impact on quality of life in individuals.

The ANIMA project features among others the development of a new mobile application, called ANIMA Research, nicknamed simply AnimApp (see Fig. 11.8). Using the application, we carry out an ESM study about the impact of the sound-scape and landscape of the surrounding environment on people's perception of the environment and their quality of life around airports.

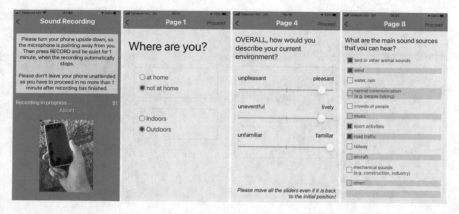

Fig. 11.8 Example screenshots of the Anima Research application 'AnimApp'

Global Structure of the Application

Explanations and Permissions

After installing the app, during the first start, the test procedure is explained, permissions are asked for location and microphone usage, and notification sending. In addition, settings have to be reviewed and adjusted at will, so the frequency and time-span of the test suits the participant's lifestyle. Then the user exits the app. The application is designed as self-operating. Once installed, notifications will prompt for assessments at random intervals. The selection of momentary, weekly or the final questionnaire as well as the replacement of missed notifications are all automatically done without further intervention by the participants or the research team.

Momentary Assessments

During weekdays, from 7 A.M. until 11 P.M. (when not shortened by the participant), once around each full hour the app sends a notification (Fig. 11.9):

Depending on the user's preference 2 to 4 assessment notifications come a day. Hours of measurement (i.e. 7 A.M, 8 A.M, etc.) are randomised for the whole duration of the test, so the user doesn't know when the next measurement request will be prompted for. Each hour of the day is tested within the adjusted interval. The total duration of the study adapts correspondingly.

The user has to respond to a notification in 20 min and to start the momentary assessment consisting of sound recording and questionnaire filling.

During weekend days, the participants respond to the same questionnaire, however in a shorter time frame, i.e. from 10 A.M. till 10 P.M. and just every second hour only.

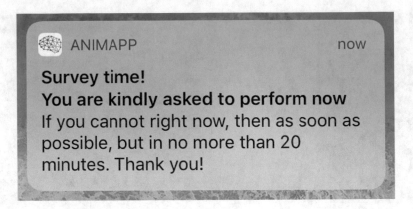

Fig. 11.9 Notification calling to perform an assessment

End-Of-Week Assessments

At the end of the work week, i.e. on Friday evening, a short end-of-week questionnaire has to be filled in. In addition, after the very last weekend's momentary assessment, the same end-of-week questionnaire is asked relating to all *weekend* days during the test.

Final Questionnaire

Once all week and week-end hours have been performed, a final questionnaire is presented to the user asking for noise sensitivity, and standardised questions on well-being [17, 33].

Acoustic Measurements and Selected Questions

The ANIMA project tries to depart from the classical approach by moving from the focus on average noise pollution and annoyance towards a broader view and the general notion of the perception of acute sound quality and its impact on quality of life. The broadened, more open content of judgments is combined with its assessment in acute, specific moments making the response more independent from memory bias and from biases that may come along with the noise-attributing wording (so-called demand characteristics) of the standardised long-term annoyance questions.

Regarding the acoustic metric describing the sound environment, it is worth trying to find better acoustic metrics, which are closer related to people's sound perception than the day-evening-night sound level L_{den}, which summarises and weights noise events over a 24 h period of the day. In order to be able to calculate most indicators that are proposed in the literature, the spectrum of the recorded sound (third octave band, each second) is stored in the AnimApp.

At each moment when a notification is received by a participant, a 1-min acoustic measurement has to be performed and then a series of questions appears. The questions that have been selected are inspired from the so-called *soundscape* questionnaires [2, 5], which capture all relevant dimensions that can explain the impact of sound environment on people and which have been recently standardised [18, 19]. The first dimension is the pleasantness of the sound, followed by the eventfulness, and the familiarity with the environment. The acoustic environment should also be described with the types of sound sources, which are present in the environment. The context is not limited to the location (which is captured by the smartphone), but should also concern the activity of the participant at the moment of the evaluation. The context also has to include visual data. Actually, it has been shown that the quality of the landscape has an influence on the perceived pleasantness of a soundscape when people are outside: the greener the landscape, the higher quality the soundscape [31, 32, 45]. When individuals are inside, the natural elements people could see through their windows reduced the negative effect due to noise [43]. We examine this further by asking participants to answer what they see through their windows, if they have a view of the outside. In the frame of our approach, we also want to question the rating of long-term annoyance by means of single items: people feel disturbed at different moments of the day, or evening, or even night, but participants could have difficulties to produce a valid annoyance rating over a longer period time (e.g. 12 months) [9, 40, 41, 44]. To examine how people add up all the different experiences deriving from their perception - at least for a one week period, we decided to ask 3 questions on the environment (overall impression on the sound pleasantness, landscape pleasantness, and representativeness of the week) at the end of each week.

Of course, the unexplained variance of noise annoyance could partly derive from personal dispositions. Accordingly, we assess the mood at each notification. Furthermore, the individual noise sensitivity and the perceived quality of life is assessed in the final questionnaire.

Development of the Mobile Application

AnimApp was developed for the two operating systems Android and iOS. This allows a widespread use of the application on the vast majority of modern smartphones in use [39] keeping participation requirements on low threshold. For the use of AnimApp on these two operating systems, several operation system specific adaptations and adjustments were applied.

In regular field studies, the procedure of the study is explained in detail to the participants and once they agree to participate, they tend to comply well, which can be enforced by offering an expense allowance. But with mobile applications, long text-based explanations, which might demotivate participants to continue participating have to be avoided. Specific effort has to be deployed in formulating instructions precisely and shortly at the same time.

Another difficulty is how to bring our test through, when the user often disregards the notifications to do an assessment. After a pre-test phase, we decided to allow 20 min delay between notifications and possible answers, and to send reminder notifications each 5 min to the user.

The regularity of data provision also needs attention: on the one hand, apps should avoid running all the time in the background (and thus draining battery), but on the other hand they must make sure to send assessment data to us, once a measurement is completed. In our case, when connectivity is not available after fulfilling a measurement, there is no other option than to schedule data sending for later time. However, we ensured the battery will be drained as little as possible and even if the application is killed by the user the schedule for the upload of data still persists.

Finally, tracking of participants' location imposes further potential problems. In our study we want to know the position of the user at the time of the assessment, so that we can estimate the aircraft noise exposure for the respective positions afterwards. Additionally, we ask our participants to allow tracking of their position all the time, so we have an impression how airport residents move during the day (i.e. to know—based on noise maps—how much they are exposed to aircraft versus other noise). This option needs consent of the user (see below; paragraph on data security and privacy).

Randomisation

For all field studies, from the point of view of later statistical analysis, the randomness of sample collection is very important. Therefore, to assure good randomisation among assessment hours and among participants, for AnimApp it has been decided (a) to let the participants perform an assessment at randomly selected hours (but along the test, each hour will be assessed just once), each day 2–4 times depending what he/she set up in the settings, (b) to define a 10 min time-frame around full hours and then randomly select the exact time in the resulting time span (e.g. between 7:55 and 8:05).

Data Security, Privacy

AnimApp made several steps to respect people's privacy and to be fully compliant to GDPR:

- It is not necessary for the users to enter any personal data to register in the study, they simply get automatically the next free user ID, thus users remain anonymous, and their answers too.
- The sound recording is right on the phone transferred into a series of 3rd octave band spectra, one for each second, and only this is transmitted to the server. This keeps privacy as the original audio recording cannot be reconstructed from these acoustic data.

- The user has explicitly to agree to constant location tracking, and can also refuse if preferred. Also, positions are rounded to a grid of 100 * 100 m on the user's phone and only then sent to the server.
- For any cases, during first use, a user must explicitly agree to our privacy policy, including the agreement that we collect/store/process data from the participant with his/her consent.

Experiences Around Two Airports

A first version of the application has been tested during winter 2018, and a feedback questionnaire has then been proposed to the "beta" testers. Based on these feedbacks, the test procedure has been refined and the final version (which has been described in this Chapter) will be used for the actual study. Two different sized airports will be observed. The application will be experimented during spring/summer 2021 around Ljubljana Airport in Slovenia, and London Heathrow Airport in the UK. Of course, the application has been translated into Slovenian for being used in Slovenia. Generally, instructions and indications are technically provided through separated libraries that are easing the adaptation to a wide range of other languages. Results should show whether such an approach could be used for more airports, and more suitable periods (more traffic for tourism, outside of a sanitary crisis like COVID-19).

Using Twitter as a Survey Tool: Understanding people's Opinions of Quality of Life Around Airports

Context

Social media has increasingly become a space where people meet to discuss, express opinions and debate over a wide range of subjects ranging from global politics to everyday life and from political and ideological opinions to advertisement of products and services. A specific part of the discussions about everyday life is the focus of this work, done as part of the ANIMA project but having wider applicability: this part concerns the understanding and subsequent classification of people's annoyance when they live, work or socialise around airports. In order to do this, we need to analyse discussions over an extended period of time which are somewhat localised since we need the involved people to either live or work around airports or to show a significant presence that would allow them to be considered as directly affected by the generated impact from the airport operation.

Actually, surveys assessing the impact of the operation of an airport on the population living in the adjacent areas have been carried out for a long time and have received particular focus (and a lot of scrutiny) in cases of airports' expansions.

These surveys have taken place in traditional ways, mainly through the selection of a representative part of the population which is contacted either by person (door-to-door), by phone or through post in order to fill a predetermined questionnaire. More recently, such surveys are being contacted either online or through mobile apps, where the invited subjects download and install an app on their mobile phones and then use the app to answer specific questions and/or allow it to monitor (parts of) their everyday life. Main disadvantages in both cases are the difficulty to extract big enough samples and to guarantee the participation of the users during the duration of the survey, since quality of life issues cannot be assessed in one-off answering. Moreover, as in all surveys, the usage of mobile apps raises various privacy concerns, which can, of course, be mitigated by extra developer effort, as it is the case in the ANIMA mobile app.

Compared to these methods, surveys based on social media research and analysis exhibit various advantages and disadvantages. On the advantages side, we can put the infrastructure for capturing social media posts once and then monitor the discussions for an extended period of time with no extra cost (besides the cost of processing and storage of the posts). Additionally, social media platforms like Twitter or Facebook can provide easy access to thousands or millions of users and millions or billions of relevant tweets (depending on the airport, the area, etc.) so as to extend the sample that "participates" in the survey. The users are actually taking part in online discussions based on their own interest, with no strict requirements. On the disadvantage side, online social media brings its own biases, for example it is well known that people from older generations use them very little or only for purposes of communication with family and friends. One more problem is that posts do not necessarily carry location information, so sometimes localising a discussion is not possible or becomes a costly operation by itself. Finally, discussions on social media are directly affected by whatever captures the public's eye as well as from the actual reality, for example during the recent COVID-19 crisis discussions on social media are overwhelmingly dominated by this and the lack of actual flights mitigated the issues and the discussions. The richness of information in social media can also be a curse: not all discussions are relevant to the specific subject. In that respect, we need first to extract the relevant posts or discussions, which is not a trivial subject by itself. Additionally, in the case of Twitter and other microblogging services the imposed limit on the number of characters for each post forces people to express themselves in unique and sometimes difficult to understand ways. Nevertheless, the number of posts that can be captured and the number of users that participate make it a viable alternative that—with the necessary scientific precautions—can provide valuable insights on the opinions and sentiments over quality-of-life issues around airports.

As part of the ANIMA project, we develop a set of reusable tools and methodologies that allow capturing the necessary relevant social media posts, extracting the topics of discussions around quality of life and classifying the sentiments around these topics as positive, neutral or negative trying to depict a qualitative assessment of the opinions of people from the area around airports. For the purposes of the project, we focus on the area around Heathrow airport in London, UK (one of the busiest airports in a very densely populated area) but the methodology and principles described can be easily applied in any other similar case.

Scientific Background

The core of the work in this task is the extraction of opinions and the analysis of sentiments contained within those opinions that would allow us to classify those opinions as positive, neutral and negative.

Sentiment classification is a hard challenge that faces several challenges such as dealing with trivial posts, incomplete sentences, misspelling and abbreviation due to size restrictions, dealing with specific meanings such as irony or humour and the use of emotional expressions. Sentiment classification approaches can be classified into three main categories: (i) machine learning, (ii) lexicon based [22, 23, 47] and (iii) hybrid approach.

Machine learning-based sentiment analysis consists in predicting the polarity of sentiments by training a machine learning model with examples of emotions in text to automatically learn how to detect sentiment without human input. In the literature, one can find works that use emoticons [35], slang language and acronyms [16], words in text and their respective part-of-speech (POS). Other elements to consider are intensifiers such as all caps and characters' repetitions (e.g., happpyyy) [21], punctuation marks, n-grams [21] and negation marks [29] or (all possible) combinations [21] as features of the analysed tweets. In [35] and [6], authors performed a 2-way classification (i.e., positive or negative) on data with emoticons and applied respectively SVM (Support Vector Machine) and NB (Naive Bayes) algorithms that were able to achieve more than 70% accuracy.

Recent works tried neural networks with word embeddings for the representation of tweets and showed that they achieved much better performance in sentiment analysis [20, 27, 36, 42]. Word embeddings represent words by dense vectors with much lower dimensionality. Each word is positioned via its vector value into a multi-dimensional space (embedding space) which helps to consider their semantics (i.e., synonyms are geometrically close, antonyms are far from each other). Mathematical operations can also be applied on vectors and produces semantically correct results, e.g., the sum of the word embeddings of *king* and *female* produces the word embedding of *queen*. Ren et al. [36] has used a context-based convolutional neural network (CNN) to apply sentiment classification on Twitter corpus. Tang et al. [42] encodes sentiment information of texts (e.g., sentences and words) together with contexts of words in sentiment embeddings. They showed that sentiment embeddings consistently outperform context-based embeddings in tasks such as word-level sentiment analysis, sentence level sentiment classification and building sentiment lexicons.

Besides machine learning approaches for sentiment classification, in the literature we find lexicon-based approaches. While finding or constructing those lexicons is not always an easy task and the difficulty might vary depending on the language, these methods allow us to use a list of words (dictionary of subjective words) [22], where each word is associated with a specific sentiment; emoticons are used the same way as well. Yadollahi et al. [48] discuss the ability to use more than one dataset to take into account multiple subjective perspectives of the word and to modify the existing dictionary in order to satisfy the topic sentiment characteristics. Asghar et al. [1] proposed to classify sentiments in reviews by combining an emoticon classifier, a modifier and negation classifier and a classifier based on the opinion lexicon SentiWordNetwork (SWNC) in a sequential, then input the text to a domain specific classifier (DSC) that takes into account the polarity of domain specific words both existing or unknown in SWNC. Such a hybrid approach consists of the combination of both machine learning and lexicon-based approaches, which can improve the results of sentiment classification. More information on related works in the area can be found in [25].

Analysis Pipeline

For our sentiment analysis task, we propose an approach that uses data mining and machine learning for the extraction of relevant tweets, then a lexicon-based sentiment classifier to calculate their polarities and classify them to negative, positive and neutral. More specifically, we provide a processing pipeline of four distinct sequential and interdependent steps. More specifically:

Collection and Preprocessing of Tweets

Tweets are collected through the Twitter API, which provides a standard way to get (a part of) the real time stream of public tweets and filter those by keywords, location, language, users, etc. We used mainly keyword-based and location-based queries. We are extracting only English language tweets and use keywords like: "Heathrow", "LHR", "noise", "annoyance", etc. to increase the chances to get relevant messages. Also, location queries were used based on Heathrow's day, evening and night level (L_{den}) noise contours in order to bound the area of interest. (This led to a bounding box of 167 km wide and 73 km long, centred around Heathrow airport to be used as location filter to Twitter API).

Those tweets go through a pre-processing phase, where we firstly remove links, numbers, emoticons and Twitter specific words, then we make all words lowercase and apply tokenisation. On those tokenised words, we correct as many errors as possible (mainly spelling errors) and then we assign part-of-speech (POS) tags and lemmatise the words in order to work with a more compact and stronger set for understanding relevance. It should be noted here that the removed parts (e.g., emoticons) are not deleted permanently but are passed to the next processing steps.

Relevance Classification

From the previous step we end up with a bag of words (including hashtags), so here we use these words to form unigrams, bigrams and hashtags as features and use tf-idf as a metric to represent tweets. Then the SVM algorithm is trained on a manually annotated sample of tweets, so as to be able to classify tweets as relevant or not. We filter the relevant tweets through a lexicon-based classifier in order to benefit from the domain knowledge (expressed through the lexicon), which assigns a relevance score to each tweet. We keep those tweets that exceed a threshold. This double classification provides better results compared with other methods, for more details please see [25].

Sentiment Analysis of Relevant Tweets

Based on the selection of relevant tweets, we proceed to classify those tweets as positive, negative or neutral. We do this by exploring various facets of the tweets and calculate different scores that represent each facet. At the end, we put together those scores in order to compute a single score per relevant tweet that would allow the system (based on this value) to classify it. In order to do this, we use three different facets: (a) emoticons (collected from tweets and labeled as positive or negative), (b) lexicon-based polarity of words (using dictionaries, where each word has been classified as positive or negative and given a score) and (c) the SentiWordNet, a dictionary where each word has been attributed at the same time a positive, a negative and a neutral score with the restriction that these scores add up to 1. The final weighted score is calculated based on the individual scores and is used for classifying the tweet.

The overall processing pipeline is depicted in Fig. 11.10 and has been published in more detail in [25].

Preliminary Results

At the time of writing this text, we already had some promising preliminary results, at least in the sense of capturing correctly the overall sentiment of the population involved, given the limitations discussed in the beginning. Although the complexity of the pipeline amplifies the errors we have in the processing, preliminary results show quite good accuracy in the classification of sentiments found in the relevant tweets. Moreover, the errors we calculate are equally distributed between the different classes, which shows that the method does not introduce any bias towards a specific class. Results can be visualised either as graphs or as localised data with the use of a map for the visual background.

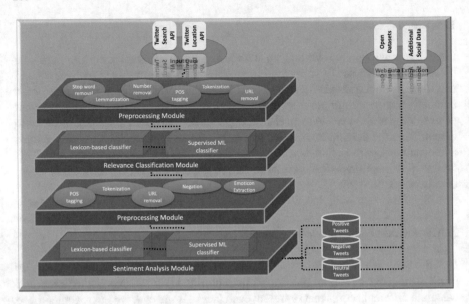

Fig. 11.10 The twitter analysis pipeline [25]

Future Work

The main effort is the large-scale application and evaluation of the proposed methodology. Given the effects of the COVID-19 pandemic on airport traffic but also on the discussions on twitter and other social media, we will rely on historical data (i.e. data recorded prior to the pandemic) to do our processing. This eases the requirements on real-time processing and allows us to apply additional methods, like embeddings, which can improve the accuracy of sentiment classification. Unlike the "bag of words" representation used by our methods so far, where the context plays a small role; these newer methods are able to detect similarities and hence classify unseen words which are similar to other words seen in the training set. Recurrent Neural Network (RNN) offers a memory factor that helps to consider the previous and the following words to better predict the sentiment of current words and hence to efficiently predict the sentiment of the whole sentence, which improves the accuracy of the classifier.

Conclusion

In this chapter, we presented how recent technological innovations could help to collect and analyse data from people who live around airports, and so improve our understanding of the adverse effect of noise on humans.

A Virtual Reality simulation made it possible to evaluate how visual settings of the aircraft (wide body, narrow body, blended wing body) and of the landscape (green park vs. urban situation) influenced sound perception. The quality of the tool has been tested, showing that 96% of the participants felt surrounded by the environment, and 78% found the virtual audio-visual environment very or extremely realistic. This application rendered highly immersive audio-visual situations, so, it has been hypothesised that it could be relevant to communicate with residents in a fair approach, showing the impact of future airport scenarios in land-use planning.

A mobile application (AnimApp) has been developed to study the impact of the audio-visual environment on sound perception and on quality of life around airports. The method of experience sampling has been chosen, because it captures subjective experiences as they are experienced in-situ in real life. Perceptual data on the sound as well as the visual environment are collected in addition to acoustic data. The final study will take place during spring/summer 2021 at two locations: one at Ljubljana Airport in Slovenia, the other at London Heathrow Airport in the UK with the aim of convincing more than 60 participants in each site. This experiment will probably suffer from the reduction of air traffic due to the COVID-19 pandemic, but this application can be also used in the future to collect valuable perceptual data synchronised with acoustic ones with more participants.

Finally, using social media as a means to survey people's opinions on various subjects, including quality of life and issues of noise around airports, seems to be promising and produces credible results. The process gives us insights based on existing online discussions and based on complex learning pipelines; it discovers, classifies and localises the opinions of the users. The complexity of the current processing architectures is significant but the results produced so far are promising for the future. Being able to combine those data with additional offline data and data from multiple sources (e.g. other opinion sites, land use, etc.) could improve the quality of the insights provided into people's responses to aircraft noise, and, thus, and allow to further refine the process of aviation noise management in airport regions.

References

1. Asghar MZ, Khan A, Ahmad S, Qasim M, Khan IA (2017) Lexicon-enhanced sentiment analysis framework using rule-based classification scheme. PLOS One 12(2):e0171649
2. Axelsson O, Nilsson M, Berglund B (2010) A principal component model of soundscape perception. J Acoust Soc Am 128(5):2836–2846
3. Bolger N, Davis A, Rafaeli E (2003) Diary methods: capturing life as it is lived. Annu Rev Psychol 54:579–616. https://doi.org/10.1146/annurev.psych.54.101601.145030
4. Brambilla G, Maffei L (2006) Responses to noise in urban parks and in rural quiet areas. Acta Acustica united with Acustica 92(6):881–886
5. Cain R, Jennings P, Poxon J (2013) The development and application of the emotional dimensions of a soundscape. Appl Acoust 74:232–239
6. Cortes C, Vapnik V (1995) Support-vector networks. Machine Learn 20(3):273–297

7. Craig A, Moore D, Knox D (2017) Experience sampling: assessing urban soundscapes using in-situ participatory methods. Appl Acoust 117:227–235. https://doi.org/10.1016/j.apacoust.2016.05.026

8. Dedieu R, Lavandier C, Aalmoes R, Legriffon I, Boullet I (2021) Studies on virtual and auralisation tools. ANIMA Deliverable D3.10

9. Fiebig A, Sottek R (2015) Contribution of peak events to overall loudness. Acta Acust united Ac 101:1116–1129

10. Fields JM, De Jong RG, Gjestland T, Flindell IH, Job RFS, Kurra S, Lercher P, Vallet M,Yano T, Guski R, Felscher-Suhr U, Schumer R (2001) Standardised general-purpose noise response questions for community noise surveys: research and a recommendation. J Sound Vib 242(4):641–679. Last visited on the 26th of March. https://www.sciencedirect.com/science/article/pii/S0022460X00933844

11. Fujiwara D, Lawton RN, MacKerron G (2017) Experience sampling in and around airports. Momentary subjective wellbeing, airports, and aviation noise in England. Transp Res Part D: Transp Environ 56:43–54. ISSN 1361–9209. https://doi.org/10.1016/j.trd.2017.07.015. (https://www.sciencedirect.com/science/article/pii/S136192091530256X)

12. Gille L-A, Marquis-Favre C (2015) Physical and perceptual characterisation of aircraft noise to better assess noise annoyance. In: Proceedings internoise, San Francisco, USA

13. Guski R (1999) Personal and social variables as co-determinants of noise annoyance. Noise Health 3:45–56

14. Guski R (2017) The increase of aircraft noise annoyance in communities. Causes and consequences, Keynote, Proc ICBEN 2017, Zurich, Switzerland, ID 4164, (2017) Available from http://www.icben.org/2017/ICBEN%202017%20Papers/Keynote04_Guski_4164.pdf

15. Guski R, Schreckenberg D, Schuemer R (2017) WHO environmental noise guidelines for the European Region: a systematic review on environmental noise and annoyance. Int J Environ Res Public Health 14(12):1539. https://doi.org/10.3390/ijerph14121539

16. Hutto C, Gilbert E, Vader A (2014) Parsimonious rule-based model for sentiment analysis of social media text. In: Proceedings of the international AAAI conference on web and social media. vol 8

17. International Wellbeing Group, Personal Wellbeing Index: 5th Edition (2013). Available from http://www.acqol.com.au/uploads/pwi-a/pwi-a-english.pdf (May 2019)

18. ISO 12913–1 (2014) Acoustics—Soundscape—Part 1: definition and conceptual framework. Geneva, Switzerland, International Organisation for Standardisation (ISO)

19. ISO 12913–2 (2018) Acoustics—Soundscape—Part 2: data collection and reporting requirements. Geneva, Switzerland, International Organisation for Standardisation (ISO)

20. Jianqiang Z, Xiaolin G, Xuejun Z (2018) Deep convolution neural networks for twitter sentiment analysis. IEEE Access 6:23253–23260. https://doi.org/10.1109/ACCESS.2017.2776930

21. Kouloumpis E, Wilson T, Moore J (2011) Twitter sentiment analysis: the good the bad and the OMG!. In: Proceedings of the international AAAI conference on web and social media. vol 5.

22. Liu B, Hu M, Cheng J (2005) Opinion observer: analysing and comparing opinions on the web. In: Proceedings of the 14th international conference on WorldWideWeb. pp 342–351

23. Loughran T, Mc DB (2011) When is a liability not a liability? textual analysis, dictionaries, and 10-ks. J Financ 66(1):35–65

24. Maisonneuve N, Matthias N (2010) Participatory noise pollution monitoring using mobile phones. Information Polity 15:51–71

25. Meddeb I, Lavandier C, Kotzinos D (2020) Using twitter streams for opinion mining: a case study on airport noise. Springer Series "Communications in Computer and Information Science". 1197:145–160

26. More SR (2011) Aircraft noise characteristics and metrics. Ph.D. Thesis, Purdue University, West Lafayette, USA

27. Ouyang X, Zhou P, Li CH, Liu L (2015) Sentiment analysis using convolutional neural network. In: IEEE international conference on computer and information technology; ubiquitous computing and communications; dependable, autonomic and secure computing. Pervasive Intelligence and Computing, pp 2359–2364. https://doi.org/10.1109/CIT/IUCC/DASC/PICOM.2015.349

28. Page JA, Hogdon KK, Hunte RP, Davis DE, Gaugler TA, Downs R et al. (2019) Quiet supersonic flights 2018 (QSF18) test: galveston, texas risk reduction for future community testing with a low-boom flight demonstration vehicle, NASA, Hapton, VA (NASA/CR–2020–220589/Volume I), (2020). Last visited on the 26th of March 2021. https://ntrs.nasa.gov/search.jsp?R=20200003223

29. Pak A, Paroubek P (2010) Twitter as a corpus for sentiment analysis and opinion mining. LREc 10:1320–1326

30. Paté A, Lavandier C, Minard A, Le Griffon I (2017) Perceived unpleasantness of aircraft flyover noise: influence of temporal parameters. Acta Acust Acust 103:34–47. https://doi.org/10.3813/AAA.919031

31. Pheasant R, Horoshenkov K, Watts G, Barrett B (2008) The acoustic and visual factors influencing the construction of tranquil space in urban and rural environments tranquil spaces-quiet places? J Acoust Soc Am 123(3):1446–1457

32. Preis A, Hafke-Dys H, Szychowska M, Kociński J, Felcyn J (2016) Audiovisual interaction of environmental noise. Noise Cont Eng J 64:36–45

33. Psychiatric Research Unit, WHO Collaborating Centre in Mental Health. WHO (Five) Well-Being Index, (1998) Available from https://www.psykiatri-regionh.dk/who-5/Documents/WHO-5%20questionaire%20-%20English.pdf Last visited on the 26th of March 2021

34. Radicchi A, Henckel D, Memmel M (2018) Citizens as smart, active sensors for a quiet and just city. The case of the 'open source soundscapes' approach to identify, assess and plan 'everyday quiet areas' in cities. Noise Mapping J 5:1–20

35. Read J (2005) Using emoticons to reduce dependency in machine learning techniques for sentiment classification. In: Proceedings of the ACL student research workshop, pp 43–48

36. Ren Y, Zhang Y, Zhang M, Ji D (2016) Context-sensitive twitter sentiment classification using neural network. In: Proceedings of the AAAI conference on artificial intelligence. vol 30

37. Ricciardi P, Delaitre P, Lavandier C, Torchia F, Aumond P (2015) Sound quality indicators for urban places in Paris cross-validated by Milan data. J Acoust Soc Am 138(4):2337–2348

38. Rizzi SA, Sullivan BM, Aumann AR (2008) Recent developments in aircraft flyover noise simulation at NASA langley research center. NATO RTO Specialist Meeting AVT-158 on Noise Issues Associated with Gas Turbine Powered Military Vehicles, 13–17 Oct, Montreal, Canada

39. Stat Counter Global Stats, Mobile Operating System Market Share in Europe (2019) Online Available from: http://gs.statcounter.com/os-market-share/mobile/europe (March 2021)

40. Steffens J, Guastavino C (2015) Trend effects in momentary and retrospective soundscape judgments. Acta Acust United Acust 101:713–722

41. Susini P, McAdams S, Smith BK (2002) Global and continuous loudness estimation of time-varying levels. Acta Acust united Ac 88:536–548

42. Tang D, Wei F, Qin B, Yang N, Liu T, Zhou M (2015) Sentiment embeddings with applications to sentiment analysis. IEEE Trans Knowl Data Eng 28(2):496–509

43. Van Renterghem T, Botteldooren D (2016) View on outdoor vegetation reduces noise annoyance for dwellers near busy roads. Landsc Urb Plan 148:203–215

44. Västfjäll D (2004) The 'end effect' in retrospective sound quality evaluation. Acoust Sci Technol 25:170–172

45. Viollon S, Lavandier C, Drake C (2002) Influence of visual setting on sound ratings in an urban environment. Appl Acoust 63(5):493–511

46. WHO Environmental Noise Guidelines for the European Region (2018) ISBN 978 92 890 5356 3. Last visited on the 26th of March 2021. https://www.euro.who.int/en/health-topics/environment-and-health/noise/publications/2018/environmental-noise-guidelines-for-the-european-region-2018

47. Wilson T, Wiebe J, Hoffmann P (2005) Recognising contextual polarity in phrase-level sentiment analysis. In: Proceedings of human language technology conference and conference on empirical methods in natural language processing, pp 347–354
48. Yadollahi A, Shahraki AG, Zaiane OR (2017) Current state of text sentiment analysis from opinion to emotion mining. ACM Comput Surv (CSUR) 50(2):1–33

Towards Mapping of Noise Impact

Explaining ANIMA Efforts to Support New Approaches for Noise Impact Management Through Noise Management Toolset, Virtual Community Tool, and Dynamic Noise Maps

Ferenc Marki , **Peter Rucz** , **Nico van Oosten, Emir Ganić** , **and Ingrid Legriffon**

Abstract Noise impact management goes hand in hand with the capability to predict the noise impact on exposed communities. Three tools to that purpose are presented in this chapter: the Noise Management Toolset (NMT), the Demo Virtual Community Tool (VCT) and Dynamic Noise Mapping. The NMT is a web-based tool giving stakeholders the opportunity to evaluate scenarios through not only noise exposure, but also noise impact, by introducing annoyance related metrics like the awakening index, with an easy-to-use interface. The VCT is the underlying research tool exploring and testing new indicators and options that might be of relevance to target audiences, such as land use planning information about location dependent activities or window insulation. The third approach, Dynamic Noise Mapping, adds the important aspect of population movement to classical noise mapping approaches where temporal changes of noise maps are tracked and included in noise exposure evaluation.

Keywords Aircraft noise · Air traffic scenarios · Noise exposure · Annoyance · Perception oriented metrics · Land use planning · Human mobility patterns · National travel survey

F. Marki (✉) · P. Rucz
Budapest University of Technology and Economics, Budapest, Hungary
e-mail: marki@vik.bme.hu

P. Rucz
e-mail: rucz@hit.bme.hu

N. van Oosten
Anotec Engineering SL, Motril, Spain
e-mail: nico@anotecengineering.com

E. Ganić
University of Belgrade—Faculty of Transport and Traffic Engineering, Belgrade, Serbia
e-mail: e.ganic@sf.bg.ac.rs

I. Legriffon
ONERA, Université Paris Saclay, Chatillon, France
e-mail: ingrid.legriffon@onera.fr

Introduction

When taking decisions with regard to land use planning, changes in air tracks etc., stakeholders need to rely on numbers. At present many tools are available that allow the user to generate noise exposure maps for airport scenarios. Although most of these tools have a rather sophisticated graphical user interface, their proper operation requires significant technical skills, usually only available at specialised consultants or at environmental departments of big airports. Due to the cost involved, this will normally limit the use of such tools to specific scenarios, required for compliance with legal requirements. It does thus usually not allow other interested stakeholders like land use planners, policymakers, airport staff, etc. to "play" with such tools to get a better understanding of the factors influencing airport noise management.

On the other hand, state-of-the-art tools usually only generate information on noise *exposure*. Although this is relevant for planning purposes, it falls short when airport noise issues need to be managed at a detailed level. As has been highlighted by the ANIMA project, an understanding of the reaction of people on interventions, aimed at reducing the noise *impact*, is required to maximise the benefits of such interventions.

Also, generating information on noise impact does not only imply knowledge of the noise sources in space and time, but also of the impacted population. Movement of people necessarily influences their exposure to noise and hence their perceived impact. Taking that varying parameter into account when estimating noise impact seems unavoidable, if it is to be done realistically.

In the following, three tools are presented. The above described shortcomings of current airport noise prediction models and mapping approaches have been addressed in ANIMA through the development of the Noise Management Toolset (NMT), the Virtual Community Tool (VCT) and Dynamic Noise Mapping. While the first one (NMT) offers a range of versions, going from a public version to a tool for aircraft noise experts, the second one (VCT) is a research tool elaborating, testing and validating new indicators, visualisations and options that can be implemented into the NMT if deemed of interest to stakeholders. The third approach, Dynamic Noise Mapping, adds the important aspect of population movement to classical noise mapping approaches.

Noise Management Toolset

Objectives of the Tool

The NMT has been developed with the aim to overcome the main shortcomings of existing airport noise tools, highlighted above. Therefore the main objectives of the tool are to:

- Reduce the required technical skills so the tool can be used by a wider range of users
- Extend the scope from noise *exposure* to noise *impact.*

To further enhance the scope of the tool, the following additional forward-looking objective has been included:

- Allow for the inclusion of advanced aircraft/powerplant concepts (so as to study their noise impact, not only exposure)

Working Principle

In the European context airport noise models have to be compatible with the methodology described in ECAC Doc29 [1]. Therefore the NMT has been developed around the SONDEO model [2], which implements the latest version of this methodology. The design of SONDEO is such that it separates the complex noise calculations required to obtain the noise levels for each individual aircraft (the so-called "single events"), from the generation of the map that reflects the noise exposure of the full fleet operation, representative for a certain scenario. After this, impact related features can be calculated. Figure 1 provides a schematic representation of the workflow of the NMT.

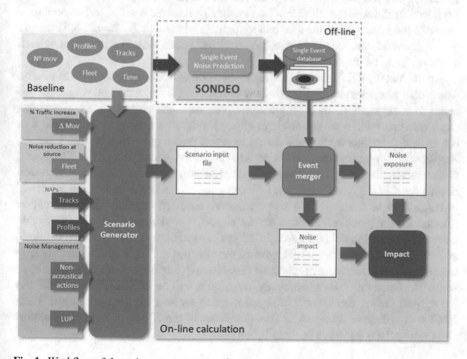

Fig. 1 Workflow of the noise management toolset

The first step (generation of the single events database) inevitably requires in-depth technical knowledge of airport noise modeling. To avoid exposing the end-user to this part, the single-event database will be generated "off-line" by specialists (here called the NMT Administrator). This database will contain the individual noise footprint of each combination of aircraft type, flight track, flight profile, etc. that may be considered later on in the scenarios. With the support of additional simulation tools, the single events for any not yet existing aircraft can also be generated here. It is noted that the single event database will need to be generated only once.

The resulting single event database will be available at the start of the on-line user experience. The user can build his own scenarios by selecting the relevant combinations of single events and in this way generate the full set of operations for which the noise results shall be calculated. Apart from the standard noise exposure metrics, relevant noise impact related metrics are also calculated.

The results of the calculations are graphically presented and the user can select which results to display and can compare the results of various scenarios.

Airports

A so-called Public Toolset (PT) is available for the general public. The PT contains a virtual airport, with the aim to illustrate the basic concepts of airport noise mapping, explained in the ANIMA Best Practice Portal. To this end a set of traffic scenarios has been included, which the user can visualise and for which relevant information on the essential components (aircraft operations, tracks, noise contours, populated areas, etc.) can be displayed.

Apart from the PT, limited to the illustration of basic concepts, a premium version (the Noise Management Toolset or NMT) has been developed that addresses the needs of users such as land use planners, policymakers, airport staff, universities, etc. These users will want to be able to generate their own scenarios at a real airport, e.g. to design an intervention and assess its effect on the noise exposure and impact. Registration is required to obtain access to this advanced version of the NMT. As a first step, an authorised person (here called the Airport Owner) shall request inclusion of an airport in the NMT database, following the procedure described on the NMT website. This person shall have the permissions required to publish the information relevant for the noise calculations at that airport. The Airport Owner shall provide a dataset with which the NMT Administrator can generate the single event database and configure the system for inclusion of the new airport. This dataset will include information on the runway(s), standard flight routes (tracks), aircraft fleet, populated areas, etc. Once this database is ready, the airport will be available in the NMT and the Airport Owner can invite additional users to register for the new airport. Each NMT user will have access to all functionalities of the NMT for the virtual airport and the airport(s) (s)he has been assigned to.

Building of Scenarios

The basic structure for the calculations to be done with the NMT is the Scenario. A Scenario represents a certain noise situation at an airport and as such consists of a specific combination of the single events available in the database. Each single event is defined by a unique combination of:

- Aircraft type
- Type of operation: Arrival or Departure
- Vertical flight procedure ("Profile")
- Distance flown (indicator for the weight at take-off)
- Runway
- Track

To define a Scenario the user should provide the following information for each single event (see Fig. 2):

- Number of operations
- Time of these operations (exact local time or period (Day, Evening, Night))

The NMT provides several options to create a new scenario. As a first option a scenario can be generated from scratch. In this case the user has to provide all the information on the aircraft operations. This can be done by manually filling in a table like the one presented in Fig. 2, or, more conveniently, by uploading a file containing the airport flight plan (similar to the time table usually managed at airports, i.e. providing the time of each individual operation) or a so-called operations file (resembling the table shown in Fig. 2, grouping the operations by period of the day, thus losing the time information of the individual operations). Templates for both are available on the NMT website. Usually the user will need to create the first scenario for an airport in one of these manners. However, once this is done, it is generally more convenient to clone the existing scenario and then change only those operations required to define the new scenario. Several smart features are available to assist the user in defining the new scenario. It is possible to move a certain amount of operations from one track to another, for all aircraft and time periods, or for specific

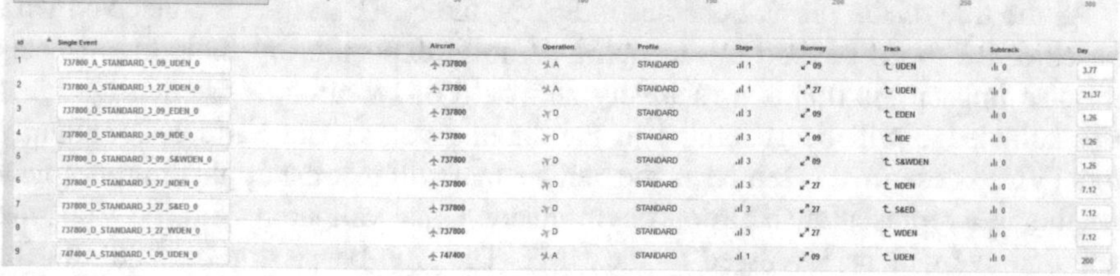

Fig. 2 Example of building a scenario, based on single events

aircraft types and/or time periods individually. Similarly, it is possible to increase the number of operations by a certain percentage to easily simulate traffic growth.

Once a scenario has been created by one of the procedures explained above, the noise calculation is invoked, which will generate the noise exposure and impact maps.

Visualisation of Results

Once a scenario has been calculated the results can be visualised. The user can select the noise/impact contours that should be shown on the map. Information on track usage, specific populated areas, etc. can also be presented.

To assess the effect of the intervention represented by a scenario, the results (noise exposure and impact contours) of that scenario can be compared with those from another scenario. A maximum of 4 scenarios can be compared at the same time.

The reader is now encouraged to get a taste about the capability of the tool by visiting:

https://anima-project.eu/noise-platform/noise-management-toolset   and   by following the path along Noise Toolset to The Public Noise Toolset.

Future Work

Although the currently available NMT addresses the most relevant needs of the targeted stakeholders, further development of the tool is envisaged, to cater for additional functionalities that are considered of added value to the users.

Within ANIMA a more comprehensive desktop toolset has been developed for use by aircraft noise experts. Some of the functionalities implemented in this tool may be migrated to the web-based NMT. As a first additional feature, it is envisaged that the web-based NMT will be extended with the calculation of emissions (CO_2 and NOx). In this manner the user will be able to obtain in a single execution both the noise and the emissions corresponding to a scenario. This will allow for the determination of the interdependencies between both environmental aspects and provide the user with means to perform trade-off studies.

As the title of the present chapter indicates, the NMT has been conceived with the objective to go beyond the mapping of noise exposure, by including aspects of noise impact. As this is a relatively new field of research, it is envisaged that new knowledge will be generated in the coming years. The Virtual Community Tool (VCT) described hereafter is the vehicle that will test the applicability of new insights in a representative airport environment. Once validated with the VCT, the new findings will be leveraged to the NMT. The web-based approach allows for an instantaneous upgrade of the NMT, allowing its users access to state-of-the-art knowledge on airport noise impact and its management.

Virtual Community Tool

Aim of the Tool

As it was seen in former chapters, noise causes annoyance, but its amount cannot be clearly related to noise levels. Decision makers have a hard time trying to protect people living around airports: on the one hand they must use *objective*, measurable quantities, but on the other hand they should use something, which reflects people's *subjective* reactions to noise. Until now, separate daily, evening, and night-time noise levels, or their combination (L_{DEN}) are calculated in most cases. But unfortunately these level metrics are normally computed for longer periods only, i.e., a month, the busiest six months, or a year, thus blurring the annoying effect of some worse days or even some hours of the day. Nevertheless, L_{DEN} could already be seen as something that is at least a bit perception oriented, because it gives a penalty weighting for the evening and night hours, taking the more adverse effect of noise during these hours into account. Unfortunately it is, by far, not "human friendly" enough, as it absolutely does not consider the nature of aircraft noise being a series of individual events in contrast to much more continuous noises. L_{DEN} can neither take critical hours into consideration, such as trying to get some rest outside after work, the time when people try to fall asleep, or when they are in a light-sleep phase soon before getting up; nor can it tell too much about sleep disturbance by the noise. (Similar findings are described in [3], which thus recommends that "supplementary Single Event metrics are routinely published by airports to better reflect the way in which noise is experienced on the ground".) Besides the metrics utilised, another issue with the current noise computation approach is the computational cost. The strategies applied currently are time consuming thus it is too costly to analyse various scenarios, like the rearrangement of flight hours or the increase of flight traffic in the future, or the renewal of the fleets, etc.

The introduction of new metrics to use as an evaluation tool on the harmful effect of air traffic is not just a difficult scientific task, it is also a hard decision, because there does not exist a universal, undisputed metric. Each of the metrics utilised nowadays emphasises one certain effect of noise, while suppressing the others. Instead of trying to invent new formulas of combined metrics, in ANIMA we try to move the current, mainly level-oriented decision approach into a direction, where more factors are considered. Therefore a tool has been developed, which is able to compute several metrics, each being strongly related to annoyance. A strong emphasis has also been set on the ability of the tool to easily change scenarios, i.e., to quickly analyse various possibilities. We don't know yet whether such a tool could be accepted by decision makers therefore we call our tool the "Demo Virtual Community Tool" (referred as "demo VCT")—as it demonstrates a new approach on evaluating aircraft noise effects.

When comparing it to the NMT, it is developed in a computer language, which allows very fast program development, so we can quickly implement whatever we think it could be useful, but it cannot correctly support user right management.

Therefore it is good as an experimentation tool to find useful features that could be later on implemented in a more commercial-like software, like the NMT.

Please note that all examples presented here serve only to demonstrate the capability and the potential of the tool. Although you'll see computed metrics around Budapest Airport, the applied schedules are fictional, so it is strongly emphasised that the reader **shall not draw any conclusions** on the situation around BUD airport.

Working Principle

Figure 3 displays the overall workflow of the tool. There are two main inputs that are required for the computations performed by the Virtual Community Tool. First, an *airport database* must be available. This database contains the ground acoustical data (i.e. acoustical footprints) of all possible flight operations of a single airport. The dataset is sampled over a geographical grid that covers a given area in the vicinity of the given airport. The airport database may also be supplemented by additional regional information, containing a demographic map of the population density, an insulation map of the buildings in the area and further auxiliary data. The second input is called a *flight schedule* and it contains a list of actual operations performed in a given frame of time. A typical schedule may contain all operations of a week or a fortnight, whereas short schedules of a single day, or long schedules containing the traffic over a whole year are also handled by the tool. Once a schedule is available, all of its flights are matched to the airport database in order to establish the connection of the actual operation with the corresponding acoustical footprint.

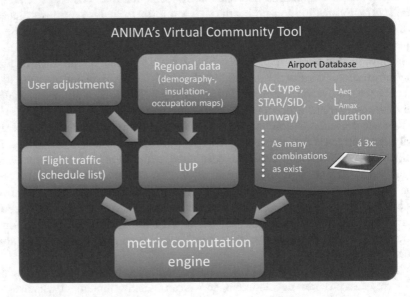

Fig. 3 Workflow of the demo VCT

By means of the metric computation engine, the tool enables the analysis of various *scenarios* for one airport, i.e., a given airport database. A scenario contains the flight schedule together with the traffic modifications defined by the user as well as the land use plan (LUP) areas and the settings of the computations. The user is able to edit the traffic composition through the graphical user interface. The main idea of the tool is to allow for adjustment of the overall properties of the composition of the air traffic instead of managing single operations one-by-one. This approach allows very easy modification/redesign of the air traffic and thus facilitates the analysis of the effects of planned interventions or expected trends, such as increasing or decreasing the amount of operations on a flight path. Moreover, land use plans (e.g. financing window insulation in a certain area, or establishing a business area, etc.) can easily be defined and modified using the interactive map visualisation window.

Once a scenario is defined, the computation of the selected acoustical and non-acoustical indicators is performed. These indicators, each being defined on the geographical grid of the airport, are referred to as *metrics*. The computation of all metrics necessitates iterating over the flight schedule and accumulating the corresponding acoustical footprints as well as calculating non-acoustical indicators at the same time. To be able to handle long schedules containing several months of traffic in a short amount of time, the computation of each metric is specialised and the possibilities to reduce the computational burden are exploited wherever possible. The computation of acoustical indicators has been validated by comparing the results provided by our tool to reference computations carried out using commercial software packages.

The main functionality of the tool is then the analysis and comparison of the metrics computed for one or several different scenarios by means of a powerful visualisation engine. Its map visualisation window (see Fig. 4) allows for displaying the data as colormaps rendered over the satellite view of the area. The colormaps can be augmented by an arbitrary number of contours that are fully customisable by the user. At the same time, the visualised data can be exported and later imported facilitating further comparisons.

Features

The most important capability of the Virtual Community Tool is that it is able to compute various acoustical and non-acoustical indicators, including standardised quantities (e.g. L_{DEN} or L_{night}) and metrics that are customisable by the user. One example of such customisation is the ability to change the typical period of sleeping hours of people which affects the calculation of the awakening indicator.

Also a key, unique feature of the tool is that it enables the user to modify the global properties of the traffic of the airport through a clean graphical user interface that is easy to handle. In particular, the following properties of air traffic can be adjusted:

Fig. 4 Map visualization window showing various metrics for one scenario. One can observe that areas defined by crucial limits of the metrics mark out somewhat different areas

- The global and the hourly amount of operations can be modified, meaning that the traffic as a whole can be increased or decreased. Furthermore, the traffic can be reorganised among the hours of the day, for example by moving a certain percentage of the traffic from one hour to one or many others. This feature is particularly useful for examining the day-evening-night balance of the noise and annoyance caused by the current or the planned air traffic.
- The usage of runways and flight paths, allowing the adjustment of the amount of traffic on them. Besides the ability of foretelling the effects of introducing new flights, this feature can be especially useful for predicting the change of indicators by such events as a renewal of a runway.
- The composition of the fleet, i.e. the relative amount of different types of aircraft, allowing the replacement of older aircraft types by newer ones, as well as completely banning operations of given aircraft types. This functionality allows for forecasting the effect of the renewal of the fleet of airlines.

Areas for various types of land use plan actions may be defined by the user by simply marking them on the map. Area types include green parks, business areas, university campuses, or areas where window insulation for the houses is funded. The areas defined by the user affect the computation of both acoustical and non-acoustical indicators.

If a demographic map of the examined area is available, the VCT imports it automatically. Once the demographic map is loaded, the acoustical and annoyance metrics can also be visualised taking the number of affected people into account, indicating the *seriousness* of possible aircraft noise related problems. Furthermore, the demographic map may be supplemented by a so-called occupancy map, which describes—on an hourly basis—what percentage of the habitants are in fact at home. This enables taking the number of affected people dynamically into account. This unique feature is also exploited in defining the land use planning for the vicinity of the airport. As an example, it can be expected that a business area or a shopping center is not inhabited during the night hours.

The user has the possibility to either perform a series of adjustments on one scenario, resulting in a *modified scenario* or to perform different adjustments on the same scenario and then to save them as separate scenarios.

The integrated map visualisation window allows for a straightforward comparison in both cases: the modified scenario to the original, hence enabling a quick overview of the effects of the planned changes or the comparison of various options for a given starting situation. Moreover, the capability to compare several scenarios also allows for comparing different schedules, i.e. worse day versus long-term average or preferred versus non-preferred oparation mode of the airport.

Scenario Show-Cases

Let us recall a statement from European Parliament [4]:

"Furthermore, the use of new metrics like Number of Events above a certain noise value are being pushed forward. As it is indicated in the WHO 2018 Environmental Noise Guidelines for the European Region *"There is additional uncertainty when characterising exposure using the acoustical description of aircraft noise by means of Lden or Lnight. Use of these average noise indicators may limit the ability to observe associations between exposure to aircraft noise and some health outcomes (such as awakening reactions); as such, noise indicators based on the number of events (such as the frequency distribution of LA,max) may be better suited. However, such indicators are not widely used"*.

There is, therefore, a proposal to start giving more priority to other noise indicators (in particular event-related metrics) as well as calculating lower noise level contours to present noise exposure, which is a challenging modification considering the way the noise effects have been studied until now.

This also supports the notion that annoyance is not just a yearly value and cannot be characterised by a single metric. More and more countries are considering various metrics simultaneously. Here, a good software comes in handy especially for "starting the journey" airports, i.e. airports with less practice in aircraft noise abatement.

In the following you will see several scenarios demonstrating the capabilities of the demo Virtual Community Tool. By showing differences in contour-sizes, we definitely don't want to give a position on what is appropriate to use. We just want

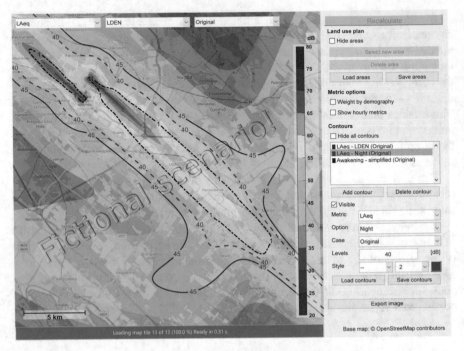

Fig. 5 Multiple metrics can be shown at the same time. Color mapping depicts L_{den}, brown continuous contour shows the limit for $L_{den} = 45$ dB, blue dashed curve shows $L_{night} = 40$ dB, while the black dash-dotted line encloses the area with > 1 additional awakening per night

decision makers to be in a better position to know what to expect. This information could be used for communication purposes or taking actions.

1. **Multi-metrics evaluations**

State-of-the art research suggests the definition of protection zones based on several properties, not just one. For aircraft noise, WHO recommends limiting the aircraft noise exposure to less than 45 dB L_{den}, and for L_{night} 40 dB [5]. Another recommendation is to keep the average additional awakenings induced by noise below 1 per night [6]. As our demo VCT is capable of computing and showing several metrics at the same time, one can clearly observe what areas should be protected to fulfill all three conditions (See Fig. 5.)

2. **Sleep-time preference**

People differ from each other. Some prefer to go to bed later and also get up later, others go earlier to sleep and get up also earlier. With the ability to flexibly change the sleeping hours of people, independently from the night period defined in regulations by each country, with our tool one can observe that "late sleepers" in a much larger area have their nights unprotected from being woken up more than once in average by the air traffic (see Fig. 6.) This example definitely shows

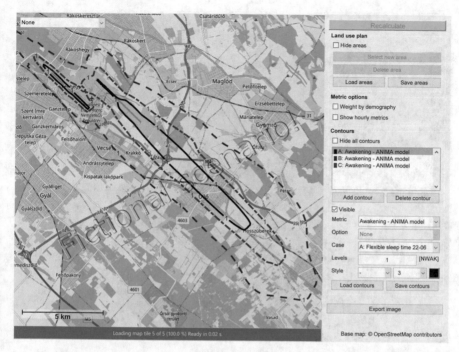

Fig. 6 More than one additional awakening per night area for people sleeping from 22–06 (black curve) and for those sleeping from 23–07 (brown dashed curve). Blue dash-dotted curve shows these latter sleep hours with the reorganisation of the morning traffic: 50% of the flights between 06–07 are moved to 07–08

that while it is not realistic to completely shift the "airport start" to one hour later, it is an option to shift at least some flights to after 7 o'clock, or in case there is budget for it, to extend window insulation programs to a larger area, or finally at least to spread simply the knowledge that "late-sleepers should not live near airports".

3. Scenarios to expect

It could be preferable to consider at the same time a long-term average and some kind of maximum operation. Especially normal, but non-preferred configurations could be cause for complaints, because published maps often present only the long-term average, lowering those less-frequently happening, inconvenient levels of areas, which receive high noise load only during non-preferred times. The effect is even stronger when preferred and non-preferred configurations are considered. It is worth explaining to people by visualisation why they sometimes feel so bad about the noise: because after a period with favourable conditions the contrast to the unfavourable is much more pronounced. However such scenarios are computed from completely different flight schedule lists. This is

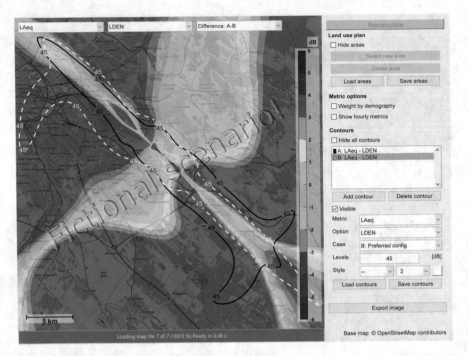

Fig. 7 The colormap shows the level difference of the non-preferred versus the preferred configuration. Although the strongly affected area (red area, > 6 dB) is huge, beyond the 45 dB contour curves the noise is overall quite low. Still, the area between the two contour lines is remarkably large. (Black curve is the non-preferred, white dashed curve is the preferred configuration.)

not a problem for the VCT, as the intelligent visualisation engine is able to present several scenarios on the same map. (See Fig. 7).

4.   Future

During Land Use Planning, it is wise to look a bit towards the future. While nobody can tell what will actually happen, most airports already have experience in evaluating their flight traffic over the years. Most probably the increase in flight operations and the renewal of airlines' fleets can be estimated. To compute such expectations is really easy for the user: just the increase in the total number of flights need to be changed, and a few replacements of some older, but frequently used aircraft types by some current ones and voilà, one can have an idea how the airport's footprint will perhaps evolve in the upcoming years. Figure 8. depicts such an estimation.

5.   Critical Hours

We know from experience that some hours are more critical than others, e.g. falling asleep is more prone to disturbance by noise events then when one is already asleep. So people could be interested in traffic during these hours. As our Virtual Community Tool performs internal computations on an hourly basis, the Map Display Window

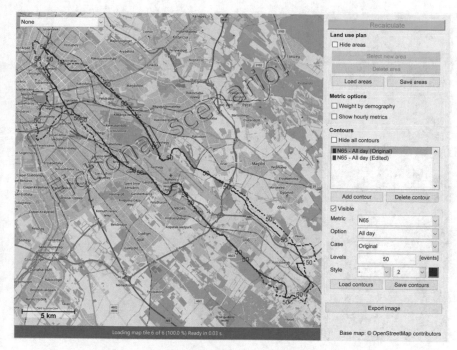

Fig. 8 Actual state (brown curve) versus 25% increased traffic with upgrading 20% of the older part of the fleet by newer aircraft (blue dash-dotted curve). Contour curves show 50 movements per day with a maximum noise level above 65 dB (N65)

allows the presentation of metrics for specific hours over the map but also the overlaying of contours for several hours. (See Fig. 9).

6. Land Use Planning

Land Use Planning is a powerful way to control noise annoyance. Building well soundproofed business-areas or shopping centers near airports could be examples for it. In these areas no population is there to be disturbed during the night time, and during the day, business areas can afford to pay for well soundproofed buildings, while in shopping centers the noise levels are usually already so high indoors that higher outside noise levels are not relevant. But also the effect of financing window insulation in a certain area is worth studying, especially if we know the typical original sound insulation quality of houses and the seasonal habit of people to close or leave their windows open during the nights. The VCT allows for an easy definition of land use planned areas by defining simply their functionality. Also a map containing the typical sound insulation quality of houses around the airport can be used by the tool, so the effect of soundproofing improvement can be easily studied. On Fig. 10 an example scenario is shown: some areas received window insulation and a business center has been established near the airport.

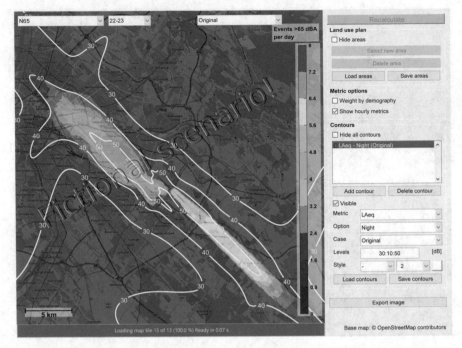

Fig. 9 This figure shows the overall night-time's noise load with white contour curves. While the average noise-load towards the city is similar to those away from it, actually the number of loud events depicted by the colormap (maximum noise level above 65 dB) between 22–23 o'clock is about half as in the opposite direction. This is a favourable effect for the densely populated city area

Dynamic Noise Maps

Background and Definition

Dynamic noise mapping represents a relatively new concept in the airport noise literature. Different authors have used this term for various purposes in the previous years. Before going into further details, it is beneficial to compare the current usage of this term and to precisely define the meaning of the "dynamic noise maps" concept in this book.

Most of the research studies have used this term to present noise in a given moment of time, i.e. to differentiate between noise maps for different period of the day (peak or off-peak hours) and for the different days of the week (weekdays and weekends) [7, 8]. Another research project [9] defines dynamic noise maps as acoustic maps that illustrate in real time the temporal change of noise levels. Such noise maps are constantly updated using algorithms and software in real time for different operating conditions (sources, traffic, and weather conditions), by detecting

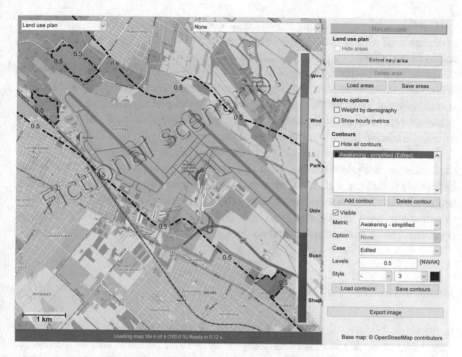

Fig. 10 Awakening is reduced in areas with LUP functionality. Purple marked area depicts a business area, while the orange ones define window insulation programs

noise and meteorological data from low-cost monitoring stations and weather sensors [10].

When real-time noise levels obtained from noise monitoring stations are used for dynamic noise mapping, measurements could be taken only at specific points due to limitations in the number of noise monitors. To obtain the existing noise levels for the rest of the area of interest some estimates are needed. This is usually done by updating the previously calculated noise levels using a reverse engineering method (based on a sound power assigned to each existing noise source and the distance to the measuring point) [11].

In several research studies, production of dynamic noise maps has been performed by including the citizens into the process of collecting the noise levels data in their surroundings instead of using noise monitoring stations. In that sense, citizens act as sensors and measure the level of noise using applications on their mobile phone or some other smart device [12–14].

Although the ultimate goal of making any noise map should be to determine the number of people exposed to noise, none of the mentioned approaches consider the dynamics of population movement. Even though such dynamic noise maps indicate different noise levels during the observed periods for which they are made, it is assumed that the population is constant in all observed locations, which is not the case in reality.

The first attempt to include the dynamics of a population's movements into the assessment of aircraft noise exposure was carried out by Ganić and Babić [15], followed by a series of papers by Ganić et al. [16–18] and Ho-Huu et al. [19]. In all these research efforts, the emphasis was on optimising the aircraft assignment to departure and arrival routes, while dynamic noise maps were created only for one-day scenarios to demonstrate the possibilities of the developed algorithm to reduce noise annoyance and fuel consumption. Furthermore, the calculations of daily population mobility included many assumptions and were based only on data from the census.

In this book, different ways of collecting human mobility patterns will be explained in more detail along with the methodology used to incorporate it into dynamic noise maps. Furthermore, a real case study conducted within the ANIMA project, based on the one-year air traffic data, will shed light on the benefits of using this new approach and demonstrate how daily movements influence the estimated population noise annoyance around an airport.

Human Mobility Patterns

In the sense of dynamic noise mapping, human mobility patterns (sometimes also referred to as population daily mobility or movement patterns) are defined as the movements of human beings (individuals as well as groups) in space and time. Motivation behind people's movements on a daily basis is manifold. While most common daily trips include commuting to and from work or school, they are also connected with the social, leisure and other activities.

During the last decade, substantial progress has been made in the study of human mobility. Not only by the significant advancement in the field of information and communication technologies enabling more accurate tracking of people's movements, but also in that the collection and processing of such data is more accessible to the general public.

While geography might be the first discipline to analyse mobility data and put forward corresponding theories to describe travel patterns [20], the study of human mobility currently spans several disciplines. It is widely used in transportation studies to describe how people plan and schedule their daily travel, as well as to provide better forecasts of future travel patterns.

A better understanding of human mobility patterns leads to more appropriate urban planning and infrastructure design, new tools to monitor health and well-being in cities, reduction of pollution, internal security and epidemic modelling, to name but a few. In this book, the special emphasis will be given on the use of human mobility patterns to estimate more accurately the number of people annoyed by aircraft noise.

Different Ways of Collecting Human Mobility Patterns

Collection of information on human mobility has a long tradition. Some of the well-known and widely used techniques to collect the data include surveys and question-naires. In particular, census data, collected periodically through national surveys, contain the questions related to the location of the workplace and school/faculty as well as the place of residence. By creating an origin–destination matrix, these data can be used to estimate daily/weekly commuting flows within a city or on a country level.

Another way of using questionnaires for collecting the mobility data is by conducting National travel surveys which have proved to be valuable for modelling and planning of transport systems. Compared to the census data, travel surveys contain more detailed information about daily activities of persons participating in the survey, including number of trips per day, origin and destination of each trip, time of the beginning of each trip, travel time and distance of each trip, purpose of a trip, mode of transport, etc. By using proven statistical methods, data collected through travel surveys allow us to simulate the movements of the whole population with great detail.

Due to technological development and the increase in usage of the internet, the methods to obtain the data have changed through the years, though the purpose mainly remained the same. Digital footprints produced by people using various digital services such as mobile phones, smartphone applications, or social networks, could provide valuable insights into their daily movement patterns.

These digital footprints can be classified as passive and active [21]. The main difference between them is whether they are left voluntary or involuntary. Many online activities such as tweeting or tagging a photo carry an information (electronic trail) about the location and timestamp that could be used to track the movement of the users when the frequency of such activities is satisfactory. Passive footprints are collected involuntary by using smart-card data, mobile phone records, GPS data, while active footprints come from the users themselves when they expose locational data in photos or messages while using social networks such as Twitter, Facebook of Foursquare and photo-sharing web sites, like Flickr or Instagram.

All these methods can be used to collect population mobility data necessary to develop dynamic noise maps. Nevertheless, numerous challenges can be encountered due to privacy issues. To protect the anonymity of individual persons or businesses, any data must be collected in accordance with the legislative framework, such as the General Data Protection Regulation (GDPR) that harmonises data privacy laws across Europe. Some of the measures can include signing a Confidentiality agreement and Declaration on Data Protection by the person who will use such sensitive data.

Methodology

To be able to assess the daily mobility patterns of the population, as a first step suggested here is development of an adequate travel model. Some broad types of models used in transportation planning include the following [22]: sketch-planning models, strategic-planning models, trip-based models, and activity-based models.

The choice of model depends on the purpose for which it will be used. Because activity-based models typically function at the level of individual persons and represent how these persons travel across the entire day, they are most suitable for dynamic noise maps. Since detailed explanation of the activity-based travel models goes beyond the scope of this book, only some brief description of the main concept will be given herein. For more details, interested readers may refer to [22].

Activity-Based Travel Models

Activity-based models consider activity and travel choices for each person throughout the entire day, taking into account different types and priorities of the activities that individuals are participating in [22]. The structure of activity-based models varies in the literature. Nevertheless, as shown in Fig. 11, between the model inputs and outputs, most activity-based models include the following major types of components [22]: synthetic population, longer-term and mobility choices, daily activity patterns, tour and trip details, and trip assignment.

The first component, synthetic population, represents a basis for predicting the behaviour of the households and persons in the modelled area. It contains anonymous microdata with the appropriate variables and granularity, statistically equivalent to the actual population of interest, to serve as input for micro-simulation agent-based models.

Next component involves modelling of choices that are made on a less frequent, longer-term basis, such as where to live, work or study. In addition, decisions whether to own a car, driving license, bicycles or transit passes also belong to longer-term and mobility choices that can significantly influence the availability and attractiveness of different location, mode, and scheduling choices that create daily activity and travel patterns. These choices are simulated for each agent in the synthetic population.

Conditional upon predicted longer-term and mobility choices, all travels during the day are simulated using day-pattern, tour-level, and trip-level models. The main

Fig. 11 Activity-based travel model components [22]

output of the daily activity pattern model is the exact number of tours that each individual makes for each of a number of different activity and tour purposes. Scheduling, location choice, and mode choice are all relevant at both the tour level and the trip level.

Scheduling model within the tour predicts the departure and arrival times for mandatory purposes (such as work or school), while the trip-level models simulate additional stops within the tour, for example on the way home, for other non-mandatory purposes (such as shopping, social visits, recreation, etc.).

Apart from the usual home and work/school locations that are predicted within the longer-term and mobility models, location model is used to predict the location of any intermediate stops along the tour, as well as tour origin and the tour destination.

Mode choice model predicts the use of mode at both the tour level and the trip level. In most cases, people are inclined to use the same mode for an entire tour, where the selected mode is the same for each trip within the tour. Nevertheless, there are also infrequent cases of multimodal tours which could include carpooling or park-and-ride concepts.

The final step in the activity-based travel modelling is to assign simulated trips to the networks. The whole process could be iterated to recalculate the travel times or some other parameters, if needed.

Calculation of Dynamic Noise Maps

After obtaining daily mobility patterns of population, the next step is to extract the distribution of people at desired spatial and temporal resolution. The most detailed spatial resolution would include every single location where people spend time. Nevertheless, such a detailed approach is not practical nor needed for airport noise impact studies since the aircraft noise levels do not differ significantly among closely located points. Another approach is to aggregate points into grid cells (e.g. 500 × 500 m) and to calculate noise levels only for grid cell centroid which will then represent all the points within that cell.

On the other hand, temporal resolution will depend on the change in number of people at different locations and frequency of activities in the observed model. The minimum temporal detail should include at least four or five time periods in the day, as opposed to some models that use continuous time (e.g., 1,440 one-minute periods in the day). Furthermore, temporal resolution could be observed separately for working and nonworking days since population daily mobility patterns can be completely different from each other.

The noise metric that needs to be calculated for each location is LAeq,T or the A-weighted, equivalent continuous sound level determined over time period T. After calculating LAeq noise levels, the next step is to match the number of people exposed to those noise levels at each location (spatial resolution) during each time period (temporal resolution). As a result, cumulative noise impact for each person for the whole day could be calculated.

For example, if the model uses 24 time periods in the study, LAeq,1 h will be used, where 1 h denotes the one-hour time period over which the fluctuating sound levels need to be averaged based on yearly average traffic for each hour. The overall traffic for the whole year should be divided into hourly periods, separately for working days (Monday to Friday) and non-working days (Saturday and Sunday) and then averaged (divided by number of days which will be different for working and non-working days). As a result, 48 different scenarios will need to be calculated.

How Do Human Mobility Patterns Influence the Population Noise Exposure Around an Airport?

To demonstrate to what extent human mobility patterns could influence population noise exposure, let us observe an example of Heathrow airport and work-related commuting patterns to and from one local authority. London Borough of Hounslow, located east of the airport, is used here as an example since the largest part of its territory is situated within the Heathrow airport L_{den} 55 dB noise contours.

Figure 12 shows work-related commuting from and to (number shown in brackets) Hounslow for each local authority around the airport. According to the 2011 UK Census, there are 102,654 residents aged 16 and over in employment living in Hounslow. By analysing the location of usual residence and place of work, it is detected

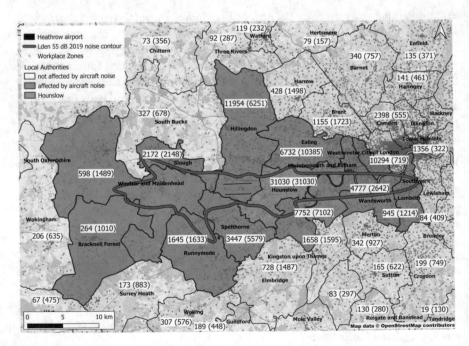

Fig. 12 Work-related commuting from (to) Hounslow

that only 30% of them (31,030) work in Hounslow, while the rest of the residents work in other local authorities. This implies that for 70% of the working population within this area (71,624 residents), the noise exposure could be incorrectly estimated (probably overestimated) since they will be spending a large portion of the day (at least while during their working hours) outside of their usual residence. For example, 11,954 residents of Hounslow (11.64%) work in London Borough of Hillingdon, while 10,294 (10.03%) of them work in City of Westminster and City of London which are outside of the L_{den} 55 dB noise contours. There are 43,730 residents who commute from Hounslow to 12 local authorities that are also, at least partially, affected by aircraft noise. This indicates that more than half of Hounslow employees work further away from the airport, in the areas not affected by aircraft noise, thus being exposed to less noise than anticipated based only on the residential location.

On the other hand, Fig. 12 also shows that, out of 105,007 employees working in Hounslow, 70% of them (73,977) live outside this local authority. Most of these employees (59.2%) travel from local authorities where the aircraft noise levels are considered insignificant, since only 42,831 of them (40.2%) are residents of 12 local authorities affected by aircraft noise, such as nearby Ealing (10,385) and Richmond upon Thames (7102).

When combining the employees traveling to and from Hounslow, this work-related commuting results in the presence of additional 2353 people within this area during the working hours compared to the number of residents. Nevertheless, the main change in individual noise exposure comes from the fact that 70% of the residents will experience noise levels different from the expected one at their residential locations. The biggest difference between inflow and outflow of commuters is observed for the City of Westminster and City of London where 10,294 of residents commute from Hounslow to work there, while only 719 of their employees work in Hounslow. This is easily explained having in mind that the highest number and concentration of workplace zones is in this part of London.

The similar conclusion can be drawn from the commuting patterns of high school pupils and students, while the proximity of elementary schools and kindergartens to residential locations makes the commuting patterns of the youngest members of the society irrelevant for this kind of analysis. Apart from the trips from and to work or school, which are regarded as main or mandatory activities, there are many additional non-mandatory activities with various trip purposes that are also relevant for dynamic noise mapping. Timing when these activities occur as well as their duration also affect the noise exposure of individuals performing these activities. Results from London Travel Demand Survey, conducted by Transport for London, shown in Fig. 13, indicate that travel patterns are steady throughout the years in terms of the time of the beginning of the trip. The highest peaks, when most of the trips start, are observed from 7 to 9 AM and again from 3 to 5 PM. This is usually in correlation with the time when people leave for, and return from, work or school.

Usual time of the beginning of the trip differs significantly for different travel purposes. Figure 14 shows the results from the National Travel Survey conducted in the UK, where all the activities are combined and presented as eight different

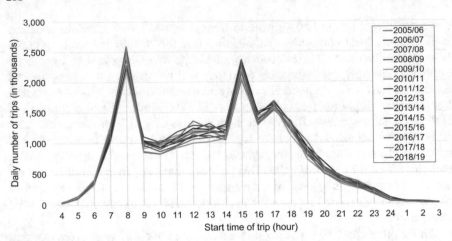

Fig. 13 Trips by start time in London (Source: London Travel Demand Survey)

purposes of trips. At a first glance, obvious differences in trip start times for different trip purposes can be observed.

Although for some trip purposes, such as business (Fig. 14b) or holiday (Fig. 14h), there is a fairly uniform distribution of trips during each hour of the day, for most of the purposes a pronounced peak can be clearly seen. For example, the largest number of employees start their journey around 6 AM, while the second peak with a much smaller number of trips occurs around 6 PM (Fig. 14a). For trips with the educational purpose (Fig. 14c) including escorting trips (Fig. 14d), there are two approximately equal peaks, around 9 AM and 4 PM. Most of the shopping activities start within the period from 11 AM to 3 PM with the moderate intensity also shown during evening hours until the midnight (Fig. 14e). As expected, only activities that include entertainment and visiting friends are dominant during late evening and early morning hours (Fig. 14g).

All these differences are considered when developing human mobility patterns for dynamic noise mapping. They influence the temporal and spatial distribution of people, thus leading to population noise exposure other than expected when the movements of residents are disregarded in the noise mapping process.

ANIMA Case Study: Dynamic Noise Maps for Ljubljana Airport

The case study that will be presented here aims to show how population movements affect the estimated number of people exposed to aircraft noise. The research was done within the ANIMA project, to demonstrate the capabilities of dynamic noise mapping compared to the traditional way of developing noise maps. More detailed

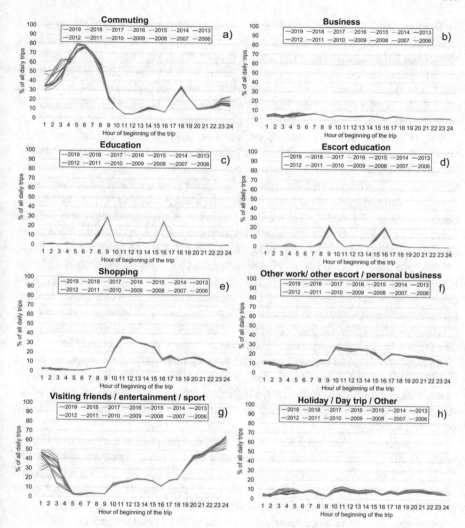

Fig. 14 Trip start time by trip purpose (source: UK National Travel Survey)

information about the methodology and input data for this study can be found in [23].

Ljubljana Jože Pučnik Airport was chosen for the case study. It is the busiest airport in Slovenia, handling more than 1.7 million passengers and approximately 31 thousand aircraft operations in 2019. The airport has a single runway 3300 m long and is located 20 km northwest of the Ljubljana capital. Even though the Ljubljana airport is not recognised as a "major airport " as defined in the Directive 2002/49/EC, proactive noise assessment actions have been undertaken by the airport authority including the development of noise contour maps and regular continuous noise monitoring in the most noise exposed areas for several years.

The year 2018 has been selected for this case study based on the availability of data regarding population mobility patterns and air traffic. The yearly traffic at Ljubljana airport for 2018 consisted of 35,512 aircraft operations, performed by 213 different aircraft types, indicating that on average, there were approximately 97 operations per day, or 4 movements per hour. There were several peaks during the day, with the most flights occurring between 5 and 6 PM (nine flights in average), while only 6% of operations were performed during the night hours.

The population mobility patterns are assessed using a dedicated national travel survey. The most recent one conducted in Slovenia was the Daily Passenger Mobility Survey (TR-MOB 2017). The data were obtained from the Statistical Office of the Republic of Slovenia through the special request for protected microdata containing detailed information about each trip on an individual level. The number of respondents in this survey was 8,842, while the number of trips conducted was 24,195. Since 1355 (15.3%) survey participants reported that they stayed at home, the average number of trips per day can be calculated as 2.7 or 3.2, depending on whether all the respondents are considered or only the ones conducting the trips. In terms of different trip purposes, leisure activities were the reasons behind most of the recorded trips (35.6%), followed by commuting trips to and from work (24.3%). Other trip purposes included education, professional and personal business, shopping, and escorting (driving/picking up/accompanying a child or other person). More information about this survey, including explanations about methodology, is contained in [24].

Once all the data had been collected, the A-weighted equivalent continuous sound level was calculated to match the number of people exposed to those noise levels at each location (500 m × 500 m grid) during each one-hour period. As a result, the cumulative noise impact for each person has been assessed and the results are presented herein.

The results lead to the conclusion that even though people live at locations enclosed in the 37 dB L_{den} noise contour (the threshold for becoming annoyed according to [25]), 10.1% of them (marked with white square symbol on Fig. 15) are not exposed to aircraft noise levels (L_{den}) above 37 dB due to daily mobility to locations away from the airport. Furthermore, apart from the 4884 people that are living within the presented noise contour (marked with red star symbol), there are additional 704 persons (14.4%) also experiencing aircraft noise exposure (marked with yellow triangles) even though they are located outside the 37 dB noise contour. This can be explained by considering that people who live outside the area affected by aircraft noise may work or study within these areas at some time during the day and are, therefore, affected by aircraft noise. The fourth group of people (marked with grey circles) resides outside this noise contour and is not affected by aircraft noise, even when the daily mobility patterns are considered.

In order to better demonstrate the dynamics behind this novel approach, four different noise maps with estimated number of people at each location are presented in Fig. 16 describing: (a) night (02–03 h), (b) morning (08–09h), (c) afternoon (14–15h) and (d) evening (20–21h) periods. This figure illustrates the temporal and spatial variation in population around Ljubljana airport. In addition, the change in the number of operations between each hour is clearly visible through the different shapes and areas

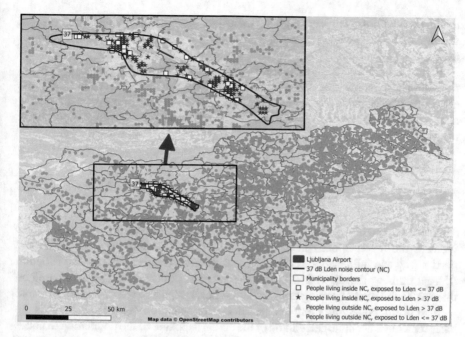

Fig. 15 Dynamic noise map for Ljubljana airport

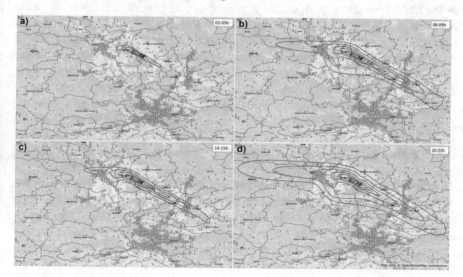

Fig. 16 Simulation of dynamic noise maps

that calculated noise contours take. As expected, the smallest changes are observed during night hours, when the traffic frequency is low as well as the movements of the people.

It should be borne in mind that this case study only considers L_{den} as a factor for estimating annoyance. As shown within the ANIMA project, aside from noise exposure, non-acoustic factors have a considerable influence on perceived annoyance, and they should also be taken into account in future research on dynamic noise mapping.

Further Developments

There are several directions for the future development of dynamic noise maps. This includes the involvement and contribution of several stakeholders and interested parties who can influence the collection and quality of necessary data, the adoption of legal frameworks, as well as the management of airport operations, all based on the new knowledge regarding the population noise exposure due to daily migration.

First of all, collecting more detailed data on population daily movements is the basis for a more accurate calculation of the actual exposure of the population to noise. Therefore, encouraging active public involvement, especially of the residents living in the vicinity of the airport, would significantly improve the quality of dynamic noise maps. One of the steps towards achieving this goal is to motivate citizens to contribute through participation in daily mobility surveys or by using dedicated applications that can collect data on people's movements. In that sense, it is pivotal to educate the citizens such that they could understand more clearly the needs of the airport and be more willing to participate in such endeavours. It goes without saying that by giving the necessary and truthful information about their movement patterns, residents are helping the airport to better solve the noise problem they could be faced with. On the other hand, it is also advisable to educate the airport authorities to adopt and take advantage of the new technologies and approaches through which dynamic noise maps can be implemented in practice more easily.

Although the above-mentioned surveys are mainly organised and conducted by the National Statistical Institutes, usually on a country level, one of the future directions may stimulate airports to also embrace such activities. In that regard, the airport authorities could conduct more detailed and more frequent surveys on a smaller sample, that could primarily include settlements around the airport that are most affected by aircraft noise. In that way, the movement habits of the local residents could be determined more accurately. Estimation of the number of people highly annoyed by aircraft noise, with such new data, could give the airport authorities a new perspective on the noise issue around the airport. Through such dedicated questionnaires, residents could also express their preferences about the periods of the day when the aircraft noise bothers them the most, or vice versa to indicate to the airport when, due to the nature of their activities, noise is not an issue since they could be far away from their residences. If applicable, airports could use such information when negotiating with the airlines about the seasonal schedule in order to reduce annoyance. Certainly, this approach is in line with the modern aspirations of every airport that has a noise issue since the current focus is primarily on improving communication with the residents, sharing information, and providing transparent reporting about the airport noise to the general public.

Air navigation service providers (ANSPs) could also benefit from dynamic noise maps and use them to reduce the adverse impact of noise on the population. The number of people in an area is a vital indicator for the noise impact analysis and should therefore be considered when making decisions regarding air traffic assignment that influence the noise allocation. Population noise exposure reduction can be achieved by optimising the distribution of aircraft on arrival and departure routes, by considering spatial and temporal variations in the number of inhabitants in the settlements around the airport, since these data are available in dynamic noise maps. One of the future developments of this approach could lead to the inclusion of dynamic noise maps into a decision support tool that could help air traffic controllers in their activities either on tactical, pre-tactical or strategic level. There are several ongoing research efforts in this direction which will allow the ANSPs to minimise the number of people exposed to noise while using the benefits of dynamic noise maps.

Finally, to apply the presented dynamic noise maps approach globally, it is pivotal to involve the policymakers who have the power of setting and directing regulatory frameworks that should follow the developments in this area. All current studies conducted on the basis of a legally imputed obligation, as is the case of strategic noise maps, consider the noise level on the most exposed façade of the building. All reported numbers of people exposed to noise are attributed to all persons living in the buildings according to the census data, regardless of their actual location. Future regulatory developments should consider the inclusion of population daily mobility patterns in the noise mapping process in order to assess the impact of noise on the population more realistically. Therefore, it is key to inform the policymakers about the possibilities of dynamic noise maps and their advantages compared to traditional noise maps currently in use. More detailed research on this topic should provide guidelines to the policymakers on how to incorporate this approach into the legislation so that any airport that has a noise problem could benefit from a dynamic noise mapping approach.

Acknowledgements Figures 4, 5, 6, 7, 8, 9, 10, 12, 15 and 16 are using OpenStreetMap data. OpenStreetMap® is open data, licensed under the Open Data Commons Open Database License (ODbL) by the OpenStreetMap Foundation (OSMF). https://www.openstreetmap.org https://opendatacommons.org

References

1. Report on Standard Method of Computing Noise Contours around Civil Airports (2019) ECAC Doc 29–4th Edition (7 December 2016). Downloadable from www.ecac-ceac.org

2. van Oosten N (2004) SONDEO: a new tool for airport noise assessment. InterNoise 2004, Prague, Czech Republic, August 22–25
3. ICCAN (Independent Commission on Civil Aviation Noise), July 2020: A review of aviation noise metrics and measurement. https://iccan.gov.uk/iccan-review-aviation-noise-metrics-mea surements/
4. EU parliament (2020) Impact of aircraft noise pollution on residents of large cities. Policy Department for Citizens' Rights and Constitutional Affairs, https://www.europarl.europa.eu/RegData/etudes/STUD/2020/650787/IPOL_STU(2020)650787_EN.pdf
5. WHO (2018) Environmental Noise Guidelines for the European Region. World Health Organisation, ISBN 978 92 890 5356 3. https://www.euro.who.int/__data/assets/pdf_file/0008/383 921/noise-guidelines-eng.pdf
6. Basner M (2008) Aircraft noise effects on sleep: substantiation of the DLR protectionconcept for airport Leipzig/Halle. In: Proceedings of 9th international congress on noise as a public health problem (ICBEN) 2008, Foxwoods, CT, pp 772–779
7. Mishra RK, Nair K, Kumar K, Shukla A (2021) Dynamic noise mapping of road traffic in an urban city. Arabian J Geosci 14(2). https://doi.org/10.1007/s12517-020-06373-9
8. Kozielecki P, Czyzewski A (2008) An application for vector-based dynamic noise maps generation. In: Joint Baltic-Nordic acoustics meeting
9. DYNAMAP (2014) [Online] Available http://www.life-dynamap.eu/project/
10. Benocci R et al. (2019) Dynamic noise mapping in the suburban area of Rome (Italy). Environments 6(7). https://doi.org/10.3390/environments6070079
11. Simón Otegui L et al. (2019) Dynamic Noise Map based on permanent monitoring network and street categorisation. In: INTER-NOISE and NOISE-CON congress and conference proceedings, vol 259(2), pp 7270–7281
12. Poslončec-Petrić V, Šlabek L, Frangeš S (2016) With the crowdsourced spatial data collection to dynamic noise map of the city of Zagreb. In: International symposium on engineering geodesy SIG 2016, pp 411–423
13. D'Hondt E, Stevens M, Jacobs A (2013) Participatory noise mapping works! An evaluation of participatory sensing as an alternative to standard techniques for environmental monitoring. Pervasive Mob Comput 9(5):681–694. https://doi.org/10.1016/j.pmcj.2012.09.002
14. Drosatos G, Efraimidis PS, Athanasiadis IN, Stevens M, D'Hondt E (2014) Privacy-preserving computation of participatory noise maps in the cloud. J Syst Softw 92(1):170–183. https://doi.org/10.1016/j.jss.2014.01.035
15. Ganić E, Babić O (2017) Air traffic assignment to reduce population noise exposure: an approach incorporating human mobility patterns. In: 21st Air transport research society world conference
16. Ganić E, Babić O, Čangalović M, Stanojević M (2018) Air traffic assignment to reduce population noise exposure using activity-based approach. Transp Res Part D: Transp Environ 63:58–71. https://doi.org/10.1016/j.trd.2018.04.012
17. Ganić E, Ho-Huu V, Babić O, Hartjes S (2018) Air traffic assignment to reduce population noise exposure and fuel consumption using multi-criteria optimisation. In: Proceedings of the 26th international conference noise and vibration, pp 69–76
18. Ganić E, Babić O, Čangalović M, Stanojević M (2017) Air traffic assignment to reduce population noise exposure: an approach incorporating human mobility patterns. In: XLIV International symposium on operational research, pp 746–751
19. Ho-Huu V, Ganić E, Hartjes S, Babić O, Curran R (2019) Air traffic assignment based on daily population mobility to reduce aircraft noise effects and fuel consumption. Transp Res Part D: Transp Environ 72:127–147. https://doi.org/10.1016/j.trd.2019.04.007
20. Barbosa H et al (2018) Human mobility: models and applications. Phys Rep 734:1–74. https://doi.org/10.1016/j.physrep.2018.01.001
21. Girardin F, Calabrese F, Fiore FD, Ratti C, Blat J (2008) Digital footprinting: uncovering tourists with user-generated content. IEEE Pervasive Comput 7(4):36–43. https://doi.org/10.1109/MPRV.2008.71

22. Castiglione J, Bradley M, Gliebe J (2014) Activity-based travel demand models: a primer. Transportation Research Board, Washington, D.C.
23. Ganić E, van Oosten N, Meliveo L, Jeram S, Louf T, Ramasco JJ (2020) Dynamic noise maps for Ljubljana airport. In: 10th SESAR innovation days
24. Statistical Office of the Republic of Slovenia (2019) In: Methodological explanation daily passenger mobility. Ljubljana, Slovenia
25. European Commission (2002) Position paper on dose response relationships between transportation noise and annoyance

ANIMA Noise Platform and ANIMA Methodology: One-Stop Shop for Aviation Noise Management

Alexandra Covrig and G. Heyes(iD)

Abstract When you think about aviation noise, you might imagine an airplane taking off. When you think about decreasing aviation noise, the first thing that usually comes up in one's mind are the new silent plane engines. This makes perfect sense, but it does not fully grasp the issue of aviation noise. The ANIMA project is based on a holistic approach to aviation noise, as it focuses on non-acoustical factors as well. Annoyance, as perceived by local communities surrounding airports, also depends on non-acoustical factors, which can be situational (time of the day, day of the week, activity performed while exposed to noise) and personal (sensitivity to noise, attitudes, noise insulation).

Keywords Airport communities · Communication · Community engagement · Design thinking

How is ANIMA Different and What is its Added Value

While seeking to better understand annoyance, ANIMA observed that aviation noise is not only an engineering issue which requires reducing noise at source. Reducing annoyance only by using quieter aircraft is indeed helpful, but not enough to make an airport a good neighbour to the local residents. Therefore, ANIMA takes a different and innovative stance to aviation noise. The project carried out research from an

---

Illustrating How ANIMA is Endeavouring to Propose a One-Stop Shop Where Various Aviation Noise-Related Stakeholders Would Find Possible Solutions for the Challenges They Are Confronted with.

A. Covrig (✉)
Airport Regions Council, Rue Montoyer 21, 1000 Brussels, Belgium
e-mail: alexandra.covrig@airportregions.org

G. Heyes
Ecology and Environment Research Centre, Department of Natural Sciences, Manchester Metropolitan University, Chester Street, Manchester M1 5GD, UK
e-mail: g.heyes@mmu.ac.uk

interdisciplinary approach, by bringing together aircraft engineers, urban geographers, psychologists, sociologists, noise experts and regulatory experts. This diverse and comprehensive partnership resulted in novel approaches to aviation noise impact management and mitigation. ANIMA is not the traditional aviation noise research project, since it aimed at complementing the existing intensity-averaged noise maps by dynamic profile-dependent annoyance maps.

In addition to its novel approach to aviation noise impact management, the holistic character of ANIMA stems also from the fact that it builds on other previous projects, such as X-NOISE, SEFA (Sound Engineering For Aircraft), COSMA (Community Oriented Solutions to Minimise aircraft noise Annoyance) and TEAM PLAY. X-NOISE was a Coordination and Support Action project which focused on aircraft noise and on lowering the noise exposure of communities. The project coordinated research activities and created an aero-acoustical knowledge base. ANIMA not only maintained the legacy of X-NOISE, but it extended its role through the setup of a specific committee gathering other relevant EU project coordinators. It also developed and consolidated a scenario-based, impact-driven strategic roadmap for aviation noise research. Throughout its four years of activity, ANIMA has been successfully leading the global coordination of European research efforts on aviation noise and it encouraged the wider network of experts and stakeholders, at both European and national levels, to maintain and enrich the roadmap developed within the project. As for SEFA, it was the pioneering EU project on aircraft noise impact. It included laboratory hearing tests and started developing the Virtual Resident tool. ANIMA followed up on SEFA by further developing the tool into a new inclusive version: Virtual Community Tool. This version comprises new scenarios with more accurate and thorough behavioural reactions to aircraft noise. The tool allows users to test traffic around airports as well as possible evolutions with new aircraft or flight scenarios. After SEFA, COSMA further built on this project, as it focused on laboratory tests and field investigations on noise in order to develop engineering criteria for aircraft design and operations that help reduce annoyance. ANIMA expanded the scope of COSMA, by exploring management and community engagement, rather than looking at noise itself. Regarding TEAM PLAY, the project created a modelling framework to support the European perspective in the international policy arena. ANIMA added to this framework an augmented modelling capability related to annoyance and a noise management toolset designed to enable use by the wider audience.

Looking at the development process of ANIMA, it can be noticed that there are two underlying elements at the core of the project: non-acoustical factors and communication. When addressing annoyance, non-acoustical factors and communication are at the centre of ANIMA's unique approach.

Community Engagement in ANIMA Project

Noise management in airport areas can only be successful if all parties, including the ones contributing to noise and those who can hear the noise, are engaged in dialogue at the same table and benefit from a common understanding of what is at stake.

Interventions meant to reduce annoyance should be designed based on structured exchanges with communities. Lack of transparent communication and fair exchanges usually lead to failed interventions, since the needs and expectations of the diverse stakeholders tend to differ. ANIMA understood that different stakeholders have different needs and there is no universal solution, but open dialogue can pave the way towards consensus. Structured exchanges between airports and communities must be fair, meaning that communities must benefit from distribution of user-friendly information which avoids technical jargon and that the standpoints of all stakeholders are taken into account. To reach consensus, shared and restored trust are prerequisites.

In this sense, ANIMA engaged with several local communities, such as residents from Gava (Spain), Iasi (Romania), Brussels (Belgium) and Kranj (Slovenia). Thanks to these numerous meetings and interactions, it became clear that there are still many knowledge gaps among relevant stakeholders. From these encounters, ANIMA learnt that, when coping with aviation noise, some residents feel "helpless", "left-out" or "unaware". In these circumstances, open and fair dialogue, which ensures equally-beneficial outcomes for everyone, can only occur when everyone is equipped with the same knowledge and understanding. To discuss aviation noise, pre-existing knowledge about a broad spectrum of different fields, such as aviation noise management, airport management, avionics, noise and exposure, health impacts and human behaviour, legislation, policymaking, might be necessary. No one can be an expert in everything, so there is a need in finding and understanding the missing parts of the noise puzzle. To facilitate engagement, communication must be inclusive, transparent and most-importantly, it must be a two-way process, where all parties can add to the dialogue. If a fair, inclusive and transparent decision-making process is set up with all stakeholders, including neighbouring communities, then authorities and airports must be ready to accept and endorse the consensus reached through the process. The goal is to develop a common noise policy for impact mitigation.

The main takeaway that can be drawn from the ANIMA events is that when it comes down to noise management, prevention and proactivity are key. If legislation is not yet available at the degree of needed protection, initiatives to increase quality of life must still be taken at national, regional and municipal level. Oftentimes, the level of awareness is not the same among stakeholders, hence the importance of working collaboratively towards common noise policy which benefits all parties. Better awareness and knowledge on different noise sources and indicators would support the understanding of the impact that noise has on human health and well-being in different ways.

ANIMA Noise Platform

As previously stated, noise is not only a technical issue, but also a social, regulatory and political issue. To mitigate impact, stakeholders, ranging from policymakers to residents and from airports to manufacturers, have to work together towards solutions.

To encourage collaborative decision-making on the development of common noise policies and to ensure engagement with local communities, ANIMA project created a Noise Platform. This platform provides the medium and tools for stakeholders to address the challenge of aviation noise exposure. The Noise Platform is the result of the successful collaboration between aircraft and airport engineers, noise specialists, urban geographers, psychologists, sociologists and experts on aviation regulation, who sought to better understand the annoyance and to develop best practice solutions to alleviate this burden. The ANIMA partners carried out research and engaged in dialogue with policymakers, airports, and noise-affected communities to assess how aviation noise mitigation interventions are implemented and how affected these measures are in reducing annoyance and improving quality of life. This platform captures the results of this research, offering an overview of aviation noise regulation and how to implement it, as well as current gaps and new solutions to bridge those gaps. It also provides tools, such as a mobile application, which is meant to help airports and authorities to capture how local communities perceive annoyance. A Noise Management Toolset was also developed and is available on this platform. This toolset aims to help airports and authorities to compute noise maps and awakening indexes in order to test the impact of different scenarios with various fleet configurations and flights. In addition, the platform gives access to an enriched Aviation Noise Research Roadmap, which supports policymakers in defining future policy and research goals. Other ANIMA results are available on the platform, such as scientific publications summarising key findings on how to address aviation noise impact as well as more tools fostering community engagement and building transparent working relations across all stakeholders. The ANIMA Noise Platform is open to everybody, featuring free and user-friendly content.

The ANIMA Noise Platform was designed and built by applying and following the principles used in the ANIMA methodology of noise interventions and management, which will be covered in subsections below.

'Designing' Effective Noise Management Measures

At its core, noise management is a process of problem solving. It sees airports seeking to provide their service (facilitating air transport) in a way that causes as little noise impact as possible on residents, via a range of noise management measures (i.e. those described under the ICAO Balanced Approach elements), including 'people issues', or communication and engagement. Airports do this due to legislation that exists to protect residents from the health impacts of noise, or as part of their social-license to operate, by demonstrating to residents that they are doing all that they can to minimise noise, and thus reduce the likelihood for complaints that otherwise can constrain airport activity. In this sense, noise management actions and interventions can also be seen as services. They are things that the airport provides for the benefit of external beneficiaries, and hence the perceived success of those interventions depends very much on understanding and responding to the needs, demands, and wishes of those stakeholders. It is for this reason why, as described in Chap. 11, considering those perspectives, through communication and engagement is so crucial to the design of effective noise management strategies.

Most business activity is not actively designed (Downe, 2020). This is the case whether it be at the process or system level, the business model level, or the strategic level. Rather, the things that businesses do tend to organically evolve over time as individual problems arise and are solved in turn. The result can be a mesh of organisational processes and strategies that have never actually been holistically designed in a systematic or targeted way, instead, they essentially exist to solve problems (either for the business or for stakeholders), often with increasing levels of complexity or bureaucracy.

If most services are not designed, it poses the natural question of how design principles and processes can help to inform their development. Indeed, there is now an entire industry dedicated to concepts such as Design Thinking and Service Design, supported by a range of academic research and publications, including a focus on the management of environmental management issues such as noise. Design and design thinking concepts have also been applied in a range of aviation contexts, including challenges related to airport and air cabin design, safety, and security and by organisations such as airlines, the military, and NASA (Hall et al. 2013; Goodheart 2016; This Is DesignThinking.net, 2016; McGowan et al. 2017; Turner Donald 2017; Design For Europe 2020). In terms of noise design-led approaches are also advocated in national guidance such as the United Kingdom's Civil Aviation Authority CAP1616 document and the United States Federal Aviation Authority Program 150 document, which both use design process and principles to inform the airspace design.

Below, the concept of design is introduced and its potential role in noise management is described. We then present the ANIMA Method for designing and delivering noise management interventions, devised through a series of case studies with airports from across European Member States.

Design-Led Approaches to Noise Management

The wide range of characteristics that define each airport and the challenges that they face means that there is no ubiquitous solution to noise that can be applied to all airports. Instead, airports need to design their own tailored solutions to the distinctive challenges that they face. They need robust yet flexible approaches that can take them from the challenges that comprise a range of unknowns, to the development and implementation of solutions that solve core issues based on a deep understanding of the challenges faced. As ANIMA research has shown, and as described in Chap. 11, this process should embed concepts of stakeholder engagement and two-way dialogues that develop empathy for residents, and to develop outcomes that are perceived as fair by, where possible, incorporating elements of public participation into decision-making processes. Developing noise management interventions should also seek out measures that are able to address core problems that trigger (for instance) complaints rather than seeing complaints as the challenge to be solved. In other words, they should address the cause of noise issues rather than the symptom.

Importantly, noise management measures, to be truly effective, have to comply with three requirements. They must be:

- *Viable* in terms of complex factors such as safety, security, environmental interdependencies and legislative compliance;
- *Feasible* in terms of airport infrastructure and financial capabilities; and
- *Desirable* to industry and community stakeholders.

This sort of thinking can be incredibly helpful in empowering organisations to understand not just what is possible and best suited to their own needs, but also to those of their stakeholders. This is of critical importance for noise management where the perspectives of those for whom noise management measures are designed to benefit can have a significant influence on the perceived success of those measures. This is particularly the case considering the important role of non-acoustic factors that are referred to throughout this book. ANIMA research has shown that airports have historically performed well in terms of feasibility and viability as these are generally technical driven considerations that the industry has been managing for many decades. These criteria are complex, but they can be informed by quantitative monitoring and modelling data that can act as an evidence base to support and communicate the decision-making process to stakeholders, and to evidence the success of given measures. The desirability of noise management decisions, i.e. in the eyes of stakeholders, is however a much more challenging concept as it inherently requires the collection of qualitative data that requires a particular set of skills, can be time-consuming and involves a range of complex and conflicting viewpoints. Design-led approaches can help in this regard. 'Design thinking' is a human-centred (van der Bijl-Brouwer and Dorst 2017) and iterative step-approach to solving problems that acknowledges that there are many levels of understanding required to solve the complex challenges faced by organisations and that these perspectives must be considered for outcomes to be regarded as successful. Design thinking proposes an iterative process through which it is possible to move from a hypothetical starting position with many unknowns, towards solutions built on the needs of those who end-users by addressing the root cause of the problem at hand. It does this not by providing a rigid set of rules and actions. Instead, design thinking creates a series of spaces in which different types of activities take place (Brown 2008; Liedtka 2015), for example, empathy building, learning, brainstorming, trialling and implementation. As a flexible process, there is no agreed definition of what design thinking looks like, but generally, it is described as comprising four key phases:

- *Discovery*: A research phase in which an initial challenge or problem is explored by researchers in order to obtain a range of data (quantitative and qualitative in nature) that can be used to get a deeper understanding of the problem at hand. For noise management, this may include the monitoring and modelling of noise data, but also speaking to industry stakeholders (i.e. national airspace providers and airlines) and community stakeholders (i.e. local authority and community members) for their important and essential insight.
- *Defining*: An interpretative phase where the collected data is analysed with the aim of providing key insights into problems faced.
- *Design:* With a deeper level of insight, creative brainstorming design exercises take place to identify as many potential solutions as possible. Consultation events and the use of dialogue forums and other community groups is an increasing trend where design options and their selection are increasingly informing decision making in aviation.
- *Delivery:* The most promising solutions can be implemented through iterative processes of testing and trials, to understand the validity and likely impacts of the designed solutions, and their potential scaling up to full deployment. This is similar to approaches already undertaken in aviation where trialling of, for example, operational procedures typically takes place before full implementation.

As shown above, the aviation industry often conducts each of these phases as important activities, however, there does not exist any standardised approach that can help airports move through these processes in a systematic way that can be evidenced to stakeholders. Design thinking is one such way and has informed the development of the ANIMA Methodology described below.

The ANIMA Methodology

Design Thinking is similar to an approach described in ANIMA research to guide airports in developing noise management measures. The work posited that effective noise management follows a similar multi-step iterative process, which poses a range of questions. These steps and a range of example questions are provided below.

Identification of the Need for an Intervention

At this stage, an airport becomes aware of the need to implement a noise management measure. This means that airports should seek to learn as much about the situation as possible to help inform decision making processes and the ultimate design of the measure. The idea for an intervention may arise from a number of sources. Perhaps a new technology has become available, such as performance-based navigation, that the airport wishes to implement. It could be that communities have identified an opportunity for a change, or have been campaigning for a change in airspace design. An airport could be seeking to build new infrastructure or be responding to legislation. The key thing to consider here is to not take the initial identification at face value, but rather seek to understand as much as possible about the change. To understand, for example, if communities are complaining, what the reasons for those complaints are, and seeking to understand if the proposed changes will adequately make a positive difference. Stakeholders who have influence over the potential changes should be spoken with and engaged in a two-way dialogue—this, importantly, includes residents. Helpful questions and thinking at this stage may include:

- Do you have multi-stakeholder and independently led stakeholder engagement pathways (including community representatives) through which the requirement for an operational change could be communicated and discussed?
- Are all communities represented in such engagement activity, so that any redistributive effectives on noise exposure can be systematically addressed and consensus built as to the most socially optimal outcome(s)?
- Are such stakeholders and community groups engaged openly and transparently to establish trust? Is noise data made available on-line for those not able to attend such forums?
- Are there other avenues through which communities or other stakeholders can raise concerns with noise managers and/or make complaints?
- Are the concerns of those contacting an airport acknowledged? Are individuals provided with tailored responses relevant to their specific concern, rather than via template responses?
- Have you taken the time to question all assumptions about the need for the measure, and attempted to define it accurately? It can be helpful here to redefine the noise problem into a 'how might we question' that can more easily lead to actionable outcomes.

Design of Options

It can be helpful, when designing how the noise management measure is implemented, to conduct design approaches in two phases. A 'green-light' brainstorming phase in which there are no wrong answers can help to generate potentially novel solutions and things can be discarded too easily for being infeasible. Rather, green-light brainstorming captures as many potential opportunities as possible and instead reviews them in a secondary—'red-light' brainstorming phase, where the proposed options are evaluated more critically. It can help to do this by using an evaluation matrix that may map options against criteria such as 'ease of implementation' and 'impact'. Doing so can help not just to determine what options have potential, but it can help to identify a pathway through which different options might be implementable. Stakeholder engagement and consultation is important at this stage as it helps to ensure that, for example, resident needs and perspectives are included in the design process, and so that more desirable outcomes can be achieved that are more likely to lead to what are perceived to be more acceptable results. Including residents in the decision-making process in this way also helps to embed the types of principles linked to perceptions of control and engagement as described in Chap. 11. When designing solutions, it can be helpful to think about questions such as:

- Are all stakeholders given the possibility of designing their own solutions to the required change, or informing in the discussion of the development of other options?
- Do stakeholders have the opportunity to work in collaboration with each other in identifying potential noise mitigation solutions?
- Are designs pre-informed by a set of criteria and objectives, for example by framing them within what is logistically feasible, safe, and regulatory compliant?
- Has time been taken to consider radically innovative solutions that may have been missed?
- Have combinations of solutions been considered rather than pursuing just one pillar of the Balanced Approach?
- Has the role of communication and engagement been considered as part of the design, for instance how non-acoustic factors may be impacted?

Selection of Intervention Option

With a range of design options developed, the most appropriate should be selected. This may include the use of a decision matrix that may include elements such as impact, desirability, feasibility and viability. Different options can be scored against each and the highest scoring option is thus the one that can be taken forward. It can be helpful here to consider long term solutions to the problem that could be implemented over longer-term periods through implementation pathways. Thus can be achieved by creating an idealised vision of what the solution to the given problem might look like in the future, and then working backwards from the idealised future state towards the present, thus creating a pathway of what needs to happen in turn for that vision to be realised. Doing so can reveal actions that need to be undertaken, stakeholders who need to be consulted with or informed, and more importantly the barriers that may need to be overcome. It can be helpful to think about questions such as the below when at this stage:

- Has modelling been carried out (ideally by an independent entity) to assess the impacts of the potential design options? Does this modelling include interdependencies?
- Are these results communicated to stakeholder forums for discussion?

- Have all stakeholders been included in the discussion, even if they appear to be removed from the designed option (to help identify unintended consequences and trade-offs between communities)?
- Have the reasons why some options may not be feasible been communicated effectively?
- Have the results of any modelling, analysis and discussions been effectively disseminated to the public? So that there is a clear and transparent pathway that shows how the requirement for change was first raised, which options were considered, and why one, in particular, has been advocated.
- Have other complementary interventions been considered? For example, could an operational change be coupled with a change in land-use planning to enhance the predicted benefits?
- Have trials been carried out to verify modelling outcomes, and to perform analysis on the impacts on communities and other stakeholders?
- Do communities understand and value the metrics and dissemination tools used? Do you need to consider a different approach to communication?

Implementation

With a decision made as to how to address the problem, the solution to the problem can then be implemented. Example questions or thoughts that might be helpful at this stage might be:

- Have all stakeholders been made aware of the intervention in advance?
- Has the rationale for the chosen solution been explained to stakeholders rather than just the outcome of the decision making process itself?
- In order to demonstrate outcomes, have you considered if you need to move noise monitoring terminals, purchase new terminals, or make use of mobile terminals?
- Is regular feedback of the progress of the implementation made available to stakeholders?
- Have contingency plans been designed should the new procedure change and you need to fall back to the previous procedure?
- Do you have plans for on-going evaluation of the procedure, and plans for regular dissemination? This includes the collection of qualitative information.

Post-Evaluation

The saying 'you can't manage what you can't measure' is as true for noise management as it is for anything else. This enables performance to be assessed and for any potential changes to be made. It enables best practice to be extended to other areas and to add to the airports' knowledge of how to address noise problems in the future. Importantly, evaluation also enables the impacts of any noise impact abatement measures to be disseminated to stakeholders so that their success (or failure) can be demonstrated. Failure should not be hidden but rather embraced and made part of an on-going journey. Example helpful questions at this stage include:

- Have you committed to long term monitoring and evaluation and reporting to stakeholders?
- Do you communicate the procedure at engagement events?
- Do you have a long-term plan for the evaluation of the outcome of the intervention on non-acoustic factors, general acceptability of the decision and quality of life implications for local residents?

The above methodology is not intended to be a rigid approach to solving any noise management issue. Rather, the intention is to illustrate how this sort of design-led thinking can help to develop and deliver more effective noise management outcomes, but taking a considered approach to noise challenges. Taking the time to think about the challenge, understand and learn about it, embed communication and engagement into the decision-making process, making decisions based on empathy, and striving towards targeted outcomes that can be effectively evaluated to demonstrate success (or failure) to stakeholders, or to modify approaches in the future. In short, they help to show that the airport is not just seeking to do 'the right thing' but also to do that thing 'in the right way'.

Lessons Learnt

The implementation of ANIMA project during its four years of activity has allowed the participating partners to learn and ultimately share with the wider audience some key aspects.

First of all, the project definitively confirmed that the problem of aviation noise is not reduced only to a question of quantities of sonic pressure. Therefore, the adequate management to try to minimise noise in such a way so that it does not become, on the one hand, an issue that decreases the quality of life of many people, putting even their health at risk, and, on the other hand, an issue that limits the daily operation of airport infrastructure, must be approached jointly from different angles with the corresponding specialists and with the necessary tools, both hard and soft. It has also been found that since various disciplines as far apart as sociology and engineering participate, a "common language" is necessary—in some sorts, it could resemble an "Esperanto" for aviation noise management, which does not imply the abandonment of any of the languages already in use, but the creation of a third party that serves as a lingua franca, ensuring that all stakeholders, from top experts to local residents, can understand, at a certain level of acceptance, what aeronautical noise really implies.

Another great lesson is that, for different reasons (legislative, political, social, etc.), aviation noise is being taken very seriously by many organisations and institutions, and that many efforts are being devoted to avoid and minimise it from all possible points of view. It is precisely these efforts—led by airports, airlines and manufacturers—that, despite situations to be resolved and/or improved, there has been a large divergence between the vertiginous increase in the number of flights and the number of people affected by aviation noise. This divergence must be increased and can be increased in order to facilitate the growth of aviation and a positive relationship with airport environments and ANIMA can, hopefully, strongly help to do so by providing information, knowledge, methodology and best practice to be carried out by certain agents in the sector.

The development of the project itself—four years with multiple internal and external activities—and the in-depth knowledge of various experiences both in the investigation of new approaches and their subsequent application, has revealed that dealing with aviation noise is not only a question of shared visions, if not also of time, dedication and will. The struggle underpinning aviation noise and its impact on local communities as well as on airport growth needs to be prioritised by policymakers and addressing it should be one of the goals of the strategic policy agenda.

References

Brown T (2008) Design thinking. Harvard Business Review. https://doi.org/10.5749/minnesota/978 0816698875.003.0002

Downe L (2020) Good Services: How to design services that work, 1st edn. BIS Publishers. ISBN: 978–148 9,063,695,439

Goodheart B (2016) Design thinking for aviation safety. Aviat Bus J, 4

Hall A et al (2013) Future aircraft cabins and design thinking: optimisation vs. win-win scenarios. Propul Power Res 2(2):85–95. https://doi.org/10.1016/j.jppr.2013.04.001

Liedtka J (2015) Perspective: linking design thinking with innovation outcomes through cognitive bias reduction. J Prod Innov Manage 32:925–938. https://doi.org/10.1111/jpim.12163

McGowan AM, Bakula C, Castner R (2017) Lessons learned from applying design thinking in a nasa rapid design study in aeronautics. In: 58th AIAA/ASCE/AHS/ASC structures, structural dynamics, and materials conference, 2017

Turner Donald EIII (2017) A case study in design thinking applied through aviation mission support tactical advancements for the next generation (TANG). Calhoun. Available at: https://calhoun.nps.edu/handle/10945/56830

Overall Perspectives

Laurent Leylekian [ID]

Abstract This section sums up the main findings from ANIMA and from associated projects related to aviation noise in order to deliver some key messages for policy-makers. These messages intend both to orient future research and to propose some concrete way to implement measures that could alleviate the noise burden endured by airports' neighbouring communities. The final table gives an even more synthetic view of these messages, the institutions they target, their objectives and the best ways that are likely to support them.

Keywords Aircraft noise · Aviation noise · Noise regulations · Annoyance · Airport communities

The world that we used to know has been increasingly shaped by a worldwide phenomenon named globalisation. For more than fifty years, globalisation has pushed our societies in a direction seen by most of the people as a progress: Interweaving of nations, economies and the exchange of goods and services have been deemed the most efficient way to increase nations' wealth and to repel the prospect of confrontations by the advent of a kind of global culture.

Transport has been and is at the heart of such a conception, privileging flows over stocks. And aviation at large has largely accompanied this trend and benefited from it. Fifty years ago, aviation was for the upper class in the Western hemisphere and for elites only elsewhere. Nowadays, even if it implies only a few percents of the world population, it has become a mass transportation. Certainly, this way of life, which spread over the world, was criticised by some groups that were marginal at the beginning. But such criticisms taking ground on the increasing economic, sociological and environmental disequilibria became audible if not dominant, despite the secondary impact of aviation on both noise and pollution as constantly evidenced by the successive European

L. Leylekian (✉)
ONERA, FR-91123, BP 80100 Palaiseau Cedex, France
e-mail: laurent.leylekian@onera.fr

Environmental Reports.[1] They ultimately led to some frontal opposition such as the "flight shame" movement pinpointing the very question of the social usefulness of air transport.

The recent COVID pandemic—which is more a systemic consequence of globalisation than a regrettable accident—gives the opportunity for an in-depth revision of the air transport system and for its possible reboot on a more sustainable basis. The aviation noise issue must be now considered in this prospect and lessons learnt from ANIMA and other past European research projects could be usefully taken into account. What are these lessons and what are recommendations stemming from them?

Lesson One—Aircraft Noise Is Not Aviation Noise

So far, most of the efforts have been put on the reduction of noise at source, i.e. on aircraft noise. Associated research is certainly much needed but not directly for relieving the quality of life of people impacted by aviation noise. **Actually, if the European Union must maintain its support of aircraft noise research , it is for the sake of competitiveness.** As recalled by the European Commission itself, aeronautics is one of the EU's key high-tech sectors on the global market, providing more than 500,000 jobs and generating a turnover of close to €180 billion euros in 2019. If the EU is a world leader in the production of civil aircraft, including helicopters, aircraft engines, parts and components, other global players constantly challenge it and low aircraft noise is a key commercial argument, especially with respect to the more and more stringent associated regulations. **Therefore, research aiming at reducing the noise at source is always needed.**

Recommendation: Member States—especially those with an aircraft industry—as well as the European Commission should maintain a significant level of financial support on research intending to lower aircraft noise for the competitiveness and the leadership of the European industry.

Lesson two—As for Technology, Research On Aircraft Noise Must Favour Disruptive Concepts

Many past research projects addressed specific noise sources such as—for instance—"airframe noise", "jet noise", "turbofan noise". A decade ago, outcomes of such projects led at low Technology Readiness Levels (TRL) were progressively embarked

---

[1] Non-road transports are responsible of only 1.7% of particle matters under 10μm, 2.9% of particle matters under 2.5μm, 7.3% of the nitrogen oxide; International aviation is responsible of 3.4% of all greenhouse gas in transport; number of people affected by aviation noise in Europe is below 3 millions, 19 millions for rail transport, 7é millions for road transport

in some demonstration platforms through some large projects such as Silencer, Openair or, more recently, CleanSky. A few years ago, the European Commission fortunately decided to support again low TRL research on aircraft noise to keep on favouring the competitiveness of the European industry. **It is recommended to orient as far as possible these low TRL research toward disruptive concepts.** This recommendation takes ground on the one hand on the fact that incremental progresses on noise sources are going to saturate. There is still margin for progress but roadmaps maintained in ANIMA and through the former X-Noise networks give reasonable assessment for such margins which are limited. Recently launched projects are for instance intending to gain no more than 0.5 to 2 dB at the utmost for new concepts of landing gears or high-lift devices studied at low TRL. If ever integrated on an actual aircraft with all its operational constraints, figures could be even lower. This recommendation for disruptive concepts also takes ground on the great deal of uncertainties about what will be the future of aviation. Energetic consideration, other environmental concerns such as noise but also societal evolutions such as the already mentioned "flight shame" may dismiss traditional long-haul/mid-haul airliners based on regional hubs saturated with an overloaded demand in resources. The recent stop in producing A380—a blatant success for the European technology that was once considered as a market game changer—is very significant in this regard. Various scenarios may arise in the future—such as for instance those put forward by the Association of European Research Establishments in Aeronautics (EREA) in its vision study 2050 (https://erea.org/erea-vision-studies/). In this prospect, breakthrough noise technologies may give genuine competitive advantages to unnoticed trends or disregarded options that would eventually come true. Recent efforts put on new aircraft design—such as in the **ARTEM project**—new propulsion types, whatever hybrid or fully distributed-electric, or on UAVs for carrying people and goods is a commendable trend and **research on evaluating and lowering the noise of such new aircraft or of newer concepts of noise reduction technologies must be pursued.**

Recommendation: Member States—especially those with an aircraft industry—as well as the European Commission should orient low TRL research for aircraft noise reduction on disruptive concepts rather than on incremental research.

Lesson Three—Noise Regulations Based on Metrics Do Not Protect Against Annoyance

Noise metrics are the historical instrument to assess noise impact, and especially aircraft noise impact. There are good reasons to justify this situation: Studies of aircraft noise are first and foremost performed by engineers and specialised acousticians for whom noise metrics are the most "scientific" and "natural" tools. "Scientific" here means related to hard science and, indeed, noise metrics are very suitable, as related measures are comparable, reproducible and objective. **That is why noise metrics are and must remain privileged for aircraft certification purposes.** But they are more than questionable for assessing annoyance. Nobody ever heard a Lden level nor an EPNL one. And if one once "heard" a dB(A), this instantaneous figure alone is not representative of annoyance. There are certainly plenty of other indicators which are more suitable for this endeavour, going from "number above" up to awakening indices. However, even these indicators cannot figure out alone what annoyance endured by an individual means, and annoyance, encompassing most of the time one or several non-acoustical factors, can barely be represented by a number or a percentage.

In this regard, the usage of noise indicators in regulations dedicated to lower annoyance, for instance the Environmental Noise Directive (2002/49/EC) or the Regulation 598/2014 introducing noise-related operating restrictions at Union airports within a Balanced Approach, is only a bottom line and must be augmented on a local basis by other instruments. **Best practices exemplified in the ANIMA project could be the basis for such more elaborated instruments, especially because many local actors are in need for such guidelines.**

Recommendation: The European Commission, the EASA along with the ICAO may perform an insightful review of the ANIMA project's outcomes in order to further refine, elaborate and endorse joint and/or official guidelines on how to cope with annoyance beyond complying to noise metric-based regulations.

The European Commission could also envisage maintaining and developing further the ANIMA top-level outcomes—and especially the Best Practice Portal—as a basis of such guidelines.

Lesson Four—We Do Not Know What Annoyance Is, But We Know How It Is Mediated

Annoyance is a kind of portmanteau word under which there is an incredible wealth of concepts and issues. The annoyance endured during business hours, which prevents concentration, and which is detrimental to children's learning, is not the same as the annoyance endured on the weekend when one cannot enjoy a party outside with relatives or friends. It is also not the same as the one which is stemming from a series of nights during which one has been prevented from sleeping or has been awakened. Eventually, it is not the same one, which ends in long-term fatigue, possible anxiety or even nervous breakdown. Last, causal pathways between such noise-induced annoyance and physiological effects such as hypertension or strokes remain unclear even if these pathways are now clearly evidenced. In this regard, ANIMA confirmed the previous conclusions from the World Health Organisation (WHO) "strongly recommending" that noise levels around airports should be reduced to below 45dBA L_{den} and 40dBA L_{night} based on the percentage of highly annoyed and highly sleep disturbed people, respectively. **ANIMA also found out that sleep disturbance is the chief impact to remedy, as it is likely to mediate long-term effects and especially physiological ones.** This conclusion says nothing about the non-acoustical factors, which are also heavily affecting annoyance felt by individuals. Though some of these non-acoustical factors are now clearly understood, additional research is needed on that specific point.

- **Recommendations: In terms of actual implementation, Civil Aviation Authorities and Airports must do their best to comply with the very challenging WHO recommendations. Priority must be put on preventing awakening.**
- **In terms of research, the European Commission should keep on supporting works dedicated to deepening our understanding of the influence of non-acoustical factors on annoyance.**

Lesson Five—Do Not Presume Communities' Expectations. Be Fair With Them, Listen To Them And Empower Them

Even if all is not understood about non-acoustical factors, some points are now quite clear. In particular, a substantial part of the annoyance derives from the feeling of being deprived of any way to cope with aviation noise. Whether it is true or not, most airports' neighbours consider that they are not associated with decisions, that such decisions are taken through top-down and technocratic processes and that eventual communications are just intending to mislead or manipulate them through screening, obscure and too sophisticated concepts.

It would be far more fruitful to engage such communities in a positive way, and there are ANIMA-evidenced recipes in this regard. They are not necessarily easy to implement, but they are worth trying.

First, airports should not presume to know what communities want; for instance, they should not presume that window insulation is a necessity or that annoyance should be evaluated through the L_{den} metrics. They should rather engage in a genuine dialogue with all the stakeholders, of course including vocal activists or environmental associations but also the silent majority among which some stakeholders are sensitive to the social and economic importance of airports' infrastructure. To be effective, the engagement must be underpinned by a 'common language' that is comprehensible to all. Jargon and dominant attitudes must be banned in order to give fair access to expertise to all the audience. In particular, the various options must be fairly exposed with all their implications and with all the other considerations than noise (for instance, prohibitive safety ones or business impact). In this formatted dialogue, the rules of the game must be clear and known to all in advance. In particular, decision-making processes must be inclusive, transparent and they must allow the validity of claims to be challenged. Citizens understand that tough decisions, not always in their interest, must be made, but if the process is unclear, even good decisions will be questioned. As a prerequisite, it also means that the instigators of such processes must be ready to accept and endorse their conclusions otherwise, it would be even more counterproductive than doing nothing.

Recommendations: Airports must locally engage their community to reach a consensus between all the stakeholders, including the silent majority, and then to endorse conclusions reached by the group.

Lesson Six—Evaluate Measures Taken On A Regular Basis

This lesson may sound obvious but it is not. Very often, some airports, even those at the leading-edge, implement some ambitious plans intending to lower the annoyance but there is little or no systematic evaluation of these efforts, nor indeed their wider consequences. Further, research into the efficacy of certain forms of communication and engagement is so limited as to be of little use to airports when designing noise management interventions or more general community outreach programmes. This additional factor may explain why in many cases airport community engagement efforts do not yield the intended benefits for airports and communities alike.

Recommendations: Airports must evaluate and survey on a regular basis measures they have taken and compare them with the updated expectations of the neighbouring communities.

Lesson Seven—Do Not Wait Until It Is Too Late For Implementing Regulation 598/2014

Regulation 598/2014 supersedes Directive EC 2002/30 and establishes "*rules and procedures with regard to the introduction of noise-related operating restrictions at Union airports within a Balanced Approach*". However, applying this regulation is mandatory only for airports above 50,000 civil aircraft movements per calendar year. It appears to be too late for small but rapidly growing airports, especially when it comes to land-use planning issues. Around some airports "starting the journey" on noise management and mitigation—for instance, in Eastern Europe—there is an actual risk that prospects for jobs and economic growth tend to override on the short-term any societal or environmental considerations. Uncontrolled or commercially driven developments of inappropriate land uses may then occur, strongly encroaching lands around airports and preventing or threatening any further enforcement of the regulation.

By considering the adoption of Balanced Approach interventions before noise becomes a constraint (i.e. via complaints and objections to developments), airports will be better placed to manage their future. When being reactive to such pressures, such airports will be forced to act quickly, potentially at a higher cost, and potentially with the issue taken out of their hands (i.e. by national policy-makers), leading to sub-optimal outcomes. Through being proactive and developing long-term noise management strategies, these rapidly growing airports will be able to better control their ongoing development on their own terms and help to shape future policy rather than being at the behest of policy decisions made by others. Land-Use Planning is perhaps the best way through which this can be done. For instance, if rapidly growing airports are able to develop long-term noise maps based on future growth, they will be able to resist the encroachment of noise-sensitive buildings such as public residences, thus leading to fewer noise problems in the longer term.

Recommendation: The ICAO and the European Commission may develop an authoritative guidance targeting regional airports beyond the threshold of 50,000 movements per year in order to provide an incentive for starting to implement Balanced Approach interventions to pave the way for a possible smooth implementation of the directive in the future.

Lesson Eight—Providing Experience Of "Pathfinders" And "Experienced Travellers" To Airports "Starting The Journey"

Through the case studies examined, ANIMA clearly showed that not all airports have the same level of experience, understanding and achievements when it comes to noise management and mitigation. On the one hand, "pathfinders" are airports at the leading-edge of Balanced Approach implementation. They are usually large and experienced platforms known to be innovative in exploring novel approaches to noise annoyance, through leading-edge interventions involving a wide range of stakeholders and Balanced Approach elements.

In the middle, there are airports that are "experienced travellers" in applying Balanced Approach principles and interventions. Experienced travellers will require support to further advance and add value to their noise management programmes, considering that they may already be engaging with stakeholders.

On the other hand, airports "starting the journey" have little to no experience in the application of Balanced Approach principles and/or community engagement. Starting the journey, airports often lack the expertise and resources required for best practice, and may face a lack of legislative drivers to encourage the implementation of effective noise abatement interventions. Such airports may require guidance in how to progress towards best practice for their own specific circumstances, rather than copying the approaches of airports with more 'advanced' noise management strategies.

It could be a good idea to make the latter benefit from the experience of the former. In this regard, the European Commission, along with dedicated associations such as ACI or ARC, possibly as well as with training companies, may envisage exchange of personnel and dedicated experts to favour learning-by-doing of the less experienced.

This could look like the TEAMING or Erasmus work programme of the European Commission.

Recommendation: The European Commission along with private partners may favour exchange of personnel and experience between advanced and less advanced airports in a kind of "Airasmus" programme

Lesson 9—Build On And Widen The Anima Experience And Its Methods

Through its successive framework programmes for Research and Innovation, the European Commission progressively enlarged its scope, starting from the key but narrow concern of scientific Excellence to the market competitiveness of the European industry and to the return on investments for its citizens and taxpayers. In this evolution, ANIMA has been a landmark with—maybe for the first time—associations of scholars from universities, research centres, industries' primes but also of experts in Human sciences and end-users that are airports. Led within Horizon 2020, ANIMA arose high expectations and the involved teams, at least, deem the project a success: on the one hand, the cross-fertilisation of up-to-then siloed disciplines not only led to effective solutions or methods, but it also partook to build up a common language, shared concerns and a comprehensive culture for the involved stakeholders. This methodology is worth extending far beyond the single aviation noise issue. It is noticeable that it is perfectly fitting new orientations put forward by Horizon Europe through its highly transversal calls for proposals, notably an increased and quick impact on citizens and taxpayers as well as the decarbonisation of European economies.

On the other hand, ANIMA not only maintained the European Strategic Roadmap on Aviation noise that was initiated by the successive X-Noise networks but it also substantially enlarged it to Airport Noise Management and Indicators and to Impacts Understanding. It also took this opportunity to enlarge the network of related national focal points (NFPs) and therefore to amplify the EU influence on regions up-to-then poorly connected to the European research ecosystem, such as for instance the Western Balkans. It is believed that such a roadmap and such a network are very valuable European assets which are not only allowing the Continent to weigh in International bodies such as the ICAO but also to act as a major pole of reference for its neighbourhood, whether Eastern or Southern.

Recommendation: The European Commission could take advantage of the ANIMA experience, on the one hand by elaborating tailored calls for proposals that would allow part of the consortium to embark with newcomers on enlarged topics with strong interdependencies, for instance related to the highly complex issue of decarbonising the air transport. It should also set conditions to maintain the unique European Strategic Roadmap on Aviation noise and the associated

network of experts as a major asset for leadership and for influence toward its Eastern and Southern neighbours.

Summary of Recommendations

#	To whom	What	How	Objective
1	Member States and the European Commission	Maintain a significant level of financial support on research intending to lower aircraft noise	Nat'l research projects & RIAs	Competitiveness of EU industry
2	Member States and the European Commission	Orient low TRL research for aircraft noise reduction on disruptive concepts rather than on incremental research	Nat'l research projects & RIAs	Competitiveness of EU industry
3	European Commission, EASA & ICAO	Elaborate and endorse guidelines on how to cope with annoyance beyond complying to noise metric-based regulations	Based on the WHO and ANIMA findings	Social acceptance of air transport
4	European Commission	Maintain and Develop further the ANIMA top-level outcomes —and especially the Best Practice Portal—as a basis of such guidelines	Internal achievement, call for tender	Dissemination of knowledge, Impact of Horizon Europe
5	Civil Aviation Authorities & Airports	Comply with the WHO recommendations, especially in order to prevent awakening	Based on the WHO and ANIMA findings	Citizens health and quality of life
6	European Commission	Keep supporting research on non-acoustical factors	RIAs & CSAs	Social acceptance of air transport and citizens health
7	Airports	Engage communities to reach a consensus between all the stakeholders, including the silent majority, and then to endorse conclusions reached by the group	ANIMA recipes in the best practice portal	Social acceptance of air transport

(continued)

(continued)

#	To whom	What	How	Objective
8	Airports	Evaluate and survey on a regular basis, take measures and compare them with the updated expectations of the neighbouring communities	Polling	Social acceptance of air transport
9	ICAO European Commission	Develop a guidance targeting regional airports below the threshold of 50,000 movements per year for convincing then to start implementing balanced approach interventions	Communication policy	Social acceptance of air transport and quality of life
10	European Commission	Favour exchange of personnel and experience between advanced and less advanced airports in a kind of "Airasmus" programme	IA, Other financial instruments	Training and knowledge of experts Quality of life
11	European Commission	Elaborate calls for proposals that would take benefit from the ANIMA experience, for instance on decarbonising the air transport	RIA	Impact of Horizon Europe
12	European Commission	Maintain the European Strategic Roadmap on Aviation noise and the associated network of experts	CSA	Leadership of the EU, Influence on neighbourhood

Glossary

ACARE Advisory Council for Aeronautics Research in Europe
ANCAT Abatement of Nuisances Caused by Air Transport Expert Group
ANERS 3AF/AIAA Aircraft Noise and Emissions Reduction Symposium
ATM Air Traffic Management
CAP Committee for Aviation Environmental Protection
CROR Counter rotating open rotor engine
CSA "Coordination and Support Action" EU funding instrument
DB *Decibel, a relative unit of sound strength*
dB(A) *decibel with A weighting*
ECAC European Civil Aviation Conference
END Directive 2002/49/EC "Assessment and Management of Environmental Noise"
EPNL, EPNdB *Effective Perceived Noise Level in dB*
ICAO *International Civil Aviation Organisation*
LUP Land Use Planning
NAP Noise Abatement Procedure
NRT Noise Reduction Technologies
UHBR Ultra High Bypass Ratio engine

Printed in the United States
by Baker & Taylor Publisher Services